Reviews of Physiology, Biochemistry and Pharmacology

Volume 183

The highly successful Reviews of Physiology, Biochemistry and Pharmacology continue to offer high-quality, in-depth reviews covering the full range of modern physiology, biochemistry and pharmacology. Leading researchers are specially invited to provide a complete understanding of the key topics in these archetypal multidisciplinary fields. In a form immediately useful to scientists, this periodical aims to filter, highlight and review the latest developments in these rapidly advancing fields.

2020 Impact Factor: 5.545, 5-Year Impact Factor: 5.645

2020 Eigenfaktor Score: 0.00036, Article Influence Score: 1.227

Christian Stock · Luis A. Pardo
Editors

Targets of Cancer Diagnosis and Treatment

Ion Transport in Tumor Biology

Springer

Editors
Christian Stock
Medizinische Hochschule Hannover
Zentrum für Innere Medizin
Hannover, Germany

Luis A. Pardo
Max-Planck-Institut für Multidisziplinäre
Naturwissenschaften
Göttingen, Germany

ISSN 0303-4240 ISSN 1617-5786 (electronic)
Reviews of Physiology, Biochemistry and Pharmacology
ISBN 978-3-031-03993-5 ISBN 978-3-031-03994-2 (eBook)
https://doi.org/10.1007/978-3-031-03994-2

This Springer imprint is published by the registered company Springer Nature Switzerland AG
The registered company address is: Gewerbestrasse 11, 6330 Cham, Switzerland

Preface

The first volume in the three-volume series "Ion Transport in Tumor Biology" provided detailed information about the ion channels and transporters found to be dysregulated in virtually all epidemiologically relevant tumor entities. The second volume dealt with how ion transport molecules help cancer cells survive in a hostile environment such as the intravascular milieu and how (reciprocal) interactions between these ion transport proteins and the cancer cells' microenvironment further tumor progression and metastasis.

A deep understanding of tumor biology, relevant in its own right, also represents the starting point for knowledge to be translated into tangible benefits for patients' survival and quality of life. Thus, in the present third volume entitled "Targets of Cancer Diagnosis and Treatment", the authors describe and discuss the transportome as a target for cancer management. In particular, Na^+, K^+, Ca^{2+} and Cl^- channels as well as organic solute carriers are critically examined regarding both their suitability as therapeutic targets, on the one hand, and their contribution to the development of resistance to chemo- or radiotherapy, on the other. The volume concludes with a chapter about the progress made towards the therapeutic exploitation of the transportome, including social aspects and future prospects.

Hannover, Germany — Christian Stock
Göttingen, Germany — Luis A. Pardo

Acknowledgements

Contributions to this volume have partly been personally invited, with kind support of the series editors D.L. Barber, E. Cordat, M. Kajimura, J. Leipziger, M.E. O'Donnell, L. Pardo, N. Schmitt, C. Stock.

Contents

Rev Physiol Biochem Pharmacol (2022) 183: 1–44
https://doi.org/10.1007/112_2020_46

Published online: 1 September 2020

Targeting Ion Channels for Cancer Treatment: Current Progress and Future Challenges

Alina L. Capatina, Dimitris Lagos, and William J. Brackenbury

Contents

Abstract Ion channels are key regulators of cancer cell pathophysiology. They contribute to a variety of processes such as maintenance of cellular osmolarity and membrane potential, motility (via interactions with the cytoskeleton), invasion, signal transduction, transcriptional activity and cell cycle progression, leading to tumour progression and metastasis. Ion channels thus represent promising targets for cancer therapy. Ion channels are attractive targets because many of them are expressed at the plasma membrane and a broad range of existing inhibitors are already in clinical use for other indications. However, many of the ion channels identified in cancer cells are also active in healthy normal cells, so there is a risk that certain blockers may have off-target effects on normal physiological function. This review describes recent research advances into ion channel inhibitors as anticancer therapeutics. A growing body of evidence suggests that a range of existing and novel

A. L. Capatina
Department of Biology, University of York, York, UK

D. Lagos
Hull York Medical School, York, UK

York Biomedical Research Institute, University of York, York, UK

W. J. Brackenbury (✉)
Department of Biology, University of York, York, UK

York Biomedical Research Institute, University of York, York, UK
e-mail: william.brackenbury@york.ac.uk

Na^+, K^+, Ca^{2+} and Cl^- channel inhibitors may be effective for suppressing cancer cell proliferation, migration and invasion, as well as enhancing apoptosis, leading to suppression of tumour growth and metastasis, either alone or in combination with standard-of-care therapies. The majority of evidence to date is based on preclinical in vitro and in vivo studies, although there are several examples of ion channel-targeting strategies now reaching early phase clinical trials. Given the strong links between ion channel function and regulation of tumour growth, metastasis and chemotherapy resistance, it is likely that further work in this area will facilitate the development of new therapeutic approaches which will reach the clinic in the future.

Keywords Calcium · Cancer · Chloride · Inhibitors · Ion channels · Immunotherapy · Potassium · Sodium

1 Introduction

Traditional chemotherapeutic approaches have been successfully used as cancer treatments for decades, partially due to their generalised, anti-proliferative and cytotoxic activity (DeVita and Chu 2008). However, the lack of specificity of chemotherapy is a limiting factor in the treatment of more advanced tumours and acquired resistance. This has driven the development of targeted therapies, such as monoclonal antibodies, small molecule pathway inhibitors and immune checkpoint inhibitors, and emerging cellular therapies (Baudino 2015). The limitations of targeted treatments can come from their specificity, making their effectiveness tumour- or antigen-dependent and thus potentially only applicable to a relatively small proportion of the population. A relatively underexplored area in cancer research is represented by the therapeutic targeting of ion channels and transporters (Oosterwijk and Gillies 2014). Plasma membrane ion channels have been shown to contribute to a variety of cellular processes in addition to their role in maintaining membrane potential (V_m) and cellular osmolarity (Yang and Brackenbury 2013; Djamgoz et al. 2014; Leslie et al. 2019). For example, as discussed in detail elsewhere in this series of Special Issues, alterations in ion flux can contribute to cellular motility, cytoskeletal rearrangements and signal transduction underpinning cellular migration (Schwab et al. 2012; Yang et al. 2020), growth and cell cycle progression (Blackiston et al. 2009; Urrego et al. 2014; Humeau et al. 2018) and gene expression (Mycielska et al. 2005; Popov et al. 2012), as well as defining the extracellular environment (e.g. pH regulation (Parks et al. 2013; Wu et al. 2017)). In the tumour microenvironment, higher levels of K^+ and Na^+ have been reported, accompanied by a relatively decreased pH and hypoxic environment compared to healthy tissue (Ouwerkerk et al. 2007; Eil et al. 2016; Leslie et al. 2019). Elevated expression of a wide range of ion channels has also been associated with metastasis, reviewed extensively elsewhere (Pardo and Stuhmer 2014; Brackenbury 2016;

Djamgoz et al. 2019). Together, these findings suggest that ion channels could serve as potential targets for anticancer therapies, particularly given the tumour-specific expression of certain channel types. Ion channels, particularly those at the plasma membrane, present potentially attractive therapeutic targets due to their location and the fact that a broad range of existing inhibitors are already in clinical use. Given that many blockers of plasma membrane ion channels can act extracellularly, they can be screened relatively easily using electrophysiological approaches. Intracellular ion channels have also been shown to be important regulators of cancer cell metabolism, apoptosis and gene expression (Leanza et al. 2013a; Jang et al. 2015; Peruzzo and Szabo 2019); these could similarly represent attractive targets for therapeutic inhibition.

On the other hand, given that many of the ion channels identified in cancer cells are expressed in healthy normal cells, there is a risk that these blockers may have off-target effects on normal physiological function. This review describes recent research advances into ion channel inhibitors for cancer treatment. Key Na^+, K^+, Ca^{2+} and Cl^- channel inhibitors are covered, followed by details on their use and effectiveness in cancer, as well as considering combining such inhibitors with standard-of-care therapies (Fig. 1).

2 Na^+ Channel Inhibitors

Several classes of Na^+ channels have been shown to be aberrantly expressed in cancer cells where they regulate cell proliferation, migration, invasion and metastasis (Leslie et al. 2019). In particular, voltage-gated Na^+ channels (VGSCs) are upregulated in tumour cells where their activity regulates V_m, morphological changes and metastatic behaviour (Grimes et al. 1995; Roger et al. 2003; Fraser et al. 2005; Nelson et al. 2014, 2015; Yang et al. 2020). VGSCs have thus been studied as potential cancer targets (Table 1). VGSCs are important clinical targets for the treatment of epilepsy and cardiac arrhythmia (George 2005; Mantegazza et al. 2010). Various Class 1B antiarrhythmic drugs, antiepileptic drugs and local anaesthetics have been studied in preclinical in vitro and in vivo cancer models (Martin et al. 2015). For example, the anticonvulsant phenytoin inhibits breast cancer cell migration, tumour growth, invasion and metastasis (Yang et al. 2012; Nelson et al. 2015). Phenytoin also inhibits migration and secretory activity in prostate and lung cancer cells (Abdul and Hoosein 2001; Fraser et al. 2003b; Onganer and Djamgoz 2005). These results are generally supported by other studies using different VGSC-inhibiting drugs in breast cancer and other cancer types, including carbamazepine, riluzole, ranolazine and ropivacaine (Abdul and Hoosein 2001, 2002b; Yip et al. 2009; Speyer et al. 2012; Djamgoz and Onkal 2013; Baptista-Hon et al. 2014; Driffort et al. 2014; Bugan et al. 2019; Guzel et al. 2019). It should be noted, however, that some compounds may elicit their anticancer effects through other mechanisms in addition to VGSC inhibition. For example, riluzole may prevent migration or promote apoptosis and cell cycle arrest (shown in glioma,

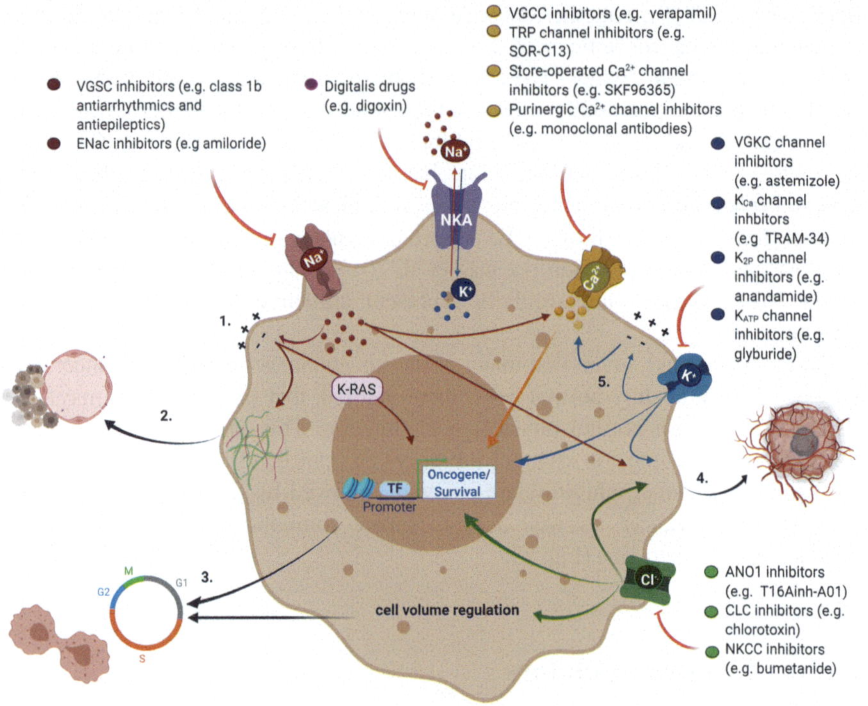

Fig. 1 Potential anticancer utility of Na^+, K^+, Ca^{2+} and Cl^- channel blockers. Principal consequences of ion channel function in cancer cells: (1) membrane potential depolarisation (Yang and Brackenbury 2013), (2) invasion and metastasis (Besson et al. 2015), (3) cell cycle progression and proliferation (Becchetti 2011), (4) angiogenesis (Fiorio Pla et al. 2012), (5) Ca^{2+} signalling in response to altered K^+ channel activity (Illek et al. 1992). Na^+, Ca^{2+}, K^+ and Cl^- channels are represented on the plasma membrane for clarity, but some functions are performed by intracellularly located channels (details in main text). Key inhibitor classes and examples of widely studied compounds are included (see tables for a complete list). *ENaC* epithelial Na^+ channel, K_{Ca} Ca^{2+}-dependent K^+ channels, *K-RAS* Kirsten rat sarcoma, *NKA* Na^+/K^+ ATPase, *NKCC* $Na^+/K^+/Cl^-$ cotransporter, *TF* transcription factors, *VGCC* voltage-gated Ca^{2+} channel, *VGKC* voltage-gated K^+ channel, *VGSC* voltage-gated Na^+ channel

neuroblastoma, lung, colon and prostate cancer) and inhibit autophagy (shown in pancreatic cancer) at least in part via its function as a non-competitive inhibitor of the metabolic glutamate receptor 1 (Akamatsu et al. 2009; Zhang et al. 2015; Seol et al. 2016; Lemieszek et al. 2018; Sun et al. 2019). Nonetheless, a number of studies now show that VGSC-inhibiting drugs suppress proliferation, promote apoptosis and reduce migration, invasion and metastasis (Martin et al. 2015).

A key advantage of Class 1B antiarrhythmic drugs is that they display state-dependent binding and preferentially block VGSCs in the inactivated state (Clare et al. 2000). Accumulating evidence suggests that cancer cells have a relatively depolarised V_m, which would mean that VGSCs present in these cells are predominantly in their inactivated state (Yang and Brackenbury 2013). Importantly, studies

Table 1 Na^+ channel/transporter inhibitors in cancer

Compound	Cancer target	Type of cancer
Amiloride	ENaC, NHE1	In vitro: multiple myeloma (Rojas et al. 2017), trophoblasts (Del Monaco et al. 2009). In vivo: hepatocellular carcinoma breast, gastric, colon, pancreatic (Matthews et al. 2011; Sparks et al. 1983)
Bupivacaine	VGSC, K_v11.1	In vitro: breast (Chang et al. 2014; Li et al. 2018), colon (Li et al. 2019), ovarian, prostate (Xuan et al. 2016)
Carbamazepine	VGSC	In vitro: prostate (Abdul and Hoosein 2001), breast (Teichmann et al. 2014; Meng et al. 2011), neuroblastoma (Lang et al. 1993)
Casein kinase 1 inhibitor IC261	VGSC	In vitro and in vivo: pancreatic cancer (Brockschmidt et al. 2008)
Desipramine	VGSC	In vitro: hepatocellular carcinoma (Yang and Kim 2017), colon (Arimochi and Morita 2008), multiple myeloma (Biber et al. 2018)
Diclofenac	VGSC	In vitro and in vivo: colon, ovarian, neuroblastoma fibrosarcoma (Pantziarka et al. 2016) In vivo: breast, lung, connective tissue tumours, prostate, pancreatic (clinical trials) (Pantziarka et al. 2016)
Digitalis drugs (ouabain, digoxin, bufalin)	Na^+/K^+ ATPase	In vitro: lung (Pongrakhananon et al. 2013; Lin et al. 2015), breast, colon, prostate, hepatocellular (Gould et al. 2018; Khajah et al. 2018; Shen et al. 2020; Zhang et al. 2008), osteosarcoma (Menger et al. 2012) In vivo: breast (Gould et al. 2018), lymphoma, leukaemia (Zhang et al. 2008; Haux et al. 2001), fibrosarcoma, colon, hepatocellular, head and neck (Menger et al. 2012)
Disopyramide	VGSC	In vitro: breast (Fraser et al. 2005)
Dronedarone	VGSC	In vitro: ovarian (Meléndez et al. 2020) In vitro and in vivo: breast (Elliott et al. 2018)
Imipramine/clomipramine/derivatives	VGSC, K_v10.1, K_v11.1	In vitro: acute myeloid leukaemia (Xia et al. 1999; Metts et al. 2017), colon (Arimochi and Morita 2006), melanoma (Gavrilova-Ruch et al. 2002; Parker et al. 2012), multiple myeloma (Biber et al. 2018) In vivo: breast (Rajamanickam et al. 2016)
Lamotrigine	VGSC	In vitro: neuroblastoma (Lang et al. 1993) In vivo: prostate (Stettner et al. 2012), breast (Pellegrino et al. 2018)
Levobupivacaine	VGSC, K_v11.1	In vitro: colon (Li et al. 2019), breast (Li et al. 2018), prostate (Jose et al. 2018)
Lidocaine	VGSC	In vitro: breast (Yoon et al. 2011; Chang et al. 2014), colon (Siekmann et al. 2019), lung

(continued)

Table 1 (continued)

Compound	Cancer target	Type of cancer
		(Onganer and Djamgoz 2005) In vivo: breast (Chang et al. 2014)
Mexiletine/ RS100642	VGSC, $K_v11.1$	In vitro: breast (Fraser et al. 2005)
Nortriptyline	VGSC, Kv11.1	In vitro: melanoma (Parker et al. 2012), multiple myeloma (Biber et al. 2018) In vivo: bladder (Yuan et al. 2015)
Phenytoin + analogues	VGSC	In vitro: breast (Yang et al. 2012), prostate (Abdul and Hoosein 2001; Anderson et al. 2003; Fraser et al. 2003b), lung (Onganer and Djamgoz 2005), neuroblastoma (Lang et al. 1993) In vivo: breast (Nelson et al. 2015)
NESOpAb	Neonatal $Na_v1.5$	In vitro: breast (Brackenbury et al. 2007; Chioni et al. 2005)
Propranolol	VGSC	In vitro: breast (Lee et al. 2019a)
Protriptyline	VGSC	In vitro: osteosarcoma (Su et al. 2016), prostate (Chang et al. 2015)
Quinidine	VGSC, VGKC, K_{ATP}	In vitro: glioma (Ru et al. 2015), breast (Wonderlin et al. 1995) In vivo: breast (Raderer et al. 1993)
Ranolazine	VGSC	In vitro: breast (Driffort et al. 2014; Lee et al. 2019a), colon (Guzel et al. 2019) In vivo: prostate (Bugan et al. 2019), breast (Driffort et al. 2014)
Riluzole	VGSC, metabotropic glutamate receptor 1, $K_v11.1$, K_{2P}	In vitro: prostate (Abdul and Hoosein 2002b; Akamatsu et al. 2009; Uzun et al. 2017), pancreatic (Sun et al. 2019), neuroblastoma, glioma, lung, colon, leukaemia, myeloma (Lemieszek et al. 2018; Benavides-Serrato et al. 2020; Pillozzi et al. 2018; Poupon et al. 2018) In vivo: breast (Speyer et al. 2012), melanoma (Yip et al. 2009), glioma (Zhang et al. 2015), glioblastoma (Benavides-Serrato et al. 2020) hepatocellular carcinoma (Seol et al. 2016)
Ropivacaine	VGSC, $K_v11.1$	In vitro: colon (Baptista-Hon et al. 2014), breast (Li et al. 2018)
Tarantula peptide toxin HNTX-III	VGSC	In vitro: prostate (Chen et al. 2019)
Tetracaine	VGSC	In vitro: breast (Yoon et al. 2011)
Tetrodotoxin	VGSC	In vitro: prostate (Grimes et al. 1995; Grimes and Djamgoz 1998) In vivo: prostate (Yildirim et al. 2012)
Topiramate	VGSC	In vitro: ovarian (Xu et al. 2018) In vivo: lung (Ma et al. 2011)

(continued)

Table 1 (continued)

Compound	Cancer target	Type of cancer
Valproic acid	VGSC	In vitro: prostate (Abdul and Hoosein 2001; Angelucci et al. 2006) breast (Olsen et al. 2004) In vivo: colon, prostate, gastro-oesophageal (Wheler et al. 2014)
ω-3 polyunsaturated docosahexaenoic acid	VGSC, NHE1	In vitro: breast (Isbilen et al. 2006; Gillet et al. 2011; Wannous et al. 2015)

ENaC epithelial Na^+ channel, *NHE1* Na^+/H^+ exchanger-1, *VGKC* voltage-gated K^+ channel, *VGSC* voltage-gated Na^+ channel

have shown that VGSCs expressed in cancer cells, including $Na_V1.5$, carry a small persistent Na^+ current in the inactivated state which depolarises the V_m further and permits cytosolic Na^+ accumulation (Gillet et al. 2009; Brisson et al. 2011; Yang et al. 2012; Campbell et al. 2013; Yang et al. 2020). Further evidence suggests that this persistent Na^+ current is critical for promoting metastatic cell behaviour (Driffort et al. 2014; Nelson et al. 2015). Therefore, state-dependent VGSC blockers which preferentially bind to VGSCs in the inactivated state are likely to selectively target tumour-expressing VGSCs whilst leaving VGSCs in other cells, e.g. cardiomyocytes and neurons, unaffected. There is, however, currently a lack of clinical data in support of this hypothesis. Although the VGSC-inhibiting drugs valproate and quinidine have been studied in clinical trials, their mode of action via Na^+ current suppression was not investigated (Raderer et al. 1993; Wheler et al. 2014). The therapeutic value of VGSC inhibitors in the context of cancer has been studied retrospectively in several observational cohort data studies (Walker et al. 2011; Fairhurst et al. 2014, 2015, 2016; Reddy et al. 2015; Takada et al. 2016). However, the results are inconsistent, with several studies demonstrating positive associations (Exadaktylos et al. 2006; Biki et al. 2008; Walker et al. 2011; Reddy et al. 2015; Takada et al. 2016) and another study showing a negative association, although the possibility of confounding by indication cannot be excluded (Fairhurst et al. 2015). Thus, prospective clinical trials are required to establish the utility of VGSC inhibition in cancer patients (Djamgoz et al. 2019).

Novel compounds have also been investigated as potential inhibitors of VGSC function in cancer cells. Novel α-hydroxy-α-phenylamide analogues of phenytoin have been developed in order to improve VGSC subtype specificity, and some of these have been shown to inhibit prostate cancer cell proliferation (Anderson et al. 2003; Lenkowski et al. 2004). Additional small molecule VGSC inhibitors have been developed with the aim of increasing selectivity for the neonatal splice variant of $Na_V1.5$ expressed in breast cancer cells, and these have been shown to inhibit both Na^+ current and invasion (Dutta et al. 2018). The casein kinase 1 inhibitor IC261, which induces cell cycle arrest and apoptosis in cancer cell lines, has also been shown to inhibit $Na_V1.5$ currents, suggesting that IC261 may elicit its antitumour effects partially through VGSC inhibition (Brockschmidt et al. 2008; Föhr et al. 2017). The mexiletine analogue RS100642, targeted at tetrodotoxin-resistant VGSCs, inhibits oxidative stress induced by tumour development in the DMBA rat breast cancer model (Batcioglu et al. 2012). ω-3 polyunsaturated docosahexaenoic acid, which has been shown to improve breast cancer outcomes, inhibits $Na_v1.5$ expression and activity in breast cancer cells via peroxisome proliferator-activated receptor β (PPARβ) (Isbilen et al. 2006; Gillet et al. 2011; Wannous et al. 2015).

Numerous peptide toxins bind to and inhibit VGSCs, and several of these have been explored in the context of cancer treatment. Local injection of the pan-specific VGSC-inhibiting toxin tetrodotoxin directly into subcutaneous prostate tumours in rats significantly reduces lung metastasis, improving survival (Yildirim et al. 2012). Treatment of prostate cancer cells with the tarantula peptide toxin HNTX-III downregulates $Na_v1.7$, decreases RhoA/Rac1 protein expression and inhibits cellular migration, raising the possibility that such isoform-specific toxins may have utility as anti-motility drugs (Chen et al. 2019). However, a potential issue with peptide toxins such as tetrodotoxin is that unlike Class 1B antiarrhythmic drugs, they do not display state-dependent binding. Thus, it would not be possible to administer such agents systemically without toxic side effects. Nonetheless, chemical modification of these toxins to aid tumour-specific targeting may be possible. One further issue with the use of VGSC inhibitors in general, including state-dependent blockers, is that they may also inhibit VGSCs present on immune cells, potentially reducing a desirable antitumour immune response. For example, $Na_v1.5$ is expressed on $CD4^+$ T cells where it plays a role in positive selection (Lo et al. 2012).

There has also been interest in developing ion channel-targeting monoclonal antibodies. This has proven to be relatively challenging given the complex structure of ion channel proteins, which makes it difficult to identify suitable epitopes, as well as due to the complexity of manufacturing antibodies, compared to small molecule design (Hutchings et al. 2019). A polyclonal antibody directed at the neonatal $Na_v1.5$-specific D1:S3/4 linker inhibits Na^+ current with high specificity for neonatal $Na_v1.5$ versus the adult splice variant (Chioni et al. 2005). Importantly, this antibody was additionally shown to inhibit migration and invasion of breast cancer cells (Brackenbury et al. 2007). Although the primary purpose of such antibodies has been to inhibit channel function, an additional possibility is that these antibodies may have utility as diagnostic tools (Yamaci et al. 2017) and/or as vehicles to target cytotoxic therapies to tumours (Arcangeli et al. 2009).

Epithelial Na^+ channels (ENaC) from the ENaC/degenerin family are also important players in metastatic cell behaviour (Yamamura et al. 2008; Bondarava et al. 2009; Del Monaco et al. 2009; Kapoor et al. 2009; Xu et al. 2016). ENaC activity promotes proliferation and inhibits apoptosis of hepatic carcinoma cells, as part of a hypertonicity-induced cationic channel complex (Sparks et al. 1983; Vila-Carriles et al. 2006; Bondarava et al. 2009). More recently, ENaC expression has been associated with increased expression of the achaete-scute homolog 1 (ASCL-1) transcription factor that mediates growth and progression of lung tumours (He et al. 2018). The exact mechanism that connects ENaC and ASCL-1 has not been fully defined, but these findings suggest that ENaC might contribute to tumour growth as a transcriptional target of ASCL-1 (He et al. 2018). Acid-sensing ion channels (ASIC), also members of the ENaC/degenerin family, can enhance invasive behaviour by activating the calcineurin/nuclear factor of activated T cell 1 (NFAT1) pathway in colorectal cancer cells, and treatment with cyclosporin A was shown to block the calcineurin pathway and ASIC2- mediated metastasis (Zhou et al. 2017). ASIC1 and ASIC3 promote epithelial to mesenchymal transition in pancreatic cancer cells in a Ca^{2+}-dependent manner (Zhu et al. 2017). The ENaC-inhibiting K^+-sparing diuretic amiloride has been shown to suppress ENaC-induced chorionic carcinoma cell migration in response to aldosterone (Del Monaco et al. 2009). Together with further studies showing antitumour and anti-metastatic effects of amiloride, these data suggest that pharmacological blockade of ENaC/ASIC channels may have therapeutic relevance (Matthews et al. 2011).

The ATP-dependent Na^+/K^+ pump (also known as the Na^+/K^+ ATPase) is an important regulator of Na^+/K^+ homeostasis in cancer cells (Zhang et al. 2008; Schneditz et al. 2019). This pump is the key membrane protein for transporting Na^+ out from the cell and maintaining a stable V_m (Post et al. 1969). Na^+/K^+ ATPase expression is elevated in breast cancer cells compared to normal epithelial cells, and its activity promotes proliferation, migration and invasion (Li et al. 2017; Khajah et al. 2018). Different Na^+/K^+ ATPase α-subunit isoforms have been associated with cancer malignancy: α1 mostly correlates with early stages of cancer (including prostate, lung and renal tumours), whilst α3 associates with advanced disease (Felippe Goncalves-de-Albuquerque et al. 2017). Cardiac glycoside digitalis drugs, e.g. ouabain and digoxin, which are potent inhibitors of the Na^+/K^+ ATPase (Post et al. 1969; Laursen et al. 2013), have been shown to inhibit proliferation, migration, invasion, inflammation and tumour growth and promote lysis of cancer cells (Zhang et al. 2008; Kepp et al. 2012; Gould et al. 2018; Khajah et al. 2018), reduce risk of certain cancers (Haux et al. 2001) and improve survival (Menger et al. 2012). Inhibition of the Na^+/K^+ ATPase by digitalis drugs leads to intracellular accumulation of Na^+ and subsequent reverse mode operation of the Na^+/Ca^{2+} exchanger (NCX) causing an increase in intracellular Ca^{2+}, which may then impact on cell cycle progression and survival (Chen et al. 2014). Interestingly, studies have shown that whilst the ion transport function of the Na^+/K^+ ATPase is inhibited by cardiac glycosides, these can enhance signalling via the pump, inducing activation of an associated kinase – Src – which transactivates the epidermal growth factor receptor (EGFR), forming a signalling complex that induces activation of mitogen-activated kinase (MAPK), thus indicating a complex function of the Na^+/K^+ pump in maintaining cellular physiology (Haas et al. 2002). Although the Na^+/K^+ ATPase was long thought to be the only target of ouabain and other digitalis drugs, there is emerging evidence suggesting that these compounds modulate additional targets, such as the X-hepatic receptor (Campia et al. 2012) and the steroid receptor co-activators (SRC) 1 and 3 (Wang et al. 2014). Thus, effects on cancer cells following treatment with these drugs might be derived from their impact on other targets.

In summary, a growing body of evidence suggests that pharmacological inhibition of various classes of Na^+ channels and transporters in cancer cells can inhibit proliferative and invasive capacity and may promote cell death. Further work is required to fully delineate the mechanism(s) of action of a number of these compounds in cancer cells and their potential clinical value.

3 K^+ Channel Inhibitors

Many plasma membrane K^+ channels are aberrantly expressed in cancer cells where their expression is often associated with increased proliferative capacity, and a number of these have been explored as therapeutic targets (Table 2) (Huang and Jan 2014; Pardo and Stuhmer 2014). For example, the $K_v1.3$ voltage-gated K^+ channel (VGKC) plays a role in regulation of proliferation in brain cell progenitors but also in cancer cells (Fraser et al. 2000, 2003a; Chittajallu et al. 2002). In prostate cancer cells, $K_v1.3$ was shown to be sensitive to several drugs, including dequalinium, glyburide and amiodarone, which induced growth inhibition and cell death (Abdul and Hoosein 2002a). In addition, $K_v1.3$ currents in prostate cancer cells are sensitive to verapamil, margatoxin, charybdotoxin, 4-aminopyridine and tetraethylammonium (Fraser et al. 2003a). In melanoma cells, treatment with the $K_v1.3$ inhibitors tetraethylammonium, verapamil and fampridine was shown to disrupt the interaction between $K_v1.3$ channels and β1 integrin, suggesting that integration of VGKCs in macromolecular protein complexes might provide a role in tumour cell adhesion and invasion (Artym and Petty 2002).

Another VGKC widely studied in the context of cancer is Eag1 ($K_v10.1$) (Ouadid-Ahidouch et al. 2016). $K_v10.1$ is upregulated in a number of tumour types (Ding et al. 2007; Ousingsawat et al. 2007). $K_v10.1$ promotes proliferation in cancer cell lines, and overexpression in Chinese hamster ovary cells causes tumorigenesis in vivo (Pardo et al. 1999; Ouadid-Ahidouch et al. 2001). Direct targeting of $K_v10.1$ with monoclonal antibodies inhibits K^+ currents and has an antiproliferative effect, reducing tumour growth in vivo (Gómez-Varela et al. 2007). Furthermore, development of a bifunctional antibody carrying a human $K_v10.1$ recognition site and a human TNF-related apoptosis-inducing ligand (TRAIL) domain was shown to induce selective apoptosis in prostate cancer cells sensitised with cytotoxic drugs (Hartung et al. 2011). Other well-established inhibitors of human Eag1 include astemizole and imipramine (Garcia-Ferreiro et al. 2004). Imipramine was shown to inhibit proliferation and induce apoptotic behaviour in ovarian cancer cells (Asher et al. 2011). Astemizole has also been shown to inhibit proliferation in vitro and tumour growth in vivo (Garcia-Quiroz and Camacho 2011; de Guadalupe Chavez-Lopez et al. 2015; Bernal-Ramos et al. 2017). In addition, the sea anemone toxin APETx4 inhibits $K_v10.1$, although it is cytotoxic in both cancer and non-cancer cell lines (Moreels et al. 2017b). Several novel purpurealidin analogues were also found to inhibit $K_v10.1$ current and increase cell death (Moreels et al. 2017a).

Table 2 K^+ channel inhibitors in cancer

Compound	Cancer target	Type of cancer
4-aminopyrimidine	VGKC, $K_{Ca}2.3$	In vitro: breast (Potier et al. 2006), prostate (Fraser et al. 2003a), melanoma (Artym and Petty 2002), cervical, ovarian (Han et al. 2007)
Amiodarone	$K_v1.3$, $K_v10.1$, VGSCs	In vitro: prostate (Abdul and Hoosein 2002a), breast (Abdul et al. 2003), glioma (Kim et al. 2011; Chang et al. 2018) In vivo: breast (Lee et al. 2015)
Anandamide	$K_{2P}3.1$, $K_v1.2$	In vitro: lung (Leithner et al. 2016), breast (De Petrocellis et al. 1998; Laezza et al. 2012), hepatocellular (Xie et al. 2012) In vivo: breast (Grimaldi et al. 2006)
Antibodies	$K_v10.1$, $K_v11.1$, $K_{2P}9.1$	In vitro: ovarian, neuroblastoma (Gómez-Varela et al. 2007), prostate (Hartung et al. 2011), pancreatic (Sette et al. 2013; Duranti et al. 2018), breast, colon (Duranti et al. 2018), B-cell lymphoma (Wang et al. 2007), lung (Sun et al. 2016) In vivo: breast and pancreatic (Gómez-Varela et al. 2007), pancreatic (Duranti et al. 2018), lung (Sun et al. 2016)
Apamin	$K_{Ca}2.3$	In vitro: breast (Potier et al. 2006)
APETx4	$K_v10.1$	In vitro: neuroblastoma, melanoma, prostate (Moreels et al. 2017b)
Astemizole	$K_v10.1$, $K_v11.1$	In vitro: lung (Chavez-Lopez et al. 2017), prostate (Bernal-Ramos et al. 2017), breast, hepatocellular (García-Quiroz et al. 2012; de Guadalupe Chavez-Lopez et al. 2015) In vivo: breast (García-Quiroz et al. 2014), hepatocellular (de Guadalupe Chavez-Lopez et al. 2015)
Bicalutamide	$K_{Ca}1.1$	In vitro: breast (Khatun et al. 2018)
Calcitriol/calcipotriol	$K_{Ca}1.1$, $K_v10.1$	In vitro: breast (García-Quiroz et al. 2012, Khatun et al. 2016; Khatun et al. 2018) hepatocellular (García-Quiroz et al. 2012) In vivo: breast (García-Quiroz et al. 2014)
Charybdotoxin	$K_v1.3$	In vitro: prostate (Fraser et al. 2003a)
Cisapride	$K_v11.1$	In vitro: gastric (Shao et al. 2005)
Clofazimine	$K_v1.3$	In vitro: melanoma, lymphocytes, (Leanza et al. 2012, 2013b) In vivo: melanoma (Leanza et al. 2012)
Clotrimazole	$K_{Ca}3.1$	In vitro: colon (De Marchi et al. 2009), breast (Zhang et al. 2016), pancreatic (Bonito et al. 2016)
Dequalinium	$K_v1.3$	In vitro: prostate (Abdul and Hoosein 2002a)
Ergtoxin	$K_v11.1$	In vitro: ovarian (Asher et al. 2011)
E4031 and Way123,398	$K_v11.1$	In vitro: breast (Lansu and Gentile 2013), ovarian (Asher et al. 2011), acute lymphoblastic leukaemia (Pillozzi et al. 2011), gastric (Crociani et al. 2014), colon (Crociani et al. 2013) In vivo: acute lymphoblastic leukaemia (Pillozzi et al.

(continued)

Table 2 (continued)

Compound	Cancer target	Type of cancer
		2011) gastric (Crociani et al. 2014), colon (Crociani et al. 2013)
Enzalutamide	$K_{Ca}1.1$	In vitro: breast (Khatun et al. 2018)
Glyburide	K_{ATP}, $K_v1.3$	In vitro: prostate (Abdul and Hoosein 2002a)
Iberiotoxin	$K_{Ca}1.1$	In vitro: cervical, ovarian (Han et al. 2007), glioma (Weaver et al. 2004)
Imipramine	$K_v10.1$	In vitro: ovarian (Asher et al. 2011)
Macrolide antibiotics	$K_v11.1$	In vitro and in vivo: leukaemia (Pillozzi et al. 2016)
Margatoxin	$K_v1.3$	In vitro: prostate (Fraser et al. 2003a)
Methanandamide	$K_{2P}9.1$	In vitro: ovarian (Innamaa et al. 2013)
Psora-4	$K_v1.3$	In vitro: melanoma, lymphocytes (Leanza et al. 2012, 2013b)
5-(4-phenoxybutoxy) psoralen	$K_v1.3$	In vitro: melanoma, lymphocytes (Leanza et al. 2012, 2013b)
Purpurealidin analogues	$K_v10.1$	In vitro: neuroblastoma, prostate, melanoma (Moreels et al. 2017a)
Ruthenium red	$K_{2P}9.1$	In vitro: lung (Leithner et al. 2016)
Tamoxifen	$K_v11.1$	In vitro and in vivo: breast (Luveta et al. 2020)
Tetraethylammonium	$K_{Ca}2.3$, $K_v1.3$	In vitro: breast (Potier et al. 2006), prostate (Fraser et al. 2003a), melanoma (Artym and Petty 2002), cervical, ovarian (Han et al. 2007)
TRAM-34	$K_{Ca}3.1$	In vitro: lymphoma (Wang et al. 2007), breast (Zhang et al. 2016), pancreatic (Zhang et al. 2016), glioma (Turner et al. 2014), colon (De Marchi et al. 2009), melanoma (Quast et al. 2012)
Verapamil	$K_v1.3$	In vitro: melanoma (Artym and Petty 2002), prostate (Fraser et al. 2003a)

VGKC voltage-gated K^+ channel

HERG ($K_v11.1$) is primarily associated with cardiac arrhythmias but is also upregulated in various cancers (Cherubini et al. 2000; Pillozzi et al. 2002; Lastraioli et al. 2004, 2006). As with $K_v10.1$, $K_v11.1$-mediated K^+ current has been shown to increase cancer cell proliferation (Bianchi et al. 1998; Wang et al. 2002; Arcangeli 2005). In addition, $K_v11.1$ interacts with β1 integrin, promoting adhesion interactions and adhesion-dependent signalling to regulate cancer cell survival, migration, invasion and chemoresistance (Cherubini et al. 2005; Arcangeli and Becchetti 2006; Pillozzi et al. 2007, 2011; Crociani et al. 2013). $K_v11.1$ has been studied as a potential cancer target and anticancer agents, for example, tamoxifen, have been shown to have an inhibitory effect on channel function (Thomas et al. 2003). The $K_v11.1$ inhibitor cisapride has also been shown to inhibit proliferation of gastric cancer cells (Shao et al. 2005). In addition, the $K_v11.1$ inhibitor E4031 reduced infiltration of acute lymphoblastic leukaemia cells in a mouse model, increasing survival (Pillozzi et al. 2011). E4031 and another $K_v11.1$ inhibitor (WAY123,398) also suppress gastric and colorectal cancer growth, angiogenesis (by PI3K/β1 integrin-mediated Akt activation leading to vascular endothelial growth factor (VEGF)-A transcription) and metastasis in mice (Crociani et al. 2013, 2014). E4031 was also shown to inhibit colon cancer cell proliferation, and $K_v11.1$ was identified as a biomarker of colon cancer in patient samples (Dolderer et al. 2010). E4031 and a second $K_v11.1$ inhibitor, ergtoxin, were shown to inhibit proliferation of ovarian cancer cells by inhibiting cell cycle progression, but without inducing apoptotic behaviour (Asher et al. 2011). A potential issue with the use of $K_v11.1$ blockers is the risk of off-target effects, specifically slowed cardiac repolarisation and ventricular arrhythmia (Arcangeli et al. 2009). However, this may be overcome by the use of state-dependent blockers targeting $K_v11.1$ in the open state in cancer cells whilst leaving cardiac $K_v11.1$ channels in the inactivated state unaffected (Arcangeli et al. 2009). $K_v11.1$-targeting monoclonal antibody-nanoparticle conjugates have also been explored as potential vehicles to deliver photodynamic therapies for pancreatic cancer (Sette et al. 2013), and novel recombinant anti-$K_v11.1$ single-chain fragment variable antibodies have been developed and evaluated for cancer molecular imaging (Duranti et al. 2018; Duranti and Arcangeli 2019).

The K^+ 2 pore domain (K_{2P}) channels, which contribute to setting the resting V_m, are also upregulated in a variety of cancers including breast, colon, prostate and lung tumours and have been shown to promote proliferation (Mu et al. 2003; Kim et al. 2004; Voloshyna et al. 2008). However, some members of this family appear to be downregulated in other tumour types, suggesting a complex function of K_{2P} channels in cancer progression (Williams et al. 2013). A monoclonal antibody against the extracellular domain of $K_{2P}9.1$ has been shown to inhibit tumour growth and metastasis in mice (Sun et al. 2016). Ca^{2+}-activated K^+ channels are also expressed in cancer cells (Brackenbury 2016). The large conductance $K_{Ca}1.1$ channel promotes proliferation of HeLa cervical cancer cells, and this can be inhibited by treatment with the $K_{Ca}1.1$ blocker iberiotoxin (Han et al. 2007). In addition, iberiotoxin causes cell cycle arrest and apoptosis in glioma cells (Weaver et al. 2004). The vitamin D receptor agonists calcitriol and calcipotriol and the androgen receptor antagonists bicalutamide and enzalutamide inhibit $K_{Ca}1.1$ expression in breast cancer cells, suggesting that these compounds may also elicit anti-proliferative activity via $K_{Ca}1.1$ inhibition (Khatun et al. 2016, 2018). The intermediate conductance $K_{Ca}3.1$ channel blocker TRAM-34 inhibits cell cycle progression of B lymphoma cells (Wang et al. 2007). The same study also showed that the CD20-targeting monoclonal antibody rituximab also inhibits $K_{Ca}3.1$ activity (Wang et al. 2007). Similarly, TRAM-34 inhibits proliferation and migration and promotes apoptosis of breast cancer cells (Zhang et al. 2016). TRAM-34 also inhibits $K_{Ca}3.1$-mediated glioma cell migration and invasion (Turner et al. 2014). In addition, $K_{Ca}3.1$ overexpression in breast cancer cells promotes tumour growth and metastasis (Thurber et al. 2017). However, in pancreatic cancer cells, although TRAM-34 inhibited $K_{Ca}3.1$ currents, it actually promoted migration and invasion, suggesting potential anomalous effects of this compound and/or target (Bonito et al. 2016). Inhibition of small conductance $K_{Ca}2.3$ channels with tetraethylammonium, apamin and 4-aminopyrimidine decreased breast cancer cell migration in vitro (Potier et al. 2006), and recently new lipophilic pyridine and tetrahydropyridine derivatives have been designed and synthesised which inhibit $K_{Ca}2.3$ channel activity and cellular migration (Kouba et al. 2020).

Targeting intracellular K^+ channels may also derive benefit. Mitochondrial $K_v1.3$ is widely expressed in various tissues, and a nuclear $K_v1.3$ was also identified in some breast, lung and gastric adenocarcinoma cell lines, as well as in lymphocytes and brain cells. Nuclear $K_v1.3$ functions as a regulator of gene expression by interacting with the cAMP response element-binding protein (CREB) and the c-FOS transcription factors (Jang et al. 2015). Mitochondrial $K_v1.3$ interacts with the Bcl-2 family protein, Bax, which inhibits the activity of the channel, inducing cytochrome c cytoplasmic release and subsequent apoptosis (Szabó et al. 2008, 2011). Pharmacological inhibition of intracellular $K_v1.3$ with Psora-4, clofazimine and 5-(4-Phenoxybutoxy)psoralen (PAP1) induces apoptosis in lymphocyte, fibroblast, bone and skin cancer cell lines in a Bax-/Bak-independent manner. Furthermore, the same inhibitors induce apoptosis in patient-derived leukaemia B cells, and clofazimine reduces melanoma tumour growth in vivo (Leanza et al. 2012, 2013b).

$K_v10.1$ is also expressed in the nuclear membrane of malignant brain colon and ovarian cancer cells, as well as leukaemia and fibrosarcoma (Martínez et al. 2015; Peruzzo et al. 2016). Given its location, it has been suggested that $K_v10.1$ might also impact on gene expression. However, unlike $K_v1.3$, its pro-tumorigenic function seems to occur through changes in channel conformation rather than through K^+ transport (Hegle et al. 2006; Chen et al. 2011). $K_{Ca}3.1$ and VGKCs have also been identified in mitochondria of melanoma, colon and breast cancer cells where they regulate oxidative phosphorylation and proliferation (Kovalenko et al. 2016). Combined activation of membrane and mitochondrial $K_{Ca}3.1$ is associated with breast tumour resistance to radiotherapy in vivo (Mohr et al. 2019). In addition, intracellular $K_{Ca}3.1$ is sensitive to inhibition by TRAM-34 and clotrimazole (De Marchi et al. 2009). Elevated intracellular $K_{Ca}1.1$ has been reported in the endoplasmic reticulum (ER), nucleus and Golgi of pancreatic cancer cells (Singh et al. 2012). Bax-mediated inhibition of mitochondrial $K_{Ca}1.1$ promotes apoptosis by enhancing the formation of the mitochondrial permeability transition pore (Cheng et al. 2011). The mitochondrial acid-sensing K^+ channel, TASK3, mediates survival and maintains mitochondrial integrity in melanoma cells (Kosztka et al. 2011; Nagy et al. 2014). Furthermore, inhibition of TASK3 with Zn^{2+} or methanandamide slows proliferation of ovarian cancer cells, suggesting that it might serve as a valuable target (Innamaa et al. 2013).

In summary, various classes of plasma membrane and intracellular K^+ channels are upregulated in cancer cells, and a number of studies point to pharmacological inhibition of specific subtypes as an effective approach to suppress proliferation, migration and invasion and increase apoptosis.

4 Ca^{2+} Channel Inhibitors

A number of different types of plasma membrane Ca^{2+} channel have been documented in cancer cells that could be targeted therapeutically (Table 3) (Lee et al. 2011; Prevarskaya et al. 2011; Bong and Monteith 2018; Gautier et al. 2019). Upregulation of L-type ($Ca_v1.x$) and T-type ($Ca_v3.x$) voltage-gated Ca^{2+} channels promotes differentiation, secretion of mitogenic factors, proliferation and angiogenesis (Bertolesi et al. 2002; Mariot et al. 2002; Sun et al. 2006; Gackiere et al. 2008; Lu et al. 2008). Emerging preclinical evidence suggests that repurposing Ca_v channel-inhibiting drugs to cancer may be beneficial (Buchanan and McCloskey 2016). For example, mibefradil and the $Ca_v3.x$ inhibitor NNC-55-0396 have been shown to reduce cell proliferation and induce cell apoptosis in leukaemia cell lines (Huang et al. 2015). Inhibition of $Ca_v1.3$ in endometrial carcinoma cells with nifedipine reduced proliferation and migration and induced autophagy (Bao et al. 2012). Nifedipine has also been shown to inhibit proliferation of breast cancer cells (Squecco et al. 2015). In addition, the $Ca_v3.x$ blocker KYS05090 has been shown to induce apoptosis and autophagy in lung cancer cells, although the mechanism may be channel-independent (Rim et al. 2014).

The store-operated Ca^{2+} channel proteins also play an oncogenic role (Yang et al. 2009). ORAI1 and ORAI3 heterodimerise to support Ca^{2+} influx and promote proliferation (Dubois et al. 2014). Furthermore, the store-operated Ca^{2+} channel blocker SKF96365 inhibits breast cancer metastasis in mice (Yang et al. 2009). Various transient receptor potential (TRP) channels, activated by extracellular stimuli, e.g. pH, and mechanical stimuli, are also expressed in cancer cells and can promote proliferation, survival, angiogenesis and metastasis (Thebault et al. 2006; Bidaux et al. 2007; Lehen'kyi et al. 2007; Bolanz et al. 2008; Bomben and Sontheimer 2008; Guilbert et al. 2009; Fiorio Pla et al. 2012). However, there are some exceptions, e.g. TRPM6, which is downregulated in colorectal tumours and is associated with improved survival (Xie et al. 2018) and TRPM8, which inhibits migration (Genova et al. 2017). Treatment of breast cancer cells with the TRP channel inhibitor 2-aminoethoxydiphenyl borate (2-APB) has been shown to decrease proliferation by damaging DNA (Hopkins et al. 2015). The specific TRPM7 inhibitor waixenicin A significantly decreased colon cancer cell proliferation in vitro but had no impact on aberrant crypt foci development in vivo, highlighting the importance of model selection in screening of channel inhibitors (Huang et al. 2017). In addition, given the complex involvement of various TRP channel subtypes in promoting/inhibiting cancer progression, channel inhibition will not be appropriate in certain circumstances. For example, the TRPM8 agonist WS12 suppresses endothelial cell migration and prostate cancer metastasis in mouse models (Genova et al. 2017; Grolez et al. 2019). Nonetheless, TRP channel inhibitors have been studied in the clinical setting. The TRPV6 inhibitor SOR-C13 recently went into first-in-human phase I study in patients with advanced solid tumours, and disease stabilisation in the treated cohort suggested potential antitumour activity (Fu et al. 2017).

Table 3 Ca^{2+} channel inhibitors in cancer

Compound	Cancer target	Type of cancer
2-Aminoethoxydiphenyl borate (2-APB)	TRP	In vitro: breast (Hopkins et al. 2015), glioma (Bomben and Sontheimer 2008; Bomben and Sontheimer 2010)
Bepridil	VGCC	In vitro: breast (Park et al. 2016; Nguyen et al. 2017) glioma (Kim et al. 2011) In vivo: breast (Park et al. 2016)
Cannabinoids	TRPM8 (inhibited), TRPV1 (activated)	In vitro: colon (Borrelli et al. 2014), cervical, glioma (Contassot et al. 2004a; Contassot et al. 2004b), breast (Ligresti et al. 2006), neuroblastoma (Hamtiaux et al. 2011) In vivo: colon (Borrelli et al. 2014)
Capsazepine	TRPM8	In vitro: prostate (Zhang and Barritt 2004)
Diltiazem	VGCC	In vitro: prostate (Kaddour-Djebbar et al. 2012), breast (Timar et al. 1992; Roger et al. 2004), pancreatic (Woods et al. 2015)
Felodipine	VGCC	In vitro: melanoma, breast (Honn et al. 1985) In vivo: melanoma (Honn et al. 1985)
Fendiline	VGCC	In vitro: pancreatic (Woods et al. 2015; Alhothali et al. 2019), lung, endometrial, colon (van der Hoeven et al. 2013)
Flunarizine	VGCC	In vitro: melanoma (Sezzi et al. 1985), multiple myeloma, lymphoma (Conrad et al. 2010; Schmeel et al. 2015)
Fluspirilene	VGCC	In vitro and in vivo: glioblastoma (Dong et al. 2017), hepatocellular (Shi et al. 2015)
KYS05090	VGCC	In vitro: ovarian (Jang et al. 2013) In vitro and in vivo: lung (Kang et al. 2012; Rim et al. 2014)
Mibefradil	VGCC	In vitro: leukaemia (Huang et al. 2015), breast, retinoblastoma (Bertolesi et al. 2002), colon (Dziegielewska et al. 2014), glioblastoma (Valerie et al. 2013; Zhang et al. 2017b) In vivo: glioma and glioblastoma (Holdhoff et al. 2017; Zhang et al. 2017b)
Monoclonal antibody	nfP2X7	In vivo: basal cell carcinoma (clinical trials) (Gilbert et al. 2017, 2019)
Nifedipine	VGCC	In vitro: breast (Timar et al. 1992; Roger et al. 2004; Squecco et al. 2015), melanoma (Honn et al. 1985), pancreatic (Woods et al. 2015) endometrial (Bao et al. 2012) In vivo: melanoma (Honn et al. 1985), colon (Yang and Friedlander 2001)
Nimodipine	VGCC	In vitro: melanoma, breast (Honn et al. 1984, 1985)

(continued)

Table 3 (continued)

Compound	Cancer target	Type of cancer
		In vivo: melanoma (Honn et al. 1984, 1985)
NNC-55-0396	VGCC	In vitro: leukaemia (Huang et al. 2015)
20-O-β-D-glucopyranosyl-20 (S)-protopanaxadiol, 4-chloro-m-cresol	Ryanodine receptor	In vitro: lung (Shin et al. 2018) In vivo: breast (Abdul et al. 2008)
Pimozide	VGCC	In vitro: breast, retinoblastoma (Bertolesi et al. 2002)
SKF96365 and MRS-1845	ORAI, TRPC1	In vitro: glioma (Bomben and Sontheimer 2008, 2010) In vivo: breast (Yang et al. 2009)
SOR-C13, SOR-C27	TRPV6	In vivo: solid tumours of epithelial origin (phase I clinical trial) (Fu et al. 2017), ovarian (Xue et al. 2018), prostate (Bowen et al. 2013)
Verapamil	VGCC	In vitro: pancreatic (Sato et al. 1994; Zhao et al. 2016), breast (Timar et al. 1992; Roger et al. 2004; Berzingi et al. 2016) In vivo: pancreatic (Sato et al. 1994; Zhao et al. 2016)
Waixenicin A	TRPM7	In vitro and in vivo: colon (Huang et al. 2017)

TRP transient receptor potential, *VGCC* voltage-gated Ca^{2+} channel

The purinergic P2X7 cation channel has also gained interest in the context of cancer, although conflicting results from different studies have been challenging to interpret (Roger et al. 2015). However, a monoclonal antibody targeting a unique epitope on the cancer-specific variant of P2X7 (nfP2X7) has undergone phase 1 clinical trial for basal cell carcinoma with promising results including disease stabilisation and partial and complete response (Gilbert et al. 2017, 2019).

In colon cancer cells, the cannabinoid cannabigerol suppresses proliferation and promotes reactive oxygen species (ROS) production and apoptosis via TRPM8 inhibition and slows tumour growth in vivo (Borrelli et al. 2014). In addition, upregulation of TRPV1 has been identified as a key player in cannabinoid derivative-induced apoptosis of cervical cancer and glioma cells (Contassot et al. 2004a, b). Other studies, however, propose different mechanisms for the anticancer activity of cannabinoids (Hamtiaux et al. 2011), for example, by interacting with the cannabinoid receptor 2 in addition to TRPV1 activation (Ligresti et al. 2006).

Intracellular Ca^{2+} channels may also present potential targets. For example, TRPM8 and TRPC1 both play a role in survival and proliferation of tumour cells (Zhang and Barritt 2004; Shapovalov et al. 2016). TRPM8 inhibition with capsazepine reduces survival of prostate cancer cells, and TRPM8 knockdown slows proliferation of osteosarcoma cells by interfering with Ca^{2+}-dependent Akt function (Zhang and Barritt 2004; Wang et al. 2013). TRPC1 regulates glioma cell division, and its inhibition with 2-APB, MRS-1845 and SKF96365 inhibits proliferation in vitro and reduces tumour size in mouse models (Bomben and Sontheimer 2010). Ryanodine receptors promote breast cancer cell survival, and their expression correlates with tumour grade; in addition, the ryanodine receptor inhibitor 4-chloro-m-cresol inhibits breast cancer cell proliferation in vitro (Abdul et al. 2008). Furthermore, treatment of lung cancer cells with the ryanodine receptor inhibitor 20-O-β-D-glucopyranosyl-20(S)-protopanaxadiol induces Ca^{2+}-dependent apoptosis, supporting the essential role of these channels in cancer cell survival (Shin et al. 2018).

In summary, a number of Ca^{2+} channel inhibitors have shown promise in preclinical studies, and some of these have now reached clinical trials. Several epidemiological studies show that existing Ca^{2+} channel blockers are not associated with increased cancer risk (Grimaldi-Bensouda et al. 2016; Wilson et al. 2016; Brasky et al. 2017), supporting a compelling argument for further exploration of the possibility of repurposing such drugs to treat cancer (Buchanan and McCloskey 2016). However, the complex opposing roles of some Ca^{2+} channels in cancer cells, e.g. certain TRP and P2X7 channels, highlight the importance of fully understanding their diverse physiological roles in order to permit appropriate targeting.

5 Cl^- Channel Inhibitors

Several Cl^- channels have been shown to be aberrantly expressed in cancer cells, contributing to survival and progression, and some have been explored as therapeutic targets (Table 4). The ionotropic Cl^--permeant $GABA_A$ receptor is upregulated on metastatic breast cancer cells in the brain (Neman et al. 2014), which themselves promote altered regional excitability (Simon et al. 2020). The voltage-gated Cl^- channels CLC-2 and CLC-3 are functionally active in glioma cells, and the latter is essential for facilitating mitosis and invasion by regulating cell volume (Olsen et al. 2003; Habela et al. 2008; Lui et al. 2010; Watkins and Sontheimer 2011). CLC-3 also stimulates breast cancer cell proliferation and tumour growth (Zhou et al. 2018) and promotes migration of nasopharyngeal carcinoma cells (Mao et al. 2008). However, other studies have indicated that CLC-3 can also promote apoptosis (Liu et al. 2013); thus cancer-promoting or cancer-inhibiting activity of this channel is likely finely tuned and may be context-dependent (Hong et al. 2015). CLC-3 is sensitive to non-specific Cl^- channel inhibitors such as tamoxifen and 4-5-nitro-2-(3-phenylpropylamino) benzoic acids (NPPBs) (Wang et al. 2012), inhibiting cancer cell proliferation (Shen et al. 2000). Similarly, tamoxifen was shown to only have an inhibitory effect on cancer cell migration in the presence of CLC-3, likely as a result of dysregulated cell volume management and therefore cell cycle stagnation (Mao et al. 2013).

Hydrolysis products of 4,4-diisothiocyanatostilbene-2,2′-disulfonic acid (DIDS), which are inhibitors of anion permeability, can inhibit CLC family channels, as can phenylalanine derivatives (stilbenes), clofibric acids, benzofurans, and the newer

Table 4 Cl^- channel inhibitors in cancer

Compound	Cancer target	Type of cancer
Ani9 and derivatives	ANO1	In vitro: prostate (Song et al. 2018), pancreatic, breast (Seo et al. 2018)
Bumetanide	NKCC1	In vitro & in vivo: glioma (Haas and Sontheimer 2010), colon (Malamas et al. 2015)
CaCCinh-A01	ANO1	In vitro: prostate (Song et al. 2018), colon, lung (Guan et al. 2016), breast (Britschgi et al. 2013), oesophageal and pharyngeal squamous carcinoma (Bill et al. 2014), pancreatic (Sauter et al. 2015)
Chlorotoxin	CLC-3	In vitro and in vivo (clinical trials): glioma (Deshane et al. 2003; Mamelak et al. 2006)
DIDS	Acid-induced Cl^- channels	Nasopharyngeal (Wang et al. 2012)
Digallic acid and tannic acid	ANO1	In vitro: lymphoblastoma (Bhouri et al. 2012), oesophageal and pharyngeal squamous carcinoma (Bill et al. 2014), gingival (Darvin et al. 2015), breast (Nie et al. 2016), prostate (Karakurt and Adali 2016)
Idebenone	ANO1	In vitro: pancreatic, prostate (Seo et al. 2015)
Luteolin	ANO1	Prostate (Seo et al. 2017)
NPPBs	Acid-induced Cl^- channels	In vitro: cervical (Shen et al. 2000), nasopharyngeal (Wang et al. 2012)
Tamoxifen	Acid-induced Cl^- channels (CLC-3)	In vitro: hepatocellular (Mao et al. 2013), cervical (Shen et al. 2000), nasopharyngeal (Wang et al. 2012) In vivo: breast (Luveta et al. 2020)
T16Ainh-A01	ANO1	In vitro: pancreatic (Mazzone et al. 2012), prostate (Song et al. 2018), colon, lung (Guan et al. 2016), oesophageal and pharyngeal squamous (Bill et al. 2014), pancreatic (Sauter et al. 2015)

benzimidazole derivative BIM1 (Matulef et al. 2008; Koster et al. 2018). However, the broad effect of such compounds on other Cl^- channels remains a challenge with respect to potential off-target effects (Hong et al. 2015), and applicability in cancer treatment needs to be confirmed through further studies. Specific function-blocking antibodies targeting CLC-3 have been developed (Wang et al. 2003), but their efficacy in cancer models remains to be determined. Chlorotoxin, a scorpion toxin identified as a CLC-3 inhibitor, binds to a membrane-bound matrix metalloproteinase on glioma cells (Deshane et al. 2003). Radiolabelled I^{131}-chlorotoxin has undergone a phase 1 clinical trial in adult patients with high-grade glioma with the aim of improving targeted radiation to the tumour site and demonstrated good tolerability and potential antitumoural effects (Mamelak et al. 2006).

The anoctamin Ca^{2+}-dependent Cl^- channels promote cancer cell proliferation and apoptosis of cancer cells under certain conditions (Kunzelmann et al. 2019). Inhibition of ANO1/TMEM16 with the specific inhibitor CaCCinh-A01 significantly decreased tumour progression, raising the possibility that this channel may be a potential therapeutic target (Britschgi et al. 2013). An important regulator of the Cl^- concentration within developing glial cells is the $Na^+/K^+/2Cl^-$ cotransporter

(NKCC1). NKCC1 has been proposed as a key promoter of glioma cell migration, being localised at the tip of the migratory pole of the cell, and also influences cell-cell adhesion and Cl^--dependent cell volume regulation (Habela et al. 2009; Garzon-Muvdi et al. 2012). Treatment with the NKCC1 inhibitor bumetanide inhibits migration of metastatic glioma cells both in vivo and in vitro, suggesting that NKCC1 could be a promising target in the treatment of glioma (Haas and Sontheimer 2010).

In lung cancer cells, the Cl^- intracellular channel 1 (CLIC1), which can be found both at the plasma membrane and in the cytosol, suppresses Ca^{2+} import via L-type Ca^{2+} channels, promoting survival (Lee et al. 2019b). In silico analysis showed a much higher risk of death in breast, pancreatic and liver cancer patients with high CLIC1 expression, whilst gastric cancer patients with high CLIC1 levels have a survival advantage, suggesting that its function might vary depending on tumour type (Gururaja Rao et al. 2020). Further work is required to establish the therapeutic value of CLIC1 inhibition and/or inhibition of other intracellular Cl^- channel subtypes, e.g. CLIC4 (Fernández-Salas et al. 2002; Zhong et al. 2012).

In summary, both plasma membrane and intracellular Cl^- channels are important regulators of cell cycle, proliferation and migration, making them promising targets for cancer therapies. Although some inhibitory molecules have been found effective in reducing tumour cell growth and migration, there is a strong potential for developing more specific inhibitors, targeted at both intracellular and extracellular channels.

6 Combinatorial Treatments

The fact that a number of ion channel-targeting drugs inhibit cellular functions, including proliferation, migration and invasion, and that others promote apoptosis raises the possibility that such compounds may have utility in combination with standard-of-care therapies, e.g. chemotherapy. Furthermore, perturbation of the ionic balance within tumour cells may provide favourable conditions for the intracellular partitioning of certain cytotoxic drugs, enhancing their effectiveness. For example, in triple-negative breast cancer cells, β-adrenergic receptors and $Na_v1.5$ colocalise; the β-adrenergic receptor competitive antagonist propranolol and the VGSC inhibitor ranolazine decrease Na^+ currents, migration and invasion both when administered individually and in combination (Lee et al. 2019a). Downregulation of $K_v10.1$ with shRNA or application of the $K_v10.1$ inhibitor astemizole to glioblastoma cells sensitises them to treatment with the standard-of-care chemotherapeutic temozolomide (Sales et al. 2016). Combination of astemizole with gefitinib has been shown to synergistically increase apoptosis of lung cancer cells over treatment with either agent alone (Chavez-Lopez et al. 2017). Another example is the macrolide antibiotics, which have antileukemic activity alone and in combination with chemotherapeutic drugs, and this was shown to be due to $K_v11.1$ inhibition (Pillozzi et al. 2016).

Riluzole has been shown to inhibit $K_v11.1$ and activate $K_{Ca}3.1$ in colon cancer cells, thus contributing to cisplatin uptake (Pillozzi et al. 2018). Combined administration of the $K_{Ca}3.1$ activator SKA-31 and E4031 had a similar effect, which was reproducible in mouse models, suggesting a complex interplay between $K_{Ca}3.1$ and $K_v11.1$ (Pillozzi et al. 2018). Riluzole has also been shown to activate the K_{2P} channel (TREK-1), reducing neuropathic pain and depression-like symptoms induced by treatment with oxaliplatin in colon cancer mouse models (Poupon et al. 2018). Another potentially interesting combinatorial treatment is represented by the $K_{2P}3.1$ and $K_{2P}9.1$ channel inhibitors anandamide and ruthenium red, which have been shown to additively inhibit K^+ currents in lung cancer cells, although the effect of these compounds on cell proliferation was not determined (Leithner et al. 2016).

Inhibitors of Ca^{2+} channels have also been investigated for combinatorial therapies. The $Ca_v3.x$ antagonist mibefradil has been shown to inhibit glioblastoma stem-like cell proliferation in vitro and tumour growth in a glioblastoma mouse model and sensitises tumours to treatment with temozolomide (Zhang et al. 2017b). Pharmacological inhibition of $Ca_v3.x$ channel activity with the antagonists mibefradil and pimozide also synergistically suppressed proliferation in several cancer cell lines (Bertolesi et al. 2002). Inhibition of active ion transport may also be beneficial in combinatorial treatments. For example, suppression of plasma membrane Ca^{2+} ATPase isoform 2 (PMCA2) expression was shown to inhibit proliferation of breast cancer cells on its own, as well as enhancing the cytotoxicity of doxorubicin (Peters et al. 2016). Store-operated Ca^{2+} entry induces expression of the chemotherapy resistance marker MDR1 in breast cancer cells (Babaer et al. 2018). Knockdown of ORAI1 or STIM1 thus significantly increases sensitivity to chemotherapeutic drugs including cisplatin, gentamycin and 5-fluorouracyl (Kondratska et al. 2014; Sun et al. 2017; Kischel et al. 2019). Similarly, inhibition of TRPC5, either by siRNA or by treatment with chloroquine or 3-methyladenine, increases sensitivity to doxorubicin in breast cancer cell lines (Zhang et al. 2017a). On the other hand, different studies suggest certain Ca^{2+} channels render cells more responsive to chemotherapy (Kischel et al. 2019). For example, TRPC1 expression is downregulated in drug-resistant ovarian cancer tissues compared with drug-responsive samples, and cisplatin- and carboplatin-resistant ovarian cancer cell lines were shown to also have lower levels of TRPC1 (Liu et al. 2016). TRPV2 activation with cannabidiol plays an important role in sensitising glioma cells to doxorubicin, carmustine and temozolomide (Nabissi et al. 2012). Thus, combination of certain channel-modulating drugs and chemotherapeutic drugs may have value by reducing tumour chemotherapeutic resistance, but the situation is likely channel or cell-type-dependent.

Ion channel inhibition may also be advantageous in the context of standard-of-care radiotherapy. For example, antiepileptic drug use is associated with improved overall survival of breast cancer patients with brain metastasis receiving whole brain radiotherapy (Reddy et al. 2015), raising the possibility that VGSC inhibition may radiosensitise brain metastases. In addition, TRPM8 inhibition has been shown to radiosensitise glioblastoma cells and attenuate DNA repair (Klumpp et al. 2017), and

TRPM2 inhibition enhances radiotherapy-induced cell death in leukaemia cells (Klumpp et al. 2016).

By targeting specific ion channels, certain inhibitors may enhance the capacity of the immune system to fight tumours. For example, non-small cell lung cancer patients with low serum salt levels respond poorly to immune checkpoint inhibitor therapy, illustrating potential interconnection between ionic balance and immune system-mediated tumour clearance (Fuca et al. 2018). Another study showed that an increase in extracellular K^+ caused by tumour cell necrosis has an immunosuppressive impact on effector T cells by increasing intracellular K^+. Upregulation of $K_V1.3$ in T cells resulted in K^+ export, counteracting the immunosuppressive action of the tumour-derived K^+ (Eil et al. 2016). Furthermore, high K^+ in the tumour microenvironment maintained T cells in a stem-like state capable of dividing and enhancing tumour destruction (Vodnala et al. 2019). These data suggest manipulation of K^+ flux may be effective in enhancing immunotherapeutic approaches.

In summary, considerable research has been carried out towards combining current cancer therapies and ion channel inhibitors or developing new combinatorial treatments that integrate ionic targeting. Despite significant progress, there is yet much that needs to be done to optimise and refine existing therapies, as well as to generate new and effective strategies for exploiting ionic disbalances in the tumour microenvironment.

7 Conclusions and Future Perspectives

The study of ion channel inhibitors in the context of oncology is gaining interest with time, particularly given the limitation of chemotherapy and targeted therapies and the need for new perspectives on counteracting tumour progression and metastasis. Whilst individual ion channel targeting may be effective on its own in certain circumstances, a combinatorial approach of ion channel-targeting drugs and chemotherapy, radiotherapy and/or emerging immunotherapies may derive greater benefits. However, a key obstacle remains in terms of tumour specificity, given that many of these channels are also expressed in normal cells. Therefore, the use of ion channel blockers can often be accompanied by severe side effects and might even be lethal (Vandenberg et al. 2012).

Engineering antibodies or small molecules that target tumour-specific isoforms/states of various ion channels has been a step forward towards increasing the sensitivity and specificity of ion channel-targeted therapies in cancer (Clare et al. 2000; Chioni et al. 2005; Hartung et al. 2011; Sette et al. 2013; Sun et al. 2016; Gilbert et al. 2017). Yet, the continuous dynamics of the tumour environment could limit the efficacy of these approaches through target mutations. Furthermore, antibody therapies are limited both by the size-dependent tissue penetration and by the manufacturing procedure. Nevertheless, the idea of specifically targeting tumour-associated ion channels is worth investigating for the future. The capacity to distinguish between malignant and healthy ion channels could enable more complex therapeutic approaches such as combining ion channel inhibitors that could suppress tumour growth and ion channel enhancers that would induce activation and proliferation of immune cells, enabling those to clear the malignant tissue (Chiang et al. 2017). However, before such complex strategies can be designed, a more complete understanding of tumour-specific ion channel expression, function and pharmacology is required.

In conclusion, given the strong links between ion channel function and regulation of tumour growth, metastasis and chemotherapy resistance, it is likely that further work in this area will facilitate the development of new, multilateral therapeutic approaches.

Acknowledgements This work was supported by the BBSRC (BB/M011151/1).

References

Abdul M, Hoosein N (2001) Inhibition by anticonvulsants of prostate-specific antigen and interleukin-6 secretion by human prostate cancer cells. Anticancer Res 21:2045–2048

Abdul M, Hoosein N (2002a) Expression and activity of potassium ion channels in human prostate cancer. Cancer Lett 186:99–105

Abdul M, Hoosein N (2002b) Voltage-gated sodium ion channels in prostate cancer: expression and activity. Anticancer Res 22:1727–1730

Abdul M, Santo A, Hoosein N (2003) Activity of potassium channel-blockers in breast cancer. Anticancer Res 23:3347–3351

Abdul M, Ramlal S, Hoosein N (2008) Ryanodine receptor expression correlates with tumor grade in breast cancer. Pathol Oncol Res 14:157–160

Akamatsu K, Shibata MA, Ito Y, Sohma Y, Azuma H, Otsuki Y (2009) Riluzole induces apoptotic cell death in human prostate cancer cells via endoplasmic reticulum stress. Anticancer Res 29:2195–2204

Alhothali M, Mathew M, Iyer G, Lawrence HR, Yang S, Chellappan S, Padmanabhan J (2019) Fendiline enhances the cytotoxic effects of therapeutic agents on PDAC cells by inhibiting tumor-promoting signaling events: a potential strategy to combat PDAC. Int J Mol Sci 20:2423

Anderson JD, Hansen TP, Lenkowski PW, Walls AM, Choudhury IM, Schenck HA, Friehling M, Holl GM, Patel MK, Sikes RA, Brown ML (2003) Voltage-gated sodium channel blockers as cytostatic inhibitors of the androgen-independent prostate cancer cell line PC-3. Mol Cancer Ther 2:1149–1154

Angelucci A, Valentini A, Millimaggi D, Gravina GL, Miano R, Dolo V, Vicentini C, Bologna M, Federici G, Bernardini S (2006) Valproic acid induces apoptosis in prostate carcinoma cell lines by activation of multiple death pathways. Anticancer Drugs 17:1141–1150

Arcangeli A (2005) Expression and role of hERG channels in cancer cells. Novartis Found Symp 266:225–232. discussion 232-4

Arcangeli A, Becchetti A (2006) Complex functional interaction between integrin receptors and ion channels. Trends Cell Biol 16:631–639

Arcangeli A, Crociani O, Lastraioli E, Masi A, Pillozzi S, Becchetti A (2009) Targeting ion channels in cancer: a novel frontier in antineoplastic therapy. Curr Med Chem 16:66–93

Arimochi H, Morita K (2006) Characterization of cytotoxic actions of tricyclic antidepressants on human HT29 colon carcinoma cells. Eur J Pharmacol 541:17–23

Arimochi H, Morita K (2008) Desipramine induces apoptotic cell death through nonmitochondrial and mitochondrial pathways in different types of human colon carcinoma cells. Pharmacology 81:164–172

Artym VV, Petty HR (2002) Molecular proximity of Kv1.3 voltage-gated potassium channels and β 1-integrins on the plasma membrane of melanoma cells: effects of cell adherence and channel blockers. J Gen Physiol 120:29–37

Asher V, Warren A, Shaw R, Sowter H, Bali A, Khan R (2011) The role of Eag and HERG channels in cell proliferation and apoptotic cell death in SK-OV-3 ovarian cancer cell line. Cancer Cell Int 11:6

Babaer D, Amara S, Ivy M, Zhao Y, Lammers PE, Titze JM, Tiriveedhi V (2018) High salt induces P-glycoprotein mediated treatment resistance in breast cancer cells through store operated calcium influx. Oncotarget 9:25193–25205

Bao XX, Xie BS, Li Q, Li XP, Wei LH, Wang JL (2012) Nifedipine induced autophagy through Beclin1 and mTOR pathway in endometrial carcinoma cells. Chin Med J (Engl) 125:3120–3126

Baptista-Hon DT, Robertson FM, Robertson GB, Owen SJ, Rogers GW, Lydon EL, Lee NH, Hales TG (2014) Potent inhibition by ropivacaine of metastatic colon cancer SW620 cell invasion and NaV1.5 channel function. Br J Anaesth 113(Suppl 1):i39–i48

Batcioglu K, Uyumlu AB, Satilmis B, Yildirim B, Yucel N, Demirtas H, Onkal R, Guzel RM, Djamgoz MB (2012) Oxidative stress in the in vivo DMBA rat model of breast cancer: suppression by a voltage-gated sodium channel inhibitor (RS100642). Basic Clin Pharmacol Toxicol 111:137–141

Baudino TA (2015) Targeted cancer therapy: the next generation of cancer treatment. Curr Drug Discov Technol 12:3–20

Becchetti A (2011) Ion channels and transporters in cancer. 1. Ion channels and cell proliferation in cancer. Am J Physiol Cell Physiol 301:C255–C265

Benavides-Serrato A, Saunders JT, Holmes B, Nishimura RN, Lichtenstein A, Gera J (2020) Repurposing potential of riluzole as an ITAF inhibitor in mTOR therapy resistant glioblastoma. Int J Mol Sci 21:344

Bernal-Ramos G, Hernandez-Gallegos E, Vera E, Chavez-Lopez MG, Zuniga-Garcia V, Sanchez-Perez Y, Garrido E, Camacho J (2017) Astemizole inhibits cell proliferation in human prostate tumorigenic cells expressing ether a-go-go-1 potassium channels. Cell Mol Biol (Noisy-le-Grand) 63:11–13

Bertolesi GE, Shi C, Elbaum L, Jollimore C, Rozenberg G, Barnes S, Kelly ME (2002) The Ca(2+) channel antagonists mibefradil and pimozide inhibit cell growth via different cytotoxic mechanisms. Mol Pharmacol 62:210–219

Berzingi S, Newman M, Yu H-G (2016) Altering bioelectricity on inhibition of human breast cancer cells. Cancer Cell Int 16:72–72

Besson P, Driffort V, Bon E, Gradek F, Chevalier S, Roger S (2015) How do voltage-gated sodium channels enhance migration and invasiveness in cancer cells? Biochim Biophys Acta 1848:2493–2501

Bhouri W, Boubaker J, Skandrani I, Ghedira K, Chekir Ghedira L (2012) Investigation of the apoptotic way induced by digallic acid in human lymphoblastoid TK6 cells. Cancer Cell Int 12:26–26

Bianchi L, Wible B, Arcangeli A, Taglialatela M, Morra F, Castaldo P, Crociani O, Rosati B, Faravelli L, Olivotto M, Wanke E (1998) Herg encodes a K+ current highly conserved in tumors of different histogenesis: a selective advantage for cancer cells? Cancer Res 58:815–822

Biber A, Durusu İZ, Özen C (2018) In vitro anticancer effect of tricyclic antidepressant nortriptyline on multiple myeloma. Turk J Biol 42:414–421

Bidaux G, Flourakis M, Thebault S, Zholos A, Beck B, Gkika D, Roudbaraki M, Bonnal JL, Mauroy B, Shuba Y, Skryma R, Prevarskaya N (2007) Prostate cell differentiation status determines transient receptor potential melastatin member 8 channel subcellular localization and function. J Clin Invest 117:1647–1657

Biki B, Mascha E, Moriarty DC, Fitzpatrick JM, Sessler DI, Buggy DJ (2008) Anesthetic technique for radical prostatectomy surgery affects cancer recurrence: a retrospective analysis. Anesthesiology 109:180–187

Bill A, Hall ML, Borawski J, Hodgson C, Jenkins J, Piechon P, Popa O, Rothwell C, Tranter P, Tria S, Wagner T, Whitehead L, Gaither LA (2014) Small molecule-facilitated degradation of ANO1 protein: a new targeting approach for anticancer therapeutics. J Biol Chem 289:11029–11041

Blackiston DJ, Mclaughlin KA, Levin M (2009) Bioelectric controls of cell proliferation: ion channels, membrane voltage and the cell cycle. Cell Cycle 8:3519–3528

Bolanz KA, Hediger MA, Landowski CP (2008) The role of TRPV6 in breast carcinogenesis. Mol Cancer Ther 7:271–279

Bomben VC, Sontheimer HW (2008) Inhibition of transient receptor potential canonical channels impairs cytokinesis in human malignant gliomas. Cell Prolif 41:98–121

Bomben VC, Sontheimer H (2010) Disruption of transient receptor potential canonical channel 1 causes incomplete cytokinesis and slows the growth of human malignant gliomas. Glia 58:1145–1156

Bondarava M, Li T, Endl E, Wehner F (2009) alpha-ENaC is a functional element of the hypertonicity-induced cation channel in HepG2 cells and it mediates proliferation. Pflugers Arch 458:675–687

Bong AHL, Monteith GR (2018) Calcium signaling and the therapeutic targeting of cancer cells. Biochim Biophys Acta Mol Cell Res 1865:1786–1794

Bonito B, Sauter DR, Schwab A, Djamgoz MB, Novak I (2016) KCa3.1 (IK) modulates pancreatic cancer cell migration, invasion and proliferation: anomalous effects on TRAM-34. Pflugers Arch 468:1865–1875

Borrelli F, Pagano E, Romano B, Panzera S, Maiello F, Coppola D, De Petrocellis L, Buono L, Orlando P, Izzo AA (2014) Colon carcinogenesis is inhibited by the TRPM8 antagonist cannabigerol, a Cannabis-derived non-psychotropic cannabinoid. Carcinogenesis 35:2787–2797

Bowen CV, Debay D, Ewart HS, Gallant P, Gormley S, Ilenchuk TT, Iqbal U, Lutes T, Martina M, Mealing G, Merkley N, Sperker S, Moreno MJ, Rice C, Syvitski RT, Stewart JM (2013) In vivo detection of human TRPV6-rich tumors with anti-cancer peptides derived from soricidin. PLoS One 8:e58866

Brackenbury WJ (2016) Ion channels in cancer. In: Pitt GS (ed) Ion channels in health and disease. Elsevier Inc., Amsterdam

Brackenbury WJ, Chioni AM, Diss JK, Djamgoz MB (2007) The neonatal splice variant of Nav1.5 potentiates in vitro invasive behaviour of MDA-MB-231 human breast cancer cells. Breast Cancer Res Treat 101:149–160

Brasky TM, Krok-Schoen JL, Liu J, Chlebowski RT, Freudenheim JL, Lavasani S, Margolis KL, Qi L, Reding KW, Shields PG, Simon MS, Wactawski-Wende J, Wang A, Womack C, Manson JE (2017) Use of calcium channel blockers and breast cancer risk in the women's health initiative. Cancer Epidemiol Biomarkers Prev 26:1345–1348

Brisson L, Gillet L, Calaghan S, Besson P, Le Guennec JY, Roger S, Gore J (2011) Na(V)1.5 enhances breast cancer cell invasiveness by increasing NHE1-dependent H(+) efflux in caveolae. Oncogene 30:2070–2076

Britschgi A, Bill A, Brinkhaus H, Rothwell C, Clay I, Duss S, Rebhan M, Raman P, Guy CT, Wetzel K, George E, Popa MO, Lilley S, Choudhury H, Gosling M, Wang L, Fitzgerald S, Borawski J, Baffoe J, Labow M, Gaither LA, Bentires-Alj M (2013) Calcium-activated chloride channel ANO1 promotes breast cancer progression by activating EGFR and CAMK signaling. Proc Natl Acad Sci U S A 110:E1026–E1034

Brockschmidt C, Hirner H, Huber N, Eismann T, Hillenbrand A, Giamas G, Radunsky B, Ammerpohl O, Bohm B, Henne-Bruns D, Kalthoff H, Leithauser F, Trauzold A, Knippschild U (2008) Anti-apoptotic and growth-stimulatory functions of CK1 delta and epsilon in ductal adenocarcinoma of the pancreas are inhibited by IC261 in vitro and in vivo. Gut 57:799–806

Buchanan PJ, Mccloskey KD (2016) CaV channels and cancer: canonical functions indicate benefits of repurposed drugs as cancer therapeutics. Eur Biophys J 45:621–633

Bugan I, Kucuk S, Karagoz Z, Fraser SP, Kaya H, Dodson A, Foster CS, Altun S, Djamgoz MBA (2019) Anti-metastatic effect of ranolazine in an in vivo rat model of prostate cancer, and expression of voltage-gated sodium channel protein in human prostate. Prostate Cancer Prostatic Dis 22:569–579

Campbell TM, Main MJ, Fitzgerald EM (2013) Functional expression of the voltage-gated Na(+)-channel Nav1.7 is necessary for EGF-mediated invasion in human non-small cell lung cancer cells. J Cell Sci 126:4939–4949

Campia I, Sala V, Kopecka J, Leo C, Mitro N, Costamagna C, Caruso D, Pescarmona G, Crepaldi T, Ghigo D, Bosia A, Riganti C (2012) Digoxin and ouabain induce the efflux of cholesterol via liver X receptor signalling and the synthesis of ATP in cardiomyocytes. Biochem J 447:301–311

Chang YC, Liu CL, Chen MJ, Hsu YW, Chen SN, Lin CH, Chen CM, Yang FM, Hu MC (2014) Local anesthetics induce apoptosis in human breast tumor cells. Anesth Analg 118:116–124

Chang H-T, Chou C-T, Yu C-C, Tsai J-Y, Sun T-K, Liang W-Z, Lin K-L, Tseng H-W, Kuo C-C, Chen F-A, Kuo D-H, Pan C-C, Ho C-M, Shieh P, Jan C-R (2015) The mechanism of protriptyline-induced Ca2+ movement and non-Ca2+-triggered cell death in PC3 human prostate cancer cells. J Recept Signal Transduct Res 35:429–434

Chang Y-L, Liu S-T, Wang Y-W, Lin W-S, Huang S-M (2018) Amiodarone promotes cancer cell death through elevated truncated SRSF3 and downregulation of miR-224. Oncotarget 9:13390–13406

Chavez-Lopez MG, Zuniga-Garcia V, Hernandez-Gallegos E, Vera E, Chasiquiza-Anchatuna CA, Viteri-Yanez M, Sanchez-Ramos J, Garrido E, Camacho J (2017) The combination astemizole-gefitinib as a potential therapy for human lung cancer. Onco Targets Ther 10:5795–5803

Chen Y, Sánchez A, Rubio ME, Kohl T, Pardo LA, Stühmer W (2011) Functional K(v)10.1 channels localize to the inner nuclear membrane. PLoS One 6:e19257

Chen D, Song M, Mohamad O, Yu SP (2014) Inhibition of Na+/K+-ATPase induces hybrid cell death and enhanced sensitivity to chemotherapy in human glioblastoma cells. BMC Cancer 14:716

Chen B, Zhang C, Wang Z, Chen Y, Xie H, Li S, Liu X, Liu Z, Chen P (2019) Mechanistic insights into Nav1.7-dependent regulation of rat prostate cancer cell invasiveness revealed by toxin probes and proteomic analysis. FEBS J 286:2549–2561

Cheng Y, Gulbins E, Siemen D (2011) Activation of the permeability transition pore by Bax via inhibition of the mitochondrial BK channel. Cell Physiol Biochem 27:191–200

Cherubini A, Taddei GL, Crociani O, Paglierani M, Buccoliero AM, Fontana L, Noci I, Borri P, Borrani E, Giachi M, Becchetti A, Rosati B, Wanke E, Olivotto M, Arcangeli A (2000) HERG potassium channels are more frequently expressed in human endometrial cancer as compared to non-cancerous endometrium. Br J Cancer 83:1722–1729

Cherubini A, Hofmann G, Pillozzi S, Guasti L, Crociani O, Cilia E, Di Stefano P, Degani S, Balzi M, Olivotto M, Wanke E, Becchetti A, Defilippi P, Wymore R, Arcangeli A (2005) Human ether-a-go-go-related gene 1 channels are physically linked to beta1 integrins and modulate adhesion-dependent signaling. Mol Biol Cell 16:2972–2983

Chiang EY, Li T, Jeet S, Peng I, Zhang J, Lee WP, Devoss J, Caplazi P, Chen J, Warming S, Hackos DH, Mukund S, Koth CM, Grogan JL (2017) Potassium channels Kv1.3 and KCa3.1 cooperatively and compensatorily regulate antigen-specific memory T cell functions. Nat Commun 8:14644

Chioni AM, Fraser SP, Pani F, Foran P, Wilkin GP, Diss JK, Djamgoz MB (2005) A novel polyclonal antibody specific for the Na(v)1.5 voltage-gated Na(+) channel 'neonatal' splice form. J Neurosci Methods 147:88–98

Chittajallu R, Chen Y, Wang H, Yuan X, Ghiani CA, Heckman T, Mcbain CJ, Gallo V (2002) Regulation of Kv1 subunit expression in oligodendrocyte progenitor cells and their role in G1/S phase progression of the cell cycle. Proc Natl Acad Sci U S A 99:2350–2355

Clare JJ, Tate SN, Nobbs M, Romanos MA (2000) Voltage-gated sodium channels as therapeutic targets. Drug Discov Today 5:506–520

Conrad DM, Furlong SJ, Doucette CD, West KA, Hoskin DW (2010) The Ca(2+) channel blocker flunarizine induces caspase-10-dependent apoptosis in Jurkat T-leukemia cells. Apoptosis 15:597–607

Contassot E, Tenan M, Schnüriger V, Pelte MF, Dietrich PY (2004a) Arachidonyl ethanolamide induces apoptosis of uterine cervix cancer cells via aberrantly expressed vanilloid receptor-1. Gynecol Oncol 93:182–188

Contassot E, Wilmotte R, Tenan M, Belkouch M-C, Schnüriger V, De Tribolet N, Bourkhardt K, Dietrich P-Y (2004b) Arachidonylethanolamide induces apoptosis of human glioma cells through vanilloid receptor-1. J Neuropathol Exp Neurol 63:956–963

Crociani O, Zanieri F, Pillozzi S, Lastraioli E, Stefanini M, Fiore A, Fortunato A, D'amico M, Masselli M, De Lorenzo E, Gasparoli L, Chiu M, Bussolati O, Becchetti A, Arcangeli A (2013) hERG1 channels modulate integrin signaling to trigger angiogenesis and tumor progression in colorectal cancer. Sci Rep 3:3308

Crociani O, Lastraioli E, Boni L, Pillozzi S, Romoli MR, D'amico M, Stefanini M, Crescioli S, Masi A, Taddei A, Bencini L, Bernini M, Farsi M, Beghelli S, Scarpa A, Messerini L, Tomezzoli A, Vindigni C, Morgagni P, Saragoni L, Giommoni E, Gasperoni S, Di Costanzo F, Roviello F, De Manzoni G, Bechi P, Arcangeli A (2014) hERG1 channels regulate VEGF-A secretion in human gastric cancer: clinicopathological correlations and therapeutical implications. Clin Cancer Res 20:1502–1512

Darvin P, Baeg SJ, Joung YH, Nipin SP, Kang DY, Byun HJ, Park JU, Yang YM (2015) Tannic acid inhibits the Jak2/STAT3 pathway and induces G1/S arrest and mitochondrial apoptosis in YD-38 gingival cancer cells. Int J Oncol 47:1111–1120

De Guadalupe Chavez-Lopez M, Perez-Carreon JI, Zuniga-Garcia V, Diaz-Chavez J, Herrera LA, Caro-Sanchez CH, Acuna-Macias I, Gariglio P, Hernandez-Gallegos E, Chiliquinga AJ, Camacho J (2015) Astemizole-based anticancer therapy for hepatocellular carcinoma (HCC), and Eag1 channels as potential early-stage markers of HCC. Tumour Biol 36:6149–6158

De Marchi U, Sassi N, Fioretti B, Catacuzzeno L, Cereghetti GM, Szabò I, Zoratti M (2009) Intermediate conductance Ca2+-activated potassium channel (KCa3.1) in the inner mitochondrial membrane of human colon cancer cells. Cell Calcium 45:509–516

De Petrocellis L, Melck D, Palmisano A, Bisogno T, Laezza C, Bifulco M, Di Marzo V (1998) The endogenous cannabinoid anandamide inhibits human breast cancer cell proliferation. Proc Natl Acad Sci U S A 95:8375–8380

Del Monaco SM, Marino GI, Assef YA, Damiano AE, Kotsias BA (2009) Cell migration in BeWo cells and the role of epithelial sodium channels. J Membr Biol 232:1–13

Deshane J, Garner CC, Sontheimer H (2003) Chlorotoxin inhibits glioma cell invasion via matrix metalloproteinase-2. J Biol Chem 278:4135–4144

Devita VT Jr, Chu E (2008) A history of cancer chemotherapy. Cancer Res 68:8643–8653

Ding XW, Luo HS, Jin X, Yan JJ, AI YW (2007) Aberrant expression of Eag1 potassium channels in gastric cancer patients and cell lines. Med Oncol 24:345–350

Djamgoz MB, Onkal R (2013) Persistent current blockers of voltage gated sodium channels: a clinical opportunity for controlling metastatic disease. Recent Pat Anticancer Drug Discov 8:66–84

Djamgoz MB, Coombes RC, Schwab A (2014) Ion transport and cancer: from initiation to metastasis. Philos Trans R Soc Lond B Biol Sci 369:20130092

Djamgoz MBA, Fraser SP, Brackenbury WJ (2019) In vivo evidence for voltage-gated sodium channel expression in carcinomas and potentiation of metastasis. Cancers (Basel) 11:1675

Dolderer JH, Schuldes H, Bockhorn H, Altmannsberger M, Lambers C, Von Zabern D, Jonas D, Schwegler H, Linke R, Schröder UH (2010) HERG1 gene expression as a specific tumor marker in colorectal tissues. Eur J Surg Oncol 36:72–77

Dong Y, Furuta T, Sabit H, Kitabayashi T, Jiapaer S, Kobayashi M, Ino Y, Todo T, Teng L, Hirao A, Zhao S-G, Nakada M (2017) Identification of antipsychotic drug fluspirilene as a potential anti-glioma stem cell drug. Oncotarget 8:111728–111741

Driffort V, Gillet L, Bon E, Marionneau-Lambot S, Oullier T, Joulin V, Collin C, Pages JC, Jourdan ML, Chevalier S, Bougnoux P, Le Guennec JY, Besson P, Roger S (2014) Ranolazine inhibits NaV1.5-mediated breast cancer cell invasiveness and lung colonization. Mol Cancer 13:264

Dubois C, Vanden Abeele F, Lehen'kyi V, Gkika D, Guarmit B, Lepage G, Slomianny C, Borowiec AS, Bidaux G, Benahmed M, Shuba Y, Prevarskaya N (2014) Remodeling of channel-forming ORAI proteins determines an oncogenic switch in prostate cancer. Cancer Cell 26:19–32

Duranti C, Arcangeli A (2019) Ion channel targeting with antibodies and antibody fragments for cancer diagnosis. Antibodies (Basel) 8:33

Duranti C, Carraresi L, Sette A, Stefanini M, Lottini T, Crescioli S, Crociani O, Iamele L, De Jonge H, Gherardi E, Arcangeli A (2018) Generation and characterization of novel recombinant anti-hERG1 scFv antibodies for cancer molecular imaging. Oncotarget 9:34972–34989

Dutta S, Lopez Charcas O, Tanner S, Gradek F, Driffort V, Roger S, Selander K, Velu SE, Brouillette W (2018) Discovery and evaluation of nNav1.5 sodium channel blockers with potent cell invasion inhibitory activity in breast cancer cells. Bioorg Med Chem 26:2428–2436

Dziegielewska B, Brautigan DL, Larner JM, Dziegielewski J (2014) T-type Ca2+ channel inhibition induces p53-dependent cell growth arrest and apoptosis through activation of p38-MAPK in colon cancer cells. Mol Cancer Res 12:348–358

Eil R, Vodnala SK, Clever D, Klebanoff CA, Sukumar M, Pan JH, Palmer DC, Gros A, Yamamoto TN, Patel SJ, Guittard GC, Yu Z, Carbonaro V, Okkenhaug K, Schrump DS, Linehan WM, Roychoudhuri R, Restifo NP (2016) Ionic immune suppression within the tumour microenvironment limits T cell effector function. Nature 537:539–543

Elliott MJ, Jerzak KJ, Cockburn JG, Safikhani Z, Gwynne WD, Hassell JA, Bane A, Silvester J, Thu KL, Haibe-Kains B, Mak TW, Cescon DW (2018) The antiarrhythmic drug, dronedarone, demonstrates cytotoxic effects in breast cancer independent of thyroid hormone receptor alpha 1 (THRalpha1) antagonism. Sci Rep 8:16562

Exadaktylos AK, Buggy DJ, Moriarty DC, Mascha E, Sessler DI (2006) Can anesthetic technique for primary breast cancer surgery affect recurrence or metastasis? Anesthesiology 105:660–664

Fairhurst C, Watt I, Martin F, Bland M, Brackenbury WJ (2014) Exposure to sodium channel-inhibiting drugs and cancer survival: protocol for a cohort study using the QResearch primary care database. BMJ Open 4:e006604

Fairhurst C, Watt I, Martin F, Bland M, Brackenbury WJ (2015) Sodium channel-inhibiting drugs and survival of breast, colon and prostate cancer: a population-based study. Sci Rep 5:16758

Fairhurst C, Martin F, Watt I, Doran T, Bland M, Brackenbury WJ (2016) Sodium channel-inhibiting drugs and cancer survival: protocol for a cohort study using the CPRD primary care database. BMJ Open 6:e011661

Felippe Goncalves-De-Albuquerque C, Ribeiro Silva A, Ignacio Da Silva C, Caire Castro-Faria-Neto H, Burth P (2017) Na/K pump and beyond: Na/K-ATPase as a modulator of apoptosis and autophagy. Molecules 22:578

Fernández-Salas E, Suh KS, Speransky VV, Bowers WL, Levy JM, Adams T, Pathak KR, Edwards LE, Hayes DD, Cheng C, Steven AC, Weinberg WC, Yuspa SH (2002) mtCLIC/CLIC4, an organellular chloride channel protein, is increased by DNA damage and participates in the apoptotic response to p53. Mol Cell Biol 22:3610–3620

Fiorio Pla A, Avanzato D, Munaron L, Ambudkar IS (2012) Ion channels and transporters in cancer. 6. Vascularizing the tumor: TRP channels as molecular targets. Am J Physiol Cell Physiol 302:C9-15

Föhr K, Knippschild U, Herkommer A, Fauler M, Peifer C, Georgieff M, Adolph O (2017) State-dependent block of voltage-gated sodium channels by the casein-kinase 1 inhibitor IC261. Invest New Drugs 35:277–289

Fraser SP, Grimes JA, Djamgoz MB (2000) Effects of voltage-gated ion channel modulators on rat prostatic cancer cell proliferation: comparison of strongly and weakly metastatic cell lines. Prostate 44:61–76

Fraser SP, Grimes JA, Diss JK, Stewart D, Dolly JO, Djamgoz MB (2003a) Predominant expression of Kv1.3 voltage-gated K+ channel subunit in rat prostate cancer cell lines: electrophysiological, pharmacological and molecular characterisation. Pflugers Arch 446:559–571

Fraser SP, Salvador V, Manning EA, Mizal J, Altun S, Raza M, Berridge RJ, Djamgoz MB (2003b) Contribution of functional voltage-gated Na+ channel expression to cell behaviors involved in the metastatic cascade in rat prostate cancer: I. Lateral motility. J Cell Physiol 195:479–487

Fraser SP, Diss JK, Chioni AM, Mycielska ME, Pan H, Yamaci RF, Pani F, Siwy Z, Krasowska M, Grzywna Z, Brackenbury WJ, Theodorou D, Koyuturk M, Kaya H, Battaloglu E, De Bella MT, Slade MJ, Tolhurst R, Palmieri C, Jiang J, Latchman DS, Coombes RC, Djamgoz MB (2005) Voltage-gated sodium channel expression and potentiation of human breast cancer metastasis. Clin Cancer Res 11:5381–5389

Fu S, Hirte H, Welch S, Ilenchuk TT, Lutes T, Rice C, Fields N, Nemet A, Dugourd D, Piha-Paul S, Subbiah V, Liu L, Gong J, Hong D, Stewart JM (2017) First-in-human phase I study of SOR-C13, a TRPV6 calcium channel inhibitor, in patients with advanced solid tumors. Invest New Drugs 35:324–333

Fuca G, Galli G, Poggi M, Lo Russo G, Proto C, Imbimbo M, Vitali M, Ganzinelli M, Lanti C, Molino G, Stangoni F, Zilembo N, De Braud F, Garassino MC, Signorelli D (2018) Low baseline serum sodium concentration is associated with poor clinical outcomes in metastatic non-small cell lung cancer patients treated with immunotherapy. Target Oncol 13:795–800

Gackiere F, Bidaux G, Delcourt P, Van Coppenolle F, Katsogiannou M, Dewailly E, Bavencoffe A, Van Chuoi-Mariot MT, Mauroy B, Prevarskaya N, Mariot P (2008) CaV3.2 T-type calcium channels are involved in calcium-dependent secretion of neuroendocrine prostate cancer cells. J Biol Chem 283:10162–10173

Garcia-Ferreiro RE, Kerschensteiner D, Major F, Monje F, Stuhmer W, Pardo LA (2004) Mechanism of block of hEag1 K+ channels by imipramine and astemizole. J Gen Physiol 124:301–317

Garcia-Quiroz J, Camacho J (2011) Astemizole: an old anti-histamine as a new promising anti-cancer drug. Anticancer Agents Med Chem 11:307–314

García-Quiroz J, García-Becerra R, Barrera D, Santos N, Avila E, Ordaz-Rosado D, Rivas-Suárez M, Halhali A, Rodríguez P, Gamboa-Domínguez A, Medina-Franco H, Camacho J, Larrea F, Díaz L (2012) Astemizole synergizes calcitriol antiproliferative activity by inhibiting CYP24A1 and upregulating VDR: a novel approach for breast cancer therapy. PLoS One 7: e45063

García-Quiroz J, García-Becerra R, Santos-Martínez N, Barrera D, Ordaz-Rosado D, Avila E, Halhali A, Villanueva O, Ibarra-Sánchez MJ, Esparza-López J, Gamboa-Domínguez A, Camacho J, Larrea F, Díaz L (2014) In vivo dual targeting of the oncogenic Ether-à-go-go-1 potassium channel by calcitriol and astemizole results in enhanced antineoplastic effects in breast tumors. BMC Cancer 14:745–745

Garzon-Muvdi T, Schiapparelli P, Ap Rhys C, Guerrero-Cazares H, Smith C, Kim DH, Kone L, Farber H, Lee DY, An SS, Levchenko A, Quinones-Hinojosa A (2012) Regulation of brain tumor dispersal by NKCC1 through a novel role in focal adhesion regulation. PLoS Biol 10: e1001320

Gautier M, Trebak M, Fleig A, Vandier C, Ouadid-Ahidouch H (2019) Ca(2+) channels in cancer. Cell Calcium 84:102083

Gavrilova-Ruch O, Schönherr K, Gessner G, Schönherr R, Klapperstück T, Wohlrab W, Heinemann SH (2002) Effects of imipramine on ion channels and proliferation of IGR1 melanoma cells. J Membr Biol 188:137–149

Genova T, Grolez GP, Camillo C, Bernardini M, Bokhobza A, Richard E, Scianna M, Lemonnier L, Valdembri D, Munaron L, Philips MR, Mattot V, Serini G, Prevarskaya N, Gkika D, Pla AF

(2017) TRPM8 inhibits endothelial cell migration via a non-channel function by trapping the small GTPase Rap1. J Cell Biol 216:2107–2130
George AL Jr (2005) Inherited disorders of voltage-gated sodium channels. J Clin Invest 115:1990–1999
Gilbert SM, Gidley Baird A, Glazer S, Barden JA, Glazer A, Teh LC, King J (2017) A phase I clinical trial demonstrates that nfP2X7-targeted antibodies provide a novel, safe and tolerable topical therapy for basal cell carcinoma. Br J Dermatol 177:117–124
Gilbert SM, Oliphant CJ, Hassan S, Peille AL, Bronsert P, Falzoni S, Di Virgilio F, Mcnulty S, Lara R (2019) ATP in the tumour microenvironment drives expression of nfP2X7, a key mediator of cancer cell survival. Oncogene 38:194–208
Gillet L, Roger S, Besson P, Lecaille F, Gore J, Bougnoux P, Lalmanach G, Le Guennec JY (2009) Voltage-gated sodium channel activity promotes cysteine cathepsin-dependent invasiveness and colony growth of human cancer cells. J Biol Chem 284:8680–8691
Gillet L, Roger S, Bougnoux P, Le Guennec JY, Besson P (2011) Beneficial effects of omega-3 long-chain fatty acids in breast cancer and cardiovascular diseases: voltage-gated sodium channels as a common feature? Biochimie 93:4–6
Gómez-Varela D, Zwick-Wallasch E, Knötgen H, Sánchez A, Hettmann T, Ossipov D, Weseloh R, Contreras-Jurado C, Rothe M, Stühmer W, Pardo LA (2007) Monoclonal antibody blockade of the human Eag1 potassium channel function exerts antitumor activity. Cancer Res 67:7343–7349
Gould HJ 3rd, Norleans J, Ward TD, Reid C, Paul D (2018) Selective lysis of breast carcinomas by simultaneous stimulation of sodium channels and blockade of sodium pumps. Oncotarget 9:15606–15615
Grimaldi C, Pisanti S, Laezza C, Malfitano AM, Santoro A, Vitale M, Caruso MG, Notarnicola M, Iacuzzo I, Portella G, Di Marzo V, Bifulco M (2006) Anandamide inhibits adhesion and migration of breast cancer cells. Exp Cell Res 312:363–373
Grimaldi-Bensouda L, Klungel O, Kurz X, De Groot MCH, Maciel Afonso AS, De Bruin ML, Reynolds R, Rossignol M (2016) Calcium channel blockers and cancer: a risk analysis using the UK Clinical Practice Research Datalink (CPRD). BMJ Open 6
Grimes JA, Djamgoz MBA (1998) Electrophysiological characterization of voltage-gated Na+ current expressed in the highly metastatic Mat-LyLu cell line of rat prostate cancer. J Cell Physiol 175:50–58
Grimes JA, Fraser SP, Stephens GJ, Downing JE, Laniado ME, Foster CS, Abel PD, Djamgoz MB (1995) Differential expression of voltage-activated Na^+ currents in two prostatic tumour cell lines: contribution to invasiveness in vitro. FEBS Lett 369:290–294
Grolez GP, Hammadi M, Barras A, Gordienko D, Slomianny C, Volkel P, ANGRAND PO, Pinault M, Guimaraes C, Potier-Cartereau M, Prevarskaya N, Boukherroub R, Gkika D (2019) Encapsulation of a TRPM8 agonist, WS12, in lipid nanocapsules potentiates PC3 prostate cancer cell migration inhibition through channel activation. Sci Rep 9:7926
Guan L, Song Y, Gao J, Gao J, Wang K (2016) Inhibition of calcium-activated chloride channel ANO1 suppresses proliferation and induces apoptosis of epithelium originated cancer cells. Oncotarget 7:78619–78630
Guilbert A, Gautier M, Dhennin-Duthille I, Haren N, Sevestre H, Ouadid-Ahidouch H (2009) Evidence that TRPM7 is required for breast cancer cell proliferation. Am J Physiol Cell Physiol 297:C493–C502
Gururaja Rao S, Patel NJ, Singh H (2020) Intracellular chloride channels: novel biomarkers in diseases. Front Physiol 11:96–96
Guzel RM, Ogmen K, Ilieva KM, Fraser SP, Djamgoz MBA (2019) Colorectal cancer invasiveness in vitro: predominant contribution of neonatal Nav1.5 under normoxia and hypoxia. J Cell Physiol 234:6582–6593
Haas BR, Sontheimer H (2010) Inhibition of the sodium-potassium-chloride cotransporter isoform-1 reduces glioma invasion. Cancer Res 70:5597–5606

Haas M, Wang H, Tian J, Xie Z (2002) Src-mediated inter-receptor cross-talk between the Na+/K+-ATPase and the epidermal growth factor receptor relays the signal from ouabain to mitogen-activated protein kinases. J Biol Chem 277:18694–18702

Habela CW, Olsen ML, Sontheimer H (2008) ClC3 is a critical regulator of the cell cycle in normal and malignant glial cells. J Neurosci 28:9205–9217

Habela CW, Ernest NJ, Swindall AF, Sontheimer H (2009) Chloride accumulation drives volume dynamics underlying cell proliferation and migration. J Neurophysiol 101:750–757

Hamtiaux L, Hansoulle L, Dauguet N, Muccioli GG, Gallez B, Lambert DM (2011) Increasing antiproliferative properties of endocannabinoids in N1E-115 neuroblastoma cells through inhibition of their metabolism. PLoS One 6:e26823

Han X, Wang F, Yao W, Xing H, Weng D, Song X, Chen G, Xi L, Zhu T, Zhou J, Xu G, Wang S, Meng L, Iadecola C, Wang G, Ma D (2007) Heat shock proteins and p53 play a critical role in K + channel-mediated tumor cell proliferation and apoptosis. Apoptosis 12:1837–1846

Hartung F, Stühmer W, Pardo L (2011) Tumor cell-selective apoptosis induction through targeting of KV 10.1 via bifunctional TRAIL antibody. Mol Cancer 10:109

Haux J, Klepp O, Spigset O, Tretli S (2001) Digitoxin medication and cancer; case control and internal dose-response studies. BMC Cancer 1:11

He M, Liu S, Gallolu Kankanamalage S, Borromeo MD, Girard L, Gazdar AF, Minna JD, Johnson JE, Cobb MH (2018) The epithelial sodium channel (αENaC) is a downstream therapeutic target of ASCL1 in pulmonary neuroendocrine tumors. Transl Oncol 11:292–299

Hegle AP, Marble DD, Wilson GF (2006) A voltage-driven switch for ion-independent signaling by ether-a-go-go K+ channels. Proc Natl Acad Sci U S A 103:2886–2891

Holdhoff M, Ye X, Supko JG, Nabors LB, Desai AS, Walbert T, Lesser GJ, Read WL, Lieberman FS, Lodge MA, Leal J, Fisher JD, Desideri S, Grossman SA, Wahl RL, Schiff D (2017) Timed sequential therapy of the selective T-type calcium channel blocker mibefradil and temozolomide in patients with recurrent high-grade gliomas. Neuro Oncol 19:845–852

Hong S, Bi M, Wang L, Kang Z, Ling L, Zhao C (2015) CLC-3 channels in cancer (review). Oncol Rep 33:507–514

Honn KV, Onoda JM, Diglio CA, Carufel MM, Taylor JD, Sloane BF (1984) Inhibition of tumor cell-platelet interactions and tumor metastasis by the calcium channel blocker, nimodipine. Clin Exp Metastasis 2:61–72

Honn KV, Onoda JM, Pampalona K, Battaglia M, Neagos G, Taylor JD, Diglio CA, Sloane BF (1985) Inhibition by dihydropyridine class calcium channel blockers of tumor cell-platelet-endothelial cell interactions in vitro and metastasis in vivo. Biochem Pharmacol 34:235–241

Hopkins MM, Feng X, Liu M, Parker LP, Koh DW (2015) Inhibition of the transient receptor potential melastatin-2 channel causes increased DNA damage and decreased proliferation in breast adenocarcinoma cells. Int J Oncol 46:2267–2276

Huang X, Jan LY (2014) Targeting potassium channels in cancer. J Cell Biol 206:151–162

Huang W, Lu C, Wu Y, Ouyang S, Chen Y (2015) T-type calcium channel antagonists, mibefradil and NNC-55-0396 inhibit cell proliferation and induce cell apoptosis in leukemia cell lines. J Exp Clin Cancer Res 34:54–54

Huang J, Furuya H, Faouzi M, Zhang Z, Monteilh Zoller M, Kelly Galbraith Kawabata F, Horgen D, Kawamori T, Penner R, Fleig A (2017) Inhibition of TRPM7 suppresses cell proliferation of colon adenocarcinoma in vitro and induces hypomagnesemia in vivo without affecting azoxymethane-induced early colon cancer in mice. Cell Commun Signal 15:30

Humeau J, Bravo-San Pedro JM, Vitale I, Nuñez L, Villalobos C, Kroemer G, Senovilla L (2018) Calcium signaling and cell cycle: progression or death. Cell Calcium 70:3–15

Hutchings CJ, Colussi P, Clark T (2019) Ion channels as therapeutic antibody targets. MAbs 11 (2):265–296

Illek B, Fischer H, Machen TE (1992) Intracellular Ca2+ signalling is modulated by K+ channel blockers in colonic epithelial cells (HT-29/B6). Pflugers Arch 422:48–54

Innamaa A, Jackson L, Asher V, Van Shalkwyk G, Warren A, Hay D, Bali A, Sowter H, Khan R (2013) Expression and prognostic significance of the oncogenic K2P potassium channel KCNK9 (TASK-3) in ovarian carcinoma. Anticancer Res 33:1401–1408

Isbilen B, Fraser SP, Djamgoz MB (2006) Docosahexaenoic acid (omega-3) blocks voltage-gated sodium channel activity and migration of MDA-MB-231 human breast cancer cells. Int J Biochem Cell Biol 38:2173–2182

Jang SJ, Choi HW, Choi DL, Cho S, Rim HK, Choi HE, Kim KS, Huang M, Rhim H, Lee KT, Lee JY (2013) In vitro cytotoxicity on human ovarian cancer cells by T-type calcium channel blockers. Bioorg Med Chem Lett 23:6656–6662

Jang SH, Byun JK, Jeon WI, Choi SY, Park J, Lee BH, Yang JE, Park JB, O'grady SM, Kim DY, Ryu PD, Joo SW, Lee SY (2015) Nuclear localization and functional characteristics of voltage-gated potassium channel Kv1.3. J Biol Chem 290:12547–12557

Jose C, Hebert-Chatelain E, Dias Amoedo N, Roche E, Obre E, Lacombe D, Rezvani HR, Pourquier P, Nouette-Gaulain K, Rossignol R (2018) Redox mechanism of levobupivacaine cytostatic effect on human prostate cancer cells. Redox Biol 18:33–42

Kaddour-Djebbar I, Choudhary V, Lakshmikanthan V, Shirley R, El Gaish M, Al-Shabrawey M, Al-Husein B, Zhong R, Davis M, Dong Z, Bollag WB, Kumar MV (2012) Diltiazem enhances the apoptotic effects of proteasome inhibitors to induce prostate cancer cell death. J Pharmacol Exp Ther 341:646–655

Kang HB, Rim HK, Park JY, Choi HW, Choi DL, Seo JH, Chung KS, Huh G, Kim J, Choo DJ, Lee KT, Lee JY (2012) In vivo evaluation of oral anti-tumoral effect of 3,4-dihydroquinazoline derivative on solid tumor. Bioorg Med Chem Lett 22:1198–1201

Kapoor N, Bartoszewski R, Qadri YJ, Bebok Z, Bubien JK, Fuller CM, Benos DJ (2009) Knockdown of ASIC1 and epithelial sodium channel subunits inhibits glioblastoma whole cell current and cell migration. J Biol Chem 284:24526–24541

Karakurt S, Adali O (2016) Tannic acid inhibits proliferation, migration, invasion of prostate cancer and modulates drug metabolizing and antioxidant enzymes. Anticancer Agents Med Chem 16:781–789

Kepp O, Menger L, Vacchelli E, Adjemian S, Martins I, Ma Y, Sukkurwala AQ, Michaud M, Galluzzi L, Zitvogel L, Kroemer G (2012) Anticancer activity of cardiac glycosides: at the frontier between cell-autonomous and immunological effects. Oncoimmunology 1:1640–1642

Khajah MA, Mathew PM, Luqmani YA (2018) Na+/K+ ATPase activity promotes invasion of endocrine resistant breast cancer cells. PLoS One 13:e0193779

Khatun A, Fujimoto M, Kito H, Niwa S, Suzuki T, Ohya S (2016) Down-regulation of Ca 2+-activated K + channel K Ca 1.1 in human breast cancer MDA-MB-453 cells treated with vitamin D receptor agonists. Int J Mol Sci 17:2083

Khatun A, Shimozawa M, Kito H, Kawaguchi M, Fujimoto M, Ri M, Kajikuri J, Niwa S, Fujii M, Ohya S (2018) Transcriptional repression and protein degradation of the Ca(2+)-activated K(+) channel K(Ca)1.1 by androgen receptor inhibition in human breast cancer cells. Front Physiol 9:312

Kim CJ, Cho YG, Jeong SW, Kim YS, Kim SY, Nam SW, Lee SH, Yoo NJ, Lee JY, Park WS (2004) Altered expression of KCNK9 in colorectal cancers. Apmis 112:588–594

Kim IY, Kang YJ, Yoon MJ, Kim EH, Kim SU, Kwon TK, Kim IA, Choi KS (2011) Amiodarone sensitizes human glioma cells but not astrocytes to TRAIL-induced apoptosis via CHOP-mediated DR5 upregulation. Neuro Oncol 13:267–279

Kischel P, Girault A, Rodat-Despoix L, Chamlali M, Radoslavova S, Abou Daya H, Lefebvre T, Foulon A, Rybarczyk P, Hague F, Dhennin-Duthille I, Gautier M, Ouadid-Ahidouch H (2019) Ion channels: new actors playing in chemotherapeutic resistance. Cancers (Basel) 11:376

Klumpp D, Misovic M, Szteyn K, Shumilina E, Rudner J, Huber SM (2016) Targeting TRPM2 channels impairs radiation-induced cell cycle arrest and fosters cell death of T cell leukemia cells in a Bcl-2-dependent manner. Oxid Med Cell Longev 2016:8026702

Klumpp D, Frank SC, Klumpp L, Sezgin EC, Eckert M, Edalat L, Bastmeyer M, Zips D, Ruth P, Huber SM (2017) TRPM8 is required for survival and radioresistance of glioblastoma cells. Oncotarget 8:95896–95913

Kondratska K, Kondratskyi A, Yassine M, Lemonnier L, Lepage G, Morabito A, Skryma R, Prevarskaya N (2014) Orai1 and STIM1 mediate SOCE and contribute to apoptotic resistance of pancreatic adenocarcinoma. Biochim Biophys Acta 1843:2263–2269

Koster AK, Wood CAP, Thomas-Tran R, Chavan TS, Almqvist J, Choi KH, Du Bois J, Maduke M (2018) A selective class of inhibitors for the CLC-Ka chloride ion channel. Proc Natl Acad Sci U S A 115:E4900–E4909

Kosztka L, Rusznák Z, Nagy D, Nagy Z, Fodor J, Szucs G, Telek A, Gönczi M, Ruzsnavszky O, Szentandrássy N, Csernoch L (2011) Inhibition of TASK-3 (KCNK9) channel biosynthesis changes cell morphology and decreases both DNA content and mitochondrial function of melanoma cells maintained in cell culture. Melanoma Res 21:308–322

Kouba S, Braire J, Felix R, Chantome A, Jaffres PA, Lebreton J, Dubreuil D, Pipelier M, Zhang X, Trebak M, Vandier C, Mathe-Allainmat M, Potier-Cartereau M (2020) Lipidic synthetic alkaloids as SK3 channel modulators. Synthesis and biological evaluation of 2-substituted tetrahydropyridine derivatives with potential anti-metastatic activity. Eur J Med Chem 186:111854

Kovalenko I, Glasauer A, Schöckel L, Sauter DRP, Ehrmann A, Sohler F, Hägebarth A, Novak I, Christian S (2016) Identification of KCa3.1 channel as a novel regulator of oxidative phosphorylation in a subset of pancreatic carcinoma cell lines. PLoS One 11:e0160658

Kunzelmann K, Ousingsawat J, Benedetto R, Cabrita I, Schreiber R (2019) Contribution of anoctamins to cell survival and cell death. Cancers (Basel) 11:382

Laezza C, D'alessandro A, Paladino S, Maria Malfitano A, Chiara Proto M, Gazzerro P, Pisanti S, Santoro A, Ciaglia E, Bifulco M (2012) Anandamide inhibits the Wnt/β-catenin signalling pathway in human breast cancer MDA MB 231 cells. Eur J Cancer 48:3112–3122

Lang DG, Wang CM, Cooper BR (1993) Lamotrigine, phenytoin and carbamazepine interactions on the sodium current present in N4TG1 mouse neuroblastoma cells. J Pharmacol Exp Ther 266:829–835

Lansu K, Gentile S (2013) Potassium channel activation inhibits proliferation of breast cancer cells by activating a senescence program. Cell Death Dis 4:e652

Lastraioli E, Guasti L, Crociani O, Polvani S, Hofmann G, Witchel H, Bencini L, Calistri M, Messerini L, Scatizzi M, Moretti R, Wanke E, Olivotto M, Mugnai G, Arcangeli A (2004) herg1 gene and HERG1 protein are overexpressed in colorectal cancers and regulate cell invasion of tumor cells. Cancer Res 64:606–611

Lastraioli E, Taddei A, Messerini L, Comin CE, Festini M, Giannelli M, Tomezzoli A, Paglierani M, Mugnai G, De Manzoni G, Bechi P, Arcangeli A (2006) hERG1 channels in human esophagus: evidence for their aberrant expression in the malignant progression of Barrett's esophagus. J Cell Physiol 209:398–404

Laursen M, Yatime L, Nissen P, Fedosova NU (2013) Crystal structure of the high-affinity Na+K+-ATPase-ouabain complex with Mg2+ bound in the cation binding site. Proc Natl Acad Sci U S A 110:10958–10963

Leanza L, Henry B, Sassi N, Zoratti M, Chandy KG, Gulbins E, Szabò I (2012) Inhibitors of mitochondrial Kv1.3 channels induce Bax/Bak-independent death of cancer cells. EMBO Mol Med 4:577–593

Leanza L, Biasutto L, Manago A, Gulbins E, Zoratti M, Szabò I (2013a) Intracellular ion channels and cancer. Front Physiol 4:227

Leanza L, Trentin L, Becker KA, Frezzato F, Zoratti M, Semenzato G, Gulbins E, Szabo I (2013b) Clofazimine, Psora-4 and PAP-1, inhibitors of the potassium channel Kv1.3, as a new and selective therapeutic strategy in chronic lymphocytic leukemia. Leukemia 27(8):1782–1785

Lee JM, Davis FM, Roberts-Thomson SJ, Monteith GR (2011) Ion channels and transporters in cancer. 4. Remodeling of Ca(2+) signaling in tumorigenesis: role of Ca(2+) transport. Am J Physiol Cell Physiol 301:C969–C976

Lee H-C, Su M-Y, Lo H-C, Wu C-C, Hu J-R, Lo D-M, Chao T-Y, Tsai H-J, Dai M-S (2015) Cancer metastasis and EGFR signaling is suppressed by amiodarone-induced versican V2. Oncotarget 6:42976–42987

Lee A, Fraser SP, Djamgoz MBA (2019a) Propranolol inhibits neonatal Nav1.5 activity and invasiveness of MDA-MB-231 breast cancer cells: effects of combination with ranolazine. J Cell Physiol 234:23066–23081

Lee JR, Lee JY, Kim HJ, Hahn MJ, Kang JS, Cho H (2019b) The inhibition of chloride intracellular channel 1 enhances Ca(2+) and reactive oxygen species signaling in A549 human lung cancer cells. Exp Mol Med 51:81

Lehen'kyi V, Flourakis M, Skryma R, Prevarskaya N (2007) TRPV6 channel controls prostate cancer cell proliferation via Ca(2+)/NFAT-dependent pathways. Oncogene 26:7380–7385

Leithner K, Hirschmugl B, Li Y, Tang B, Papp R, Nagaraj C, Stacher E, Stiegler P, Lindenmann J, Olschewski A, Olschewski H, Hrzenjak A (2016) TASK-1 regulates apoptosis and proliferation in a subset of non-small cell lung cancers. PLoS One 11:e0157453

Lemieszek MK, Stepulak A, Sawa-Wejksza K, Czerwonka A, Ikonomidou C, Rzeski W (2018) Riluzole inhibits proliferation, migration and cell cycle progression and induces apoptosis in tumor cells of various origins. Anticancer Agents Med Chem 18:565–572

Lenkowski PW, Ko SH, Anderson JD, Brown ML, Patel MK (2004) Block of human NaV1.5 sodium channels by novel alpha-hydroxyphenylamide analogues of phenytoin. Eur J Pharm Sci 21:635–644

Leslie TK, James AD, Zaccagna F, Grist JT, Deen S, Kennerley A, Riemer F, Kaggie JD, Gallagher FA, Gilbert FJ, Brackenbury WJ (2019) Sodium homeostasis in the tumour microenvironment. Biochim Biophys Acta Rev Cancer 1872:188304

Li L, Feng R, Xu Q, Zhang F, Liu T, Cao J, Fei S (2017) Expression of the beta3 subunit of Na(+)/K(+)-ATPase is increased in gastric cancer and regulates gastric cancer cell progression and prognosis via the PI3/AKT pathway. Oncotarget 8:84285–84299

Li R, Xiao C, Liu H, Huang Y, Dilger JP, Lin J (2018) Effects of local anesthetics on breast cancer cell viability and migration. BMC Cancer 18:666–666

Li T, Chen L, Zhao H, Wu L, Masters J, Han C, Hirota K, Ma D (2019) Both Bupivacaine and Levobupivacaine inhibit colon cancer cell growth but not melanoma cells in vitro. J Anesth 33:17–25

Ligresti A, Moriello AS, Starowicz K, Matias I, Pisanti S, De Petrocellis L, Laezza C, Portella G, Bifulco M, Di Marzo V (2006) Antitumor activity of plant cannabinoids with emphasis on the effect of cannabidiol on human breast carcinoma. J Pharmacol Exp Ther 318:1375–1387

Lin S-Y, Chang H-H, Lai Y-H, Lin C-H, Chen M-H, Chang G-C, Tsai M-F, Chen JJW (2015) Digoxin suppresses tumor malignancy through inhibiting multiple Src-related signaling pathways in non-small cell lung cancer. PLoS One 10:e0123305

Liu J, Zhang D, Li Y, Chen W, Ruan Z, Deng L, Wang L, Tian H, Yiu A, Fan C, Luo H, Liu S, Wang Y, Xiao G, Chen L, Ye W (2013) Discovery of bufadienolides as a novel class of ClC-3 chloride channel activators with antitumor activities. J Med Chem 56:5734–5743

Liu X, Zou J, Su J, Lu Y, Zhang J, Li L, Yin F (2016) Downregulation of transient receptor potential cation channel, subfamily C, member 1 contributes to drug resistance and high histological grade in ovarian cancer. Int J Oncol 48:243–252

Lo WL, Donermeyer DL, Allen PM (2012) A voltage-gated sodium channel is essential for the positive selection of CD4(+) T cells. Nat Immunol 13:880–887

Lu F, Chen H, Zhou C, Liu S, Guo M, Chen P, Zhuang H, Xie D, Wu S (2008) T-type Ca2+ channel expression in human esophageal carcinomas: a functional role in proliferation. Cell Calcium 43:49–58

Lui VC, Lung SS, Pu JK, Hung KN, Leung GK (2010) Invasion of human glioma cells is regulated by multiple chloride channels including ClC-3. Anticancer Res 30:4515–4524

Luveta J, Parks RM, Heery DM, Cheung K-L, Johnston SJ (2020) Invasive lobular breast cancer as a distinct disease: implications for therapeutic strategy. Oncol Ther 8:1–11

Ma B, Pan Y, Song Q, Tie L, Zhang Y, Xiao Y, Zhang J, Han J, Xu Y, Xiang Y, Yu H-M, Li X (2011) The effect of topiramate on tumor-related angiogenesis and on the serum proteome of mice bearing Lewis lung carcinoma. Eur J Pharmacol 663:9–16

Malamas AS, Jin E, Zhang Q, Haaga J, Lu Z-R (2015) Anti-angiogenic effects of bumetanide revealed by DCE-MRI with a biodegradable macromolecular contrast agent in a colon cancer model. Pharm Res 32:3029–3043

Mamelak AN, Rosenfeld S, Bucholz R, Raubitschek A, Nabors LB, Fiveash JB, Shen S, Khazaeli MB, Colcher D, Liu A, Osman M, Guthrie B, Schade-Bijur S, Hablitz DM, Alvarez VL, Gonda MA (2006) Phase I single-dose study of intracavitary-administered iodine-131-TM-601 in adults with recurrent high-grade glioma. J Clin Oncol 24:3644–3650

Mantegazza M, Curia G, Biagini G, Ragsdale DS, Avoli M (2010) Voltage-gated sodium channels as therapeutic targets in epilepsy and other neurological disorders. Lancet Neurol 9:413–424

Mao J, Chen L, Xu B, Wang L, Li H, Guo J, Li W, Nie S, Jacob TJ (2008) Suppression of ClC-3 channel expression reduces migration of nasopharyngeal carcinoma cells. Biochem Pharmacol 75:1706–1716

Mao J, Yuan J, Wang L, Zhang H, Jin X, Zhu J, Li H, Xu B, Chen L (2013) Tamoxifen inhibits migration of estrogen receptor-negative hepatocellular carcinoma cells by blocking the swelling-activated chloride current. J Cell Physiol 228:991–1001

Mariot P, Vanoverberghe K, Lalevee N, Rossier MF, Prevarskaya N (2002) Overexpression of an alpha 1H (Cav3.2) T-type calcium channel during neuroendocrine differentiation of human prostate cancer cells. J Biol Chem 277:10824–10833

Martin F, Ufodiama C, Watt I, Bland M, Brackenbury WJ (2015) Therapeutic value of voltage-gated sodium channel inhibitors in breast, colorectal and prostate cancer: a systematic review. Front Pharmacol 6:273

Martínez R, Stühmer W, Martin S, Schell J, Reichmann A, Rohde V, Pardo L (2015) Analysis of the expression of Kv10.1 potassium channel in patients with brain metastases and glioblastoma multiforme: impact on survival. BMC Cancer 15:839

Matthews H, Ranson M, Kelso MJ (2011) Anti-tumour/metastasis effects of the potassium-sparing diuretic amiloride: an orally active anti-cancer drug waiting for its call-of-duty? Int J Cancer 129:2051–2061

Matulef K, Howery AE, Tan L, Kobertz WR, Du Bois J, Maduke M (2008) Discovery of potent CLC chloride channel inhibitors. ACS Chem Biol 3:419–428

Mazzone A, Eisenman ST, Strege PR, Yao Z, Ordog T, Gibbons SJ, Farrugia G (2012) Inhibition of cell proliferation by a selective inhibitor of the Ca(2+)-activated Cl(-) channel, Ano1. Biochem Biophys Res Commun 427:248–253

Meléndez TA, Huanosta-Gutiérrez A, Barriga-Montoya C, González-Andrade M, Gómez-Lagunas F (2020) Dronedarone blockage of the tumor-related Kv10.1 channel: a comparison with amiodarone. Pflügers Arch 472:75–87

Meng Q, Chen X, Sun L, Zhao C, Sui G, Cai L (2011) Carbamazepine promotes Her-2 protein degradation in breast cancer cells by modulating HDAC6 activity and acetylation of Hsp90. Mol Cell Biochem 348:165–171

Menger L, Vacchelli E, Adjemian S, Martins I, Ma Y, Shen S, Yamazaki T, Sukkurwala AQ, Michaud M, Mignot G, Schlemmer F, Sulpice E, Locher C, Gidrol X, Ghiringhelli F, Modjtahedi N, Galluzzi L, André F, Zitvogel L, Kepp O, Kroemer G (2012) Cardiac glycosides exert anticancer effects by inducing immunogenic cell death. Sci Transl Med 4:143ra99

Metts J, Bradley HL, Wang Z, Shah NP, Kapur R, Arbiser JL, Bunting KD (2017) Imipramine blue sensitively and selectively targets FLT3-ITD positive acute myeloid leukemia cells. Sci Rep 7:4447

Mohr CJ, Gross D, Sezgin EC, Steudel FA, Ruth P, Huber SM, Lukowski R (2019) K(Ca)3.1 channels confer radioresistance to breast cancer cells. Cancer 11:1285

Moreels L, Bhat C, Voracova M, Peigneur S, Goovaerts H, Maki-Lohiluoma E, Zahed F, Pardo LA, Yli-Kauhaluoma J, Kiuru P, Tytgat J (2017a) Synthesis of novel purpurealidin analogs and evaluation of their effect on the cancer-relevant potassium channel KV10.1. PLoS One 12: e0188811

Moreels L, Peigneur S, Galan DT, De Pauw E, Beress L, Waelkens E, Pardo LA, Quinton L, Tytgat J (2017b) APETx4, a novel sea anemone toxin and a modulator of the cancer-relevant potassium channel KV10.1. Mar Drugs 15(9):287

Mu D, Chen L, Zhang X, See LH, Koch CM, Yen C, Tong JJ, Spiegel L, Nguyen KC, Servoss A, Peng Y, Pei L, Marks JR, Lowe S, Hoey T, Jan LY, Mccombie WR, Wigler MH, Powers S (2003) Genomic amplification and oncogenic properties of the KCNK9 potassium channel gene. Cancer Cell 3:297–302

Mycielska ME, Palmer CP, Brackenbury WJ, Djamgoz MB (2005) Expression of Na^+-dependent citrate transport in a strongly metastatic human prostate cancer PC-3M cell line: regulation by voltage-gated Na^+ channel activity. J Physiol 563:393–408

Nabissi M, Morelli MB, Santoni M, Santoni G (2012) Triggering of the TRPV2 channel by cannabidiol sensitizes glioblastoma cells to cytotoxic chemotherapeutic agents. Carcinogenesis 34:48–57

Nagy D, Gönczi M, Dienes B, Szöőr Á, Fodor J, Nagy Z, Tóth A, Fodor T, Bai P, Szücs G, Rusznák Z, Csernoch L (2014) Silencing the KCNK9 potassium channel (TASK-3) gene disturbs mitochondrial function, causes mitochondrial depolarization, and induces apoptosis of human melanoma cells. Arch Dermatol Res 306:885–902

Nelson M, Millican-Slater R, Forrest LC, Brackenbury WJ (2014) The sodium channel beta1 subunit mediates outgrowth of neurite-like processes on breast cancer cells and promotes tumour growth and metastasis. Int J Cancer 135:2338–2351

Nelson M, Yang M, Dowle AA, Thomas JR, Brackenbury WJ (2015) The sodium channel-blocking antiepileptic drug phenytoin inhibits breast tumour growth and metastasis. Mol Cancer 14:13

Neman J, Termini J, Wilczynski S, Vaidehi N, Choy C, Kowolik CM, Li H, Hambrecht AC, Roberts E, Jandial R (2014) Human breast cancer metastases to the brain display GABAergic properties in the neural niche. Proc Natl Acad Sci U S A 111:984–989

Nguyen CH, Huttary N, Atanasov AG, Chatuphonprasert W, Brenner S, Fristiohady A, Hong J, Stadler S, Holzner S, Milovanovic D, Dirsch VM, Kopp B, Saiko P, Krenn L, Jager W, Krupitza G (2017) Fenofibrate inhibits tumour intravasation by several independent mechanisms in a 3-dimensional co-culture model. Int J Oncol 50:1879–1888

Nie F, Liang Y, Jiang B, Li X, Xun H, He W, Lau HT, Ma X (2016) Apoptotic effect of tannic acid on fatty acid synthase over-expressed human breast cancer cells. Tumour Biol 37:2137–2143

Olsen ML, Schade S, Lyons SA, Amaral MD, Sontheimer H (2003) Expression of voltage-gated chloride channels in human glioma cells. J Neurosci 23:5572–5582

Olsen CM, Meussen-Elholm ET, Roste LS, Tauboll E (2004) Antiepileptic drugs inhibit cell growth in the human breast cancer cell line MCF7. Mol Cell Endocrinol 213:173–179

Onganer PU, Djamgoz MB (2005) Small-cell lung cancer (human): potentiation of endocytic membrane activity by voltage-gated Na^+ channel expression in vitro. J Membr Biol 204:67–75

Oosterwijk E, Gillies RJ (2014) Targeting ion transport in cancer. Philos Trans R Soc Lond B Biol Sci 369:20130107

Ouadid-Ahidouch H, Le Bourhis X, Roudbaraki M, Toillon RA, Delcourt P, Prevarskaya N (2001) Changes in the K+ current-density of MCF-7 cells during progression through the cell cycle: possible involvement of a h-ether.a-gogo K+ channel. Receptors Channels 7:345–356

Ouadid-Ahidouch H, Ahidouch A, Pardo LA (2016) Kv10.1 K(+) channel: from physiology to cancer. Pflugers Arch 468:751–762

Ousingsawat J, Spitzner M, Puntheeranurak S, Terracciano L, Tornillo L, BUBENDORF L, Kunzelmann K, Schreiber R (2007) Expression of voltage-gated potassium channels in human and mouse colonic carcinoma. Clin Cancer Res 13:824–831

Ouwerkerk R, Jacobs M, Macura K, Wolff A, Stearns V, Mezban S, Khouri N, Bluemke D, Bottomley P (2007) Elevated tissue sodium concentration in malignant breast lesions detected with non-invasive 23 Na MRI. Breast Cancer Res Treat 106:151–160

Pantziarka P, Sukhatme V, Bouche G, Meheus L, Sukhatme VP (2016) Repurposing drugs in oncology (ReDO)-diclofenac as an anti-cancer agent. Ecancermedicalscience 10:610–610

Pardo LA, Stuhmer W (2014) The roles of K(+) channels in cancer. Nat Rev Cancer 14:39–48

Pardo LA, Del Camino D, Sanchez A, Alves F, Bruggemann A, Beckh S, Stuhmer W (1999) Oncogenic potential of EAG K(+) channels. EMBO J 18:5540–5547

Park S-H, Chung YM, Ma J, Yang Q, Berek JS, Hu MCT (2016) Pharmacological activation of FOXO3 suppresses triple-negative breast cancer in vitro and in vivo. Oncotarget 7:42110–42125

Parker KA, Glaysher S, Hurren J, Knight LA, Mccormick D, Suovouri A, Amberger-Murphy V, Pilkington GJ, Cree IA (2012) The effect of tricyclic antidepressants on cutaneous melanoma cell lines and primary cell cultures. Anticancer Drugs 23:65–69

Parks SK, Chiche J, Pouyssegur J (2013) Disrupting proton dynamics and energy metabolism for cancer therapy. Nat Rev Cancer 13:611–623

Pellegrino M, Rizza P, Nigro A, Ceraldi R, Ricci E, Perrotta I, Aquila S, Lanzino M, Ando S, Morelli C, Sisci D (2018) FoxO3a mediates the inhibitory effects of the antiepileptic drug lamotrigine on breast cancer growth. Mol Cancer Res 16:923–934

Peruzzo R, Szabo I (2019) Contribution of mitochondrial ion channels to chemo-resistance in cancer cells. Cancer 11:761

Peruzzo R, Biasutto L, Szabò I, Leanza L (2016) Impact of intracellular ion channels on cancer development and progression. Eur Biophys J 45:685–707

Peters AA, Milevskiy MJ, Lee WC, Curry MC, Smart CE, Saunus JM, Reid L, Da Silva L, Marcial DL, Dray E, Brown MA, Lakhani SR, Roberts-Thomson SJ, Monteith GR (2016) The calcium pump plasma membrane Ca(2+)-ATPase 2 (PMCA2) regulates breast cancer cell proliferation and sensitivity to doxorubicin. Sci Rep 6:25505

Pillozzi S, Brizzi MF, Balzi M, Crociani O, Cherubini A, Guasti L, Bartolozzi B, Becchetti A, Wanke E, Bernabei PA, Olivotto M, Pegoraro L, Arcangeli A (2002) HERG potassium channels are constitutively expressed in primary human acute myeloid leukemias and regulate cell proliferation of normal and leukemic hemopoietic progenitors. Leukemia 16:1791–1798

Pillozzi S, Brizzi MF, Bernabei PA, Bartolozzi B, Caporale R, Basile V, Boddi V, Pegoraro L, Becchetti A, Arcangeli A (2007) VEGFR-1 (FLT-1), beta1 integrin, and hERG K+ channel for a macromolecular signaling complex in acute myeloid leukemia: role in cell migration and clinical outcome. Blood 110:1238–1250

Pillozzi S, Masselli M, De Lorenzo E, Accordi B, Cilia E, Crociani O, Amedei A, Veltroni M, D'amico M, Basso G, Becchetti A, Campana D, Arcangeli A (2011) Chemotherapy resistance in acute lymphoblastic leukemia requires hERG1 channels and is overcome by hERG1 blockers. Blood 117:902–914

Pillozzi S, Masselli M, Gasparoli L, D'amico M, Polletta L, Veltroni M, Favre C, Basso G, Becchetti A, Arcangeli A (2016) Macrolide antibiotics exert antileukemic effects by modulating the autophagic flux through inhibition of hERG1 potassium channels. Blood Cancer J 6:e423

Pillozzi S, D'amico M, Bartoli G, Gasparoli L, Petroni G, Crociani O, Marzo T, Guerriero A, Messori L, Severi M, Udisti R, Wulff H, Chandy KG, Becchetti A, Arcangeli A (2018) The combined activation of K(Ca)3.1 and inhibition of K(v)11.1/hERG1 currents contribute to overcome Cisplatin resistance in colorectal cancer cells. Br J Cancer 118:200–212

Pongrakhananon V, Chunhacha P, Chanvorachote P (2013) Ouabain suppresses the migratory behavior of lung cancer cells. PLoS One 8:e68623

Popov S, Venetsanou K, Chedrese P, Pinto V, Takemori H, Franco-Cereceda A, Eriksson P, Mochizuki N, Soares-Da-Silva P, Bertorello A (2012) Increases in intracellular sodium activate transcription and gene expression via the salt-inducible kinase 1 network in an atrial myocyte cell line. Am J Physiol Heart Circ Physiol 303:H57–H65

Post RL, Kume S, Tobin T, Orcutt B, Sen AK (1969) Flexibility of an active center in sodium-plus-potassium adenosine triphosphatase. J Gen Physiol 54:306–326

Potier M, Joulin V, Roger S, Besson P, Jourdan ML, Leguennec JY, Bougnoux P, Vandier C (2006) Identification of SK3 channel as a new mediator of breast cancer cell migration. Mol Cancer Ther 5:2946–2953

Poupon L, Lamoine S, Pereira V, Barriere DA, Lolignier S, Giraudet F, Aissouni Y, Meleine M, Prival L, Richard D, Kerckhove N, Authier N, Balayssac D, Eschalier A, Lazdunski M, Busserolles J (2018) Targeting the TREK-1 potassium channel via riluzole to eliminate the neuropathic and depressive-like effects of oxaliplatin. Neuropharmacology 140:43–61

Prevarskaya N, Skryma R, Shuba Y (2011) Calcium in tumour metastasis: new roles for known actors. Nat Rev Cancer 11:609–618

Quast SA, Berger A, Buttstädt N, Friebel K, Schönherr R, Eberle J (2012) General Sensitization of melanoma cells for TRAIL-induced apoptosis by the potassium channel inhibitor TRAM-34 depends on release of SMAC. PLoS One 7:e39290

Raderer M, Depisch D, Haider K, Kwasny W, Djavanmard M, Scheithauer W (1993) A phase I/II study of quinidine, a potential multidrug resistance-reversing agent, in combination with pirarubicin in patients with advanced refractory breast cancer. Oncol Res Treat 16:450–453

Rajamanickam S, Panneerdoss S, Gorthi A, Timilsina S, Onyeagucha B, Kovalskyy D, Ivanov D, Hanes MA, Vadlamudi RK, Chen Y, Bishop AJ, Arbiser JL, Rao MK (2016) Inhibition of FoxM1-mediated DNA repair by imipramine blue suppresses breast cancer growth and metastasis. Clin Cancer Res 22:3524–3536

Reddy JP, Dawood S, Mitchell M, Debeb BG, Bloom E, Gonzalez-Angulo AM, Sulman EP, Buchholz TA, Woodward WA (2015) Antiepileptic drug use improves overall survival in breast cancer patients with brain metastases in the setting of whole brain radiotherapy. Radiother Oncol 117:308–314

Rim HK, Cho S, Shin DH, Chung KS, Cho YW, Choi JH, Lee JY, Lee KT (2014) T-type Ca2+ channel blocker, KYS05090 induces autophagy and apoptosis in A549 cells through inhibiting glucose uptake. Molecules 19:9864–9875

Roger S, Besson P, Le Guennec JY (2003) Involvement of a novel fast inward sodium current in the invasion capacity of a breast cancer cell line. Biochim Biophys Acta 1616:107–111

Roger S, Le Guennec J-Y, Besson P (2004) Particular sensitivity to calcium channel blockers of the fast inward voltage-dependent sodium current involved in the invasive properties of a metastastic breast cancer cell line. Br J Pharmacol 141:610–615

Roger S, Jelassi B, Couillin I, Pelegrin P, Besson P, Jiang LH (2015) Understanding the roles of the P2X7 receptor in solid tumour progression and therapeutic perspectives. Biochim Biophys Acta 1848:2584–2602

Rojas E, Corchete L, San Segundo L, Martínez-Blanch JF, Codoñer FM, Paíno T, Puig N, García-Sanz R, Mateos MV, Ocio EM, Misiewicz-Krzeminska I, Gutiérrez NC (2017) Amiloride, an old diuretic drug, is a potential therapeutic agent for multiple myeloma. Clin Cancer Res 23 (21):6602–6615

Ru Q, Tian X, Pi MS, Chen L, Yue K, Xiong Q, Ma BM, Li CY (2015) Voltage-gated K+ channel blocker quinidine inhibits proliferation and induces apoptosis by regulating expression of microRNAs in human glioma U87MG cells. Int J Oncol 46:833–840

Sales TT, Resende FF, Chaves NL, Titze-De-Almeida SS, Bao SN, Brettas ML, Titze-De-Almeida R (2016) Suppression of the Eag1 potassium channel sensitizes glioblastoma cells to injury caused by temozolomide. Oncol Lett 12:2581–2589

Sato K, Ishizuka J, Cooper CW, Chung DH, Tsuchiya T, Uchida T, Rajaraman S, Townsend CM Jr, Thompson JC (1994) Inhibitory effect of calcium channel blockers on growth of pancreatic cancer cells. Pancreas 9:193–202

Sauter DRP, Novak I, Pedersen SF, Larsen EH, Hoffmann EK (2015) ANO1 (TMEM16A) in pancreatic ductal adenocarcinoma (PDAC). Pflugers Arch 467:1495–1508

Schmeel LC, Schmeel FC, Kim Y, Blaum-Feder S, Endo T, Schmidt-Wolf IG (2015) Flunarizine exhibits in vitro efficacy against lymphoma and multiple myeloma cells. Anticancer Res 35:1369–1376

Schneditz G, Elias JE, Pagano E, Zaeem Cader M, Saveljeva S, Long K, Mukhopadhyay S, Arasteh M, Lawley TD, Dougan G, Bassett A, Karlsen TH, Kaser A, Kaneider NC (2019) GPR35 promotes glycolysis, proliferation, and oncogenic signaling by engaging with the sodium potassium pump. Sci Signal 12(562):eaau9048

Schwab A, Fabian A, Hanley PJ, Stock C (2012) Role of ion channels and transporters in cell migration. Physiol Rev 92:1865–1913

Seo Y, Park J, Kim M, Lee HK, Kim JH, Jeong JH, Namkung W (2015) Inhibition of ANO1/TMEM16A chloride channel by idebenone and its cytotoxicity to cancer cell lines. PLoS One 10:e0133656

Seo Y, Ryu K, Park J, Jeon D-K, Jo S, Lee HK, Namkung W (2017) Inhibition of ANO1 by luteolin and its cytotoxicity in human prostate cancer PC-3 cells. PLoS One 12:e0174935

Seo Y, Kim J, Chang J, Kim SS, Namkung W, Kim I (2018) Synthesis and biological evaluation of novel Ani9 derivatives as potent and selective ANO1 inhibitors. Eur J Med Chem 160:245–255

Seol HS, Lee SE, Song JS, Lee HY, Park S, Kim I, Singh SR, Chang S, Jang SJ (2016) Glutamate release inhibitor, riluzole, inhibited proliferation of human hepatocellular carcinoma cells by elevated ROS production. Cancer Lett 382:157–165

Sette A, Spadavecchia J, Landoulsi J, Casale S, Haye B, Crociani O, Arcangeli A (2013) Development of novel anti-Kv 11.1 antibody-conjugated PEG-TiO2 nanoparticles for targeting pancreatic ductal adenocarcinoma cells. J Nanopart Res 15:2111

Sezzi ML, De Luca G, Materazzi M, Bellelli L (1985) Effects of a calcium-antagonist (flunarizine) on cancer cell movement and phagocytosis. Anticancer Res 5:265–271

Shao X-D, Wu K, Hao Z-M, Hong L, Zhang J, Fan D (2005) The potent inhibitory effects of cisapride, a specific blocker for human ether-a-go-go-related gene (HERG) channel, on gastric cancer cells. Cancer Biol Ther 4:295–301

Shapovalov G, Ritaine A, Skryma R, Prevarskaya N (2016) Role of TRP ion channels in cancer and tumorigenesis. Semin Immunopathol 38:357–369

Shen MR, Droogmans G, Eggermont J, Voets T, Ellory JC, Nilius B (2000) Differential expression of volume-regulated anion channels during cell cycle progression of human cervical cancer cells. J Physiol 529(Pt 2):385–394

Shen J-J, Zhan Y-C, Li H-Y, Wang Z (2020) Ouabain impairs cancer metabolism and activates AMPK-Src signaling pathway in human cancer cell lines. Acta Pharmacol Sin 41:110–118

Shi XN, Li H, Yao H, Liu X, Li L, Leung KS, Kung HF, Lu D, Wong MH, Lin MC (2015) In silico identification and in vitro and in vivo validation of anti-psychotic drug fluspirilene as a potential CDK2 inhibitor and a candidate anti-cancer drug. PLoS One 10:e0132072

Shin DH, Leem DG, Shin JS, Kim JI, Kim KT, Choi SY, Lee MH, Choi JH, Lee KT (2018) Compound K induced apoptosis via endoplasmic reticulum Ca(2+) release through ryanodine receptor in human lung cancer cells. J Ginseng Res 42:165–174

Siekmann W, Tina E, Von Sydow AK, Gupta A (2019) Effect of lidocaine and ropivacaine on primary (SW480) and metastatic (SW620) colon cancer cell lines. Oncol Lett 18:395–401

Simon A, Yang M, Marrison JL, James AD, Hunt MJ, O'toole PJ, Kaye PM, Whittington MA, Chawla S, Brackenbury WJ (2020) Metastatic breast cancer cells induce altered microglial morphology and electrical excitability in vivo. J Neuroinflammation 17:87

Singh H, Stefani E, Toro L (2012) Intracellular BK(Ca) (iBK(Ca)) channels. J Physiol 590:5937–5947

Song Y, Gao J, Guan L, Chen X, Gao J, Wang K (2018) Inhibition of ANO1/TMEM16A induces apoptosis in human prostate carcinoma cells by activating TNF-α signaling. Cell Death Dis 9:703

Sparks RL, Pool TB, Smith NK, Cameron IL (1983) Effects of amiloride on tumor growth and intracellular element content of tumor cells in vivo. Cancer Res 43:73–77

Speyer CL, Smith JS, Banda M, Devries JA, Mekani T, Gorski DH (2012) Metabotropic glutamate receptor-1: a potential therapeutic target for the treatment of breast cancer. Breast Cancer Res Treat 132:565–573

Squecco R, Tani A, Zecchi-Orlandini S, Formigli L, Francini F (2015) Melatonin affects voltage-dependent calcium and potassium currents in MCF-7 cell line cultured either in growth or differentiation medium. Eur J Pharmacol 758:40–52

Stettner M, Kramer G, Strauss A, Kvitkina T, Ohle S, Kieseier BC, Thelen P (2012) Long-term antiepileptic treatment with histone deacetylase inhibitors may reduce the risk of prostate cancer. Eur J Cancer Prev 21:55–64

Su C-K, Chou C-T, Lin K-L, Liang W-Z, Cheng J-S, Chang H-T, Chen IS, Lu T, Kuo C-C, Yu C-C, Shieh P, Kuo D-H, Chen F-A, Jan C-R (2016) Effect of protriptyline on [Ca2+]i and viability in MG63 human osteosarcoma cells. Toxicol Mech Methods 26:580–587

Sun YH, Gao X, Tang YJ, Xu CL, Wang LH (2006) Androgens induce increases in intracellular calcium via a G protein-coupled receptor in LNCaP prostate cancer cells. J Androl 27:671–678

Sun H, Luo L, Lal B, Ma X, Chen L, Hann CL, Fulton AM, Leahy DJ, Laterra J, Li M (2016) A monoclonal antibody against KCNK9 K(+) channel extracellular domain inhibits tumour growth and metastasis. Nat Commun 7:10339

Sun X, Wei Q, Cheng J, Bian Y, Tian C, Hu Y, Li H (2017) Enhanced Stim1 expression is associated with acquired chemo-resistance of cisplatin in osteosarcoma cells. Hum Cell 30:216–225

Sun R, He X, Jiang X, Tao H (2019) The new role of riluzole in the treatment of pancreatic cancer through the apoptosis and autophagy pathways. J Cell Biochem. https://doi.org/10.1002/jcb.29533

Szabó I, Bock J, Grassmé H, Soddemann M, Wilker B, Lang F, Zoratti M, Gulbins E (2008) Mitochondrial potassium channel Kv1.3 mediates Bax-induced apoptosis in lymphocytes. Proc Natl Acad Sci U S A 105:14861–14866

Szabò I, Soddemann M, Leanza L, Zoratti M, Gulbins E (2011) Single-point mutations of a lysine residue change function of Bax and Bcl-xL expressed in Bax- and Bak-less mouse embryonic fibroblasts: novel insights into the molecular mechanisms of Bax-induced apoptosis. Cell Death Differ 18:427–438

Takada M, Fujimoto M, Motomura H, Hosomi K (2016) Inverse association between sodium channel-blocking antiepileptic drug use and cancer: data mining of spontaneous reporting and claims databases. Int J Med Sci 13:48–59

Teichmann M, Kretschy N, Kopf S, Jarukamjorn K, Atanasov AG, Viola K, Giessrigl B, Saiko P, Szekeres T, Mikulits W, Dirsch VM, Huttary N, Krieger S, Jager W, Grusch M, Dolznig H, Krupitza G (2014) Inhibition of tumour spheroid-induced prometastatic intravasation gates in the lymph endothelial cell barrier by carbamazepine: drug testing in a 3D model. Arch Toxicol 88:691–699

Thebault S, Flourakis M, Vanoverberghe K, Vandermoere F, Roudbaraki M, Lehen'kyi V, Slomianny C, Beck B, Mariot P, Bonnal JL, Mauroy B, Shuba Y, Capiod T, Skryma R, Prevarskaya N (2006) Differential role of transient receptor potential channels in Ca2+ entry and proliferation of prostate cancer epithelial cells. Cancer Res 66:2038–2047

Thomas D, Gut B, Karsai S, Wimmer AB, Wu K, Wendt-Nordahl G, Zhang W, Kathofer S, Schoels W, Katus HA, Kiehn J, Karle CA (2003) Inhibition of cloned HERG potassium channels by the antiestrogen tamoxifen. Naunyn Schmiedebergs Arch Pharmacol 368:41–48

Thurber AE, Nelson M, Frost CL, Levin M, Brackenbury WJ, Kaplan DL (2017) IK channel activation increases tumor growth and induces differential behavioral responses in two breast epithelial cell lines. Oncotarget 8:42382–42397

Timar J, Chopra H, Rong X, Hatfield JS, Fligiel SE, Onoda JM, Taylor JD, Honn KV (1992) Calcium channel blocker treatment of tumor cells induces alterations in the cytoskeleton, mobility of the integrin alpha IIb beta 3 and tumor-cell-induced platelet aggregation. J Cancer Res Clin Oncol 118:425–434

Turner KL, Honasoge A, Robert SM, Mcferrin MM, Sontheimer H (2014) A proinvasive role for the Ca(2+) -activated K(+) channel KCa3.1 in malignant glioma. Glia 62:971–981

Urrego D, Tomczak AP, Zahed F, Stuhmer W, Pardo LA (2014) Potassium channels in cell cycle and cell proliferation. Philos Trans R Soc Lond B Biol Sci 369:20130094

Uzun S, Altun S, Bugan İ (2017) Anti-metastatic effect of riluzole on Mat-LyLu rat prostate cancer cell line. Ann Oncol 28:x3

Valerie NC, Dziegielewska B, Hosing AS, Augustin E, Gray LS, Brautigan DL, Larner JM, Dziegielewski J (2013) Inhibition of T-type calcium channels disrupts Akt signaling and promotes apoptosis in glioblastoma cells. Biochem Pharmacol 85:888–897

Van Der Hoeven D, Cho K-J, Ma X, Chigurupati S, Parton RG, Hancock JF (2013) Fendiline inhibits K-ras plasma membrane localization and blocks K-ras signal transmission. Mol Cell Biol 33:237

Vandenberg JI, Perry MD, Perrin MJ, Mann SA, Ke Y, Hill AP (2012) hERG K(+) channels: structure, function, and clinical significance. Physiol Rev 92:1393–1478

Vila-Carriles WH, Kovacs GG, Jovov B, Zhou Z-H, Pahwa AK, Colby G, Esimai O, Gillespie GY, Mapstone TB, Markert JM, Fuller CM, Bubien JK, Benos DJ (2006) Surface expression of ASIC2 inhibits the amiloride-sensitive current and migration of glioma cells. J Biol Chem 281:19220–19232

Vodnala SK, Eil R, Kishton RJ, Sukumar M, Yamamoto TN, Ha NH, Lee PH, Shin M, Patel SJ, Yu Z, Palmer DC, Kruhlak MJ, Liu X, Locasale JW, Huang J, Roychoudhuri R, Finkel T, Klebanoff CA, Restifo NP (2019) T cell stemness and dysfunction in tumors are triggered by a common mechanism. Science 363:eaau0135

Voloshyna I, Besana A, Castillo M, Matos T, Weinstein IB, Mansukhani M, Robinson RB, Cordon-Cardo C, Feinmark SJ (2008) TREK-1 is a novel molecular target in prostate cancer. Cancer Res 68:1197–1203

Walker AJ, Card T, Bates TE, Muir K (2011) Tricyclic antidepressants and the incidence of certain cancers: a study using the GPRD. Br J Cancer 104:193–197

Wang H, Zhang Y, Cao L, Han H, Wang J, Yang B, Nattel S, Wang Z (2002) HERG K+ channel, a regulator of tumor cell apoptosis and proliferation. Cancer Res 62:4843–4848

Wang GX, Hatton WJ, Wang GL, Zhong J, Yamboliev I, Duan D, Hume JR (2003) Functional effects of novel anti-ClC-3 antibodies on native volume-sensitive osmolyte and anion channels in cardiac and smooth muscle cells. Am J Physiol Heart Circ Physiol 285:H1453–H1463

Wang J, Xu YQ, Liang YY, Gongora R, Warnock DG, Ma HP (2007) An intermediate-conductance Ca(2+)-activated K (+) channel mediates B lymphoma cell cycle progression induced by serum. Pflugers Arch 454:945–956

Wang L, Ma W, Zhu L, Ye D, Li Y, Liu S, Li H, Zuo W, Li B, Ye W, Chen L (2012) ClC-3 is a candidate of the channel proteins mediating acid-activated chloride currents in nasopharyngeal carcinoma cells. Am J Physiol Cell Physiol 303:C14–C23

Wang Y, Yang Z, Meng Z, Cao H, Zhu G, Liu T, Wang X (2013) Knockdown of TRPM8 suppresses cancer malignancy and enhances epirubicin-induced apoptosis in human osteosarcoma cells. Int J Biol Sci 10:90–102

Wang Y, Lonard DM, Yu Y, Chow DC, Palzkill TG, Wang J, Qi R, Matzuk AJ, Song X, Madoux F, Hodder P, Chase P, Griffin PR, Zhou S, Liao L, Xu J, O'malley BW (2014) Bufalin is a potent small-molecule inhibitor of the steroid receptor coactivators SRC-3 and SRC-1. Cancer Res 74:1506–1517

Wannous R, Bon E, Gillet L, Chamouton J, Weber G, Brisson L, Gore J, Bougnoux P, Besson P, Roger S, Chevalier S (2015) Suppression of PPARbeta, and DHA treatment, inhibit NaV1.5 and NHE-1 pro-invasive activities. Pflugers Arch 467:1249–1259

Watkins S, Sontheimer H (2011) Hydrodynamic cellular volume changes enable glioma cell invasion. J Neurosci 31:17250–17259

Weaver AK, Liu X, Sontheimer H (2004) Role for calcium-activated potassium channels (BK) in growth control of human malignant glioma cells. J Neurosci Res 78:224–234

Wheler JJ, Janku F, Falchook GS, Jackson TL, Fu S, Naing A, Tsimberidou AM, Moulder SL, Hong DS, Yang H, Piha-Paul SA, Atkins JT, Garcia-Manero G, Kurzrock R (2014) Phase I study of anti-VEGF monoclonal antibody bevacizumab and histone deacetylase inhibitor valproic acid in patients with advanced cancers. Cancer Chemother Pharmacol 73:495–501

Williams S, Bateman A, O'kelly I (2013) Altered expression of two-pore domain potassium (K2P) channels in cancer. PLoS One 8:e74589

Wilson LE, D'aloisio AA, Sandler DP, Taylor JA (2016) Long-term use of calcium channel blocking drugs and breast cancer risk in a prospective cohort of US and Puerto Rican women. Breast Cancer Res 18:61

Wonderlin WF, Woodfork KA, Strobl JS (1995) Changes in membrane potential during the progression of MCF-7 human mammary tumor cells through the cell cycle. J Cell Physiol 165:177–185

Woods N, Trevino J, Coppola D, Chellappan S, Yang S, Padmanabhan J (2015) Fendiline inhibits proliferation and invasion of pancreatic cancer cells by interfering with ADAM10 activation and β-catenin signaling. Oncotarget 6:35931–35948

Wu Y, Gao B, Xiong Q-J, Wang Y-C, Huang D-K, Wu W-N (2017) Acid-sensing ion channels contribute to the effect of extracellular acidosis on proliferation and migration of A549 cells. Tumor Biol 39. https://doi.org/10.1177/1010428317705750

Xia Z, Bergstrand A, Depierre JW, Nassberger L (1999) The antidepressants imipramine, clomipramine, and citalopram induce apoptosis in human acute myeloid leukemia HL-60 cells via caspase-3 activation. J Biochem Mol Toxicol 13:338–347

Xie C, Liu G, Liu J, Huang Z, Wang F, Lei X, Wu X, Huang S, Zhong D, Xu X (2012) Anti-proliferative effects of anandamide in human hepatocellular carcinoma cells. Oncol Lett 4:403–407

Xie B, Zhao R, Bai B, Wu Y, Xu Y, Lu S, Fang Y, Wang Z, Maswikiti EP, Zhou X, Pan H, Han W (2018) Identification of key tumorigenesis related genes and their microRNAs in colon cancer. Oncol Rep 40:3551–3560

Xu S, Liu C, Ma Y, Ji HL, Li X (2016) Potential roles of amiloride-sensitive sodium channels in cancer development. Biomed Res Int 2016:2190216

Xu G, Fang Z, Clark LH, Sun W, Yin Y, Zhang R, Sullivan SA, Tran AQ, Kong W, Wang J, Zhou C, Bae-Jump VL (2018) Topiramate exhibits anti-tumorigenic and metastatic effects in ovarian cancer cells. Am J Transl Res 10:1663–1676

Xuan W, Zhao H, Hankin J, Chen L, Yao S, Ma D (2016) Local anesthetic bupivacaine induced ovarian and prostate cancer apoptotic cell death and underlying mechanisms in vitro. Sci Rep 6:26277–26277

Xue H, Wang Y, Maccormack TJ, Lutes T, Rice C, Davey M, Dugourd D, Ilenchuk TT, Stewart JM (2018) Inhibition of transient receptor potential vanilloid 6 channel, elevated in human ovarian cancers, reduces tumour growth in a xenograft model. J Cancer 9:3196–3207

Yamaci RF, Fraser SP, Battaloglu E, Kaya H, Erguler K, Foster CS, Djamgoz MBA (2017) Neonatal Nav1.5 protein expression in normal adult human tissues and breast cancer. Pathol Res Pract 213:900–907

Yamamura H, Ugawa S, Ueda T, Shimada S (2008) Expression analysis of the epithelial Na+ channel delta subunit in human melanoma G-361 cells. Biochem Biophys Res Commun 366:489–492

Yang M, Brackenbury WJ (2013) Membrane potential and cancer progression. Front Physiol 4:185

Yang JL, Friedlander ML (2001) Effect of nifedipine in metastatic colon cancer with DNA mismatch repair gene defect. Lancet 357(9270):1767–1768

Yang DK, Kim SJ (2017) Desipramine induces apoptosis in hepatocellular carcinoma cells. Oncol Rep 38:1029–1034

Yang S, Zhang JJ, Huang XY (2009) Orai1 and STIM1 are critical for breast tumor cell migration and metastasis. Cancer Cell 15:124–134

Yang M, Kozminski DJ, Wold LA, Modak R, Calhoun JD, Isom LL, Brackenbury WJ (2012) Therapeutic potential for phenytoin: targeting Na(v)1.5 sodium channels to reduce migration and invasion in metastatic breast cancer. Breast Cancer Res Treat 134:603–615

Yang M, James AD, Suman R, Kasprowicz R, Nelson M, O'toole PJ, Brackenbury WJ (2020) Voltage-dependent activation of Rac1 by Nav 1.5 channels promotes cell migration. J Cell Physiol 235:3950–3972

Yildirim S, Altun S, Gumushan H, Patel A, Djamgoz MBA (2012) Voltage-gated sodium channel activity promotes prostate cancer metastasis in vivo. Cancer Lett 323:58–61

Yip D, Le MN, Chan JL, Lee JH, Mehnert JA, Yudd A, Kempf J, Shih WJ, Chen S, Goydos JS (2009) A phase 0 trial of riluzole in patients with resectable stage III and IV melanoma. Clin Cancer Res 15:3896–3902

Yoon JR, Whipple RA, Balzer EM, Cho EH, Matrone MA, Peckham M, Martin SS (2011) Local anesthetics inhibit kinesin motility and microtentacle protrusions in human epithelial and breast tumor cells. Breast Cancer Res Treat 129:691–701

Yuan SY, Cheng CL, Ho HC, Wang SS, Chiu KY, Su CK, Ou YC, Lin CC (2015) Nortriptyline induces mitochondria and death receptor-mediated apoptosis in bladder cancer cells and inhibits bladder tumor growth in vivo. Eur J Pharmacol 761:309–320

Zhang L, Barritt GJ (2004) Evidence that TRPM8 is an androgen-dependent Ca2+ channel required for the survival of prostate cancer cells. Cancer Res 64:8365–8373

Zhang H, Qian DZ, Tan YS, Lee K, Gao P, Ren YR, Rey S, Hammers H, Chang D, Pili R, Dang CV, Liu JO, Semenza GL (2008) Digoxin and other cardiac glycosides inhibit HIF-1 synthesis and block tumor growth. Proc Natl Acad Sci 105:19579–19586

Zhang C, Yuan XR, Li HY, Zhao ZJ, Liao YW, Wang XY, Su J, Sang SS, Liu Q (2015) Anti-cancer effect of metabotropic glutamate receptor 1 inhibition in human glioma U87 cells: involvement of PI3K/Akt/mTOR pathway. Cell Physiol Biochem 35:419–432

Zhang P, Yang X, Yin Q, Yi J, Shen W, Zhao L, Zhu Z, Liu J (2016) Inhibition of SK4 potassium channels suppresses cell proliferation, migration and the epithelial-mesenchymal transition in triple-negative breast cancer cells. PLoS One 11:e0154471

Zhang P, Liu X, Li H, Chen Z, Yao X, Jin J, Ma X (2017a) TRPC5-induced autophagy promotes drug resistance in breast carcinoma via CaMKKbeta/AMPKalpha/mTOR pathway. Sci Rep 7:3158

Zhang Y, Cruickshanks N, Yuan F, Wang B, Pahuski M, Wulfkuhle J, Gallagher I, Koeppel AF, Hatef S, Papanicolas C, Lee J, Bar EE, Schiff D, Turner SD, Petricoin EF, Gray LS, Abounader R (2017b) Targetable T-type calcium channels drive glioblastoma. Cancer Res 77:3479–3490

Zhao L, Zhao Y, Schwarz B, Mysliwietz J, Hartig R, Camaj P, Bao Q, Jauch K-W, Guba M, Ellwart JW, Nelson PJ, Bruns CJ (2016) Verapamil inhibits tumor progression of chemotherapy-resistant pancreatic cancer side population cells. Int J Oncol 49:99–110

Zhong J, Kong X, Zhang H, Yu C, Xu Y, Kang J, Yu H, Yi H, Yang X, Sun L (2012) Inhibition of CLIC4 enhances autophagy and triggers mitochondrial and ER stress-induced apoptosis in human glioma U251 cells under starvation. PLoS One 7:e39378

Zhou Z, Song J, Li W, Liu X, Cao L, Wan L, Tan Y, Ji S, Liang Y, Gong F (2017) The acid-sensing ion channel, ASIC2, promotes invasion and metastasis of colorectal cancer under acidosis by activating the calcineurin/NFAT1 axis. J Exp Clin Cancer Res 36(1):130

Zhou FM, Huang YY, Tian T, Li XY, Tang YB (2018) Knockdown of chloride channel-3 inhibits breast cancer growth in vitro and in vivo. J Breast Cancer 21:103–111

Zhu S, Zhou HY, Deng SC, Deng SJ, He C, Li X, Chen JY, Jin Y, Hu ZL, Wang F, Wang CY, Zhao G (2017) ASIC1 and ASIC3 contribute to acidity-induced EMT of pancreatic cancer through activating Ca(2+)/RhoA pathway. Cell Death Dis 8:e2806

Rev Physiol Biochem Pharmacol (2022) 183: 45–102
https://doi.org/10.1007/112_2020_28

Published online: 27 July 2020

Novel Therapeutic Approaches of Ion Channels and Transporters in Cancer

Ana Ramírez, Janice García-Quiroz, Luis Aguilar-Eslava, Yesennia Sánchez-Pérez, and Javier Camacho

Contents

Ana Ramírez and Janice García-Quiroz contributed equally to this work.

A. Ramírez
Facultad de Ciencias Químicas e Ingeniería, Universidad Autónoma de Baja California – Campus Tijuana, Tijuana, Baja California, Mexico

J. García-Quiroz
Departamento de Biología de la Reproducción Dr. Carlos Gual Castro, Instituto Nacional de Ciencias Médicas y Nutrición Salvador Zubirán, Ciudad de México, México

L. Aguilar-Eslava and J. Camacho (✉)
Department of Pharmacology, Centro de Investigación y de Estudios Avanzados del Instituto Politécnico Nacional, Mexico City, Mexico
e-mail: fcamacho@cinvestav.mx

Y. Sánchez-Pérez
Subdirección de Investigación Básica, Instituto Nacional de Cancerología, Mexico City, Mexico

Abstract The expression and function of many ion channels and transporters in cancer cells display major differences in comparison to those from healthy cells. These differences provide the cancer cells with advantages for tumor development. Accordingly, targeting ion channels and transporters have beneficial anticancer effects including inhibition of cancer cell proliferation, migration, invasion, metastasis, tumor vascularization, and chemotherapy resistance, as well as promoting apoptosis. Some of the molecular mechanisms associating ion channels and transporters with cancer include the participation of oxidative stress, immune response, metabolic pathways, drug synergism, as well as noncanonical functions of ion channels. This diversity of mechanisms offers an exciting possibility to suggest novel and more effective therapeutic approaches to fight cancer. Here, we review and discuss most of the current knowledge suggesting novel therapeutic approaches for cancer therapy targeting ion channels and transporters. The role and regulation of ion channels and transporters in cancer provide a plethora of exceptional opportunities in drug design, as well as novel and promising therapeutic approaches that may be used for the benefit of cancer patients.

Keywords Cancer · Immunotherapy · Ion channels · Mitochondria · Oxidative stress · Toxins · Transporters

1 Introduction

Cancer is one of the most devastating diseases; it generates profound emotional, financial, and physical stress both to the patients and family members. Besides, the cost of cancer treatment has a strong impact on the economy of any country. Current treatments include surgical resection, chemotherapy, radiation, and immunotherapy. However, many patients have a very poor response to current treatments and/or acquire resistance leading to cancer relapse. Medical research has been looking for more effective and specific drugs to improve the quality of life of cancer patients (Hoelder et al. 2012). One of the most novel approaches in cancer research is to study the role of ion channels and transporters as potential therapeutic targets for anticancer therapy. Since the recognition that ion channels and transporters play an important role in the carcinogenesis process, there has been great scientific interest in discovering new treatments using these genes and proteins as novel tools in oncology. Actually, several compounds targeting ion channels and transporters demonstrate promising potential to be used in cancer patients (Arcangeli and Becchetti 2010; Litan and Langhans 2015). The anticancer potential of these compounds is enhanced when different therapeutic approaches are considered, for instance, by its combination with antineoplastic drugs, immunotherapy, or other molecules targeting essential processes in cancer development including oxidative stress or metabolic pathways. Here, we review and discuss most of the current knowledge suggesting novel therapeutic approaches for cancer therapy by targeting ion channels and transporters. The expression and activity of ion channels and transporters in cancer have been reviewed in detail in several excellent articles of these series. Thus, before going into details of the topics of this review, first we will provide a general and brief panorama of ion channels and transporters in cancer.

1.1 Ion Channels and Transporters in Cancer

Potassium (K^+) channels are some of the most studied and deregulated channels in malignancies. The voltage-gated K^+ channels Kv10.1 (EAG1) and Kv11.1 (HERG) have been implicated in the pathogenesis of various cancers (Asher et al. 2010; Serrano-Novillo et al. 2019). Kv10.1 channel ectopic expression is associated with malignant transformation, tumor development, metastasis, and poor prognosis; channel overexpression has been observed in most of the human tumors (Pardo et al. 1999; Gavrilova-Ruch et al. 2002; Gessner and Heinemann 2003; Farias et al. 2004; Camacho 2006; Hemmerlein et al. 2006; Queiroz et al. 2006; Pardo and Stühmer 2008; Garcia-Becerra et al. 2010; Asher et al. 2011; Ortiz et al. 2011; Liu et al. 2015; Martinez et al. 2015; Serrano-Novillo et al. 2019). Inhibition of either its expression or activity decreases cancer cell proliferation both in vitro and in vivo (Pardo et al. 1999; Gomez-Varela et al. 2007; Garcia-Quiroz et al. 2014; Chavez-Lopez et al. 2015). Kv11.1 channel altered expression increases cell proliferation, angiogenesis, invasiveness, migration, and lymph node dissemination and decreases cell differentiation (He et al. 2020; Jehle et al. 2011). Overexpression of Kv11.1 channels has been observed in a variety of neoplastic tissues including endometrial, colorectal, esophageal, pancreatic, gastric, ovarian, breast, thyroid, and brain cancers, as well as leukemias (Jehle et al. 2011; Lastraioli et al. 2015a, b; Iorio et al. 2018; Lastraioli et al. 2019; Iorio et al. 2020; He et al. 2020). In gastric tumors, these channels participate in the PI3K/Akt-dependent pathway that induces hypoxia-inducible factors (HIF) and vascular endothelial growth factor (VEGF) to promote cancer progression (Crociani et al. 2014). Interestingly, Kv11.1 is also aberrantly expressed in human gastric dysplasia samples, representing a potential novel marker for progression toward gastric cancer (Lastraioli et al. 2019). In pancreatic ductal adenocarcinoma (PDAC) cells, Kv11.1 activity is essential to induce cell migration by modulating the f-actin organization (Manoli et al. 2019). In addition, Kv11.1 channels may serve as prognostic factors and potential targets for cancer treatment (He et al. 2020; Lastraioli et al. 2015b). Channel blockade reduces proliferation and migration and induces apoptosis in cancer cell lines and tissues (Roy et al. 2008; Jehle et al. 2011; Lastraioli et al. 2015). Interestingly, activation of Kv11.1 also promotes anticancer effects. In SKBr3 or MDA-MB-231 mammary gland adenocarcinoma cell lines, prolonged stimulation of Kv11.1 with the diphenylurea compound NS1643 triggered a senescence-like phenotype, arresting the cell cycle in the G0/G1 phase (Lansu and Gentile 2013). Besides, NS1643 treatment (6 mg/kg) of MDA-MB-231 cell-derived breast cancer xenografts generated significantly smaller tumors, expressed lower levels of Ki67, and showed increased expression of the senescence markers $p21^{waf/cip}$ and $p16^{INK4A}$ compared with untreated mice; these NS1643-treated animals did not show cardiac function alterations (Fukushiro-Lopes et al. 2018). Likewise, NS1643 treatment of the B-RAF-dependent melanoma cell line A375 (that expresses Kv11.3 channels but not Kv11.1), significantly reduced cell proliferation. This antiproliferative effect included lowering the expression of cell cycle promoters (cyclin E, cyclin D, and phosphorylated WEE1), as well as increasing senescence markers ($p21^{waf}$ and $p16^{INK4A}$) and autophagy markers (phosphorylation of ULK1 and LC3-II), suggesting that activation of Kv11.3 generates tumor suppression (Perez-Neut et al. 2016).

Similarly, high expression of Kv1.3 channels is detected in a great number of human malignancies including breast, colon, and prostate cancer (Comes et al. 2013; Huang and Jan 2014), and blockade of these type of channels inhibits cancer cell proliferation by arresting the cell cycle in the G1 phase (Teisseyre et al. 2015). Likewise, the expression of ATP-sensitive K^+ (K_{ATP}) channels has been observed in multiple malignancies, including bladder, gastric, and cervical cancer, as well as in glioma and hepatocellular carcinoma (Monen et al. 1998; Wondergem et al. 1998; Malhi et al. 2000; Qian et al. 2008; Huang et al. 2009; Núñez et al. 2013; Vazquez-Sanchez et al. 2018). Because K^+ channels have a high potential to be targeted in cancer diagnosis and treatment, several patents have been filed concerning these channels as tools for diagnostic or therapeutic purposes in oncology (D'Amico et al. 2013).

Calcium ions participate as second messengers in cellular homeostasis like gene transcription, proliferation, migration, autophagy, and apoptosis (Bootman et al. 2001; Harr and Distelhorst 2010; Varghese et al. 2019). Some of the most studied calcium channels in cancer are those from the ORAI family and the TRP (transient receptor potential) Ca^{2+} channel superfamily. ORAI channels are located in the plasma membrane and interact with the stromal interaction molecules (STIMs) located in the endoplasmic reticulum (ER). ORAI1 and ORAI3 isoforms are overexpressed in breast cancer (Lis et al. 2007; Azimi et al. 2014); in prostate cancer, these isoforms confer apoptosis resistance (Dubois et al. 2014). The TRP family consists of seven subfamilies: TRPC (canonical), TRPV (vanilloid), TRPM (melastatin), TRPP (polycystin), TRPA (ankyrin), TRPML (mucolipin), and TRPN (NOMPC-like); they are permeable to monovalent and divalent cations and are expressed in a variety of cell types, including sensory neurons (Clapham 2003). These channels are altered in various cancers favoring carcinogenesis mainly by dysfunction in Ca^{2+} signaling pathways (Miller and Zhang 2011; Nielsen et al. 2014). Changes in the expression of several TRP channels have been implicated in prostate, breast, and lung cancer progression, as well as in ovarian cancer differentiation (Zeng et al. 2013; Azimi et al. 2014; Deliot and Constantin 2015). The SERCA-ATPase pump responsible for reloading the sarco/endoplasmic reticulum with Ca^{2+} and regulating cytosolic free Ca^{2+} has been associated with different tumors including gastric, colon, prostate, lung, and breast cancers (Denmeade and Isaacs 2005; Korošec et al. 2006; Dang and Rao 2016; Izquierdo-Torres et al. 2017). SERCA3 is downregulated or absent in colon, gastric, breast, and lung cancers (Gélébart et al. 2002; Papp and Brouland 2011; Arbabian et al. 2013), whereas SERCA2 is overexpressed in colon cancer and correlates with metastasis and decreased survival in patients (Chung et al. 2006). Voltage-gated calcium (Cav) channels are also involved in the development and progression of diverse types of cancer (Wang et al. 2015a; Martinez-Delgado and Felix 2017). These channels are organized into three subfamilies: (1) L-type, (2) P/Q-, N- and R-type, and (3) T-type channels (Gao et al. 2000, 2001; Buchanan and McCloskey 2016). Several Cav's channels are overexpressed in a variety of cancers including leukemia, sarcomas, brain, colorectal, gastric, lung, ovarian, pancreas, breast, uterus, and prostate cancer (Wang et al. 2015a; Taylor et al. 2008). Upon activation of L-type channels, gene regulation can be addressed through the activation of transcription factors such as cAMP-response-element-binding protein (CREB), nuclear factor of activated T cells (NFAT), and downstream of the regulatory element antagonist modulator (DREAM); these transcription factors favor cancer cell proliferation, invasion, and metastasis (Shankar et al. 2005; Barbado et al. 2009; Mancini and Toker 2009; Xiao et al. 2010). The blockade of T-type channel expression or activity reduces cancer cell proliferation and induces apoptosis (Bertolesi et al. 2002). Interestingly, Ca^{2+} channel blockers approved for the treatment of other conditions may be repurposed to treat some cancers (Buchanan and McCloskey 2016). Actually, the use of Ca^{2+} channel blockers for the treatment of hypertension, epilepsy, and other conditions may be inversely correlated with prostate cancer (Fitzpatrick et al. 2001; Debes et al. 2004). Lastly, voltage-gated sodium channels have been mainly associated to the metastatic potential of several cancers (Arcangeli and Becchetti 2010; Litan and Langhans 2015).

One of the major problems in cancer treatment is chemoresistance produced partly because of drug extrusion by ATP-binding cassette (ABC) transporters. Although the etiology of multidrug resistance (MDR) is multifactorial, the most common mechanism in the majority of resistant cell lines involves the overexpression of P-glycoprotein (Silva et al. 2015). Other transporters related to drug efflux are multidrug resistance-associated protein1 (MRP-1) and multixenobiotic resistance (MXR) (Xue and Liang 2012). Interestingly, some ion channels and transporters have been associated with therapy resistance by diverse mechanisms; in accordance, ion channel inhibitors restore chemotherapy sensitivity of different cancer cells (Kischel et al. 2019).

In summary, searching for high-efficacy therapies modulating the activity and/or expression of ion channels and transporters is a very active and promising field in cancer. Table 1 shows some examples of the potential therapeutic, diagnostic, and/or prognostic use of ion channels and transporters in cancer including some clinical trials in cancer patients.

A major opportunity for cancer treatment comes by taking advantage of the molecular mechanisms associating ion channels and transporters with cancer. Relevant cellular processes involved in cancer progression including oxidative stress, immune response, and mitochondrial activity, as well as chemoresistance, have been associated with the different roles of ion channels and transporters in tumor progression. Therefore, novel therapeutic approaches may be suggested by simultaneously targeting ion channels and transporters and the cell processes or molecular mechanisms involved. These approaches should provide better and potentiated effects of cancer therapies. Following, we will go into details of the novel therapeutic approaches suggested by a number of groups, based on the participation of ion channels and transporters in cancer.

2 Association of Oxidative Stress with Ion Channels and Transporters in Cancer: Friends and Foes

Reactive oxygen species (ROS) and reactive nitrogen species (RNS) participate in the regulation of metabolism, gene transcription, protein posttranslational modifications, ion transport, cell differentiation, proliferation, and migration, among other processes (Birben et al. 2012; Tochhawng et al. 2013). When ROS/RNS rises beyond their physiological levels, oxidative stress is produced (Chio and Tuveson 2017) potentially leading to DNA mutations, gene transcription alterations, protein oxidation, lipid peroxidation, ion transport alterations, mutagenesis, and cell death (Rani et al. 2016; Poprac et al. 2017). Cancer cells display increased metabolic activity that leads to ROS/RNS overproduction, to counteract this oxidative stress; they have larger pools of antioxidants (Denicola et al. 2011; Harris et al. 2015;

Table 1 Examples of the potential clinical use of ion channels and transporters in cancer

Ion channel/ transporter	Potential clinical use	Cancer type	Preclinical and clinical findings	References
SERCA	Therapeutic target	Prostate cancer Breast cancer Tumor-associated vascular endothelial cells	Blockers like thapsigargin induce cell death in prostate and breast cancer cells and decrease tumor growth in human prostate and non-prostate cancer xenografts in vivo Phase II clinical trial with mipsagargin and prior treatment with sorafenib reduced tumor blood flow in five patients with hepatocellular carcinoma	Denmeade et al. (2012), Sehgal et al. (2017), Mahalingam et al. (2019)
ORAI3	Prognostic marker	Lung cancer Breast cancer	Overexpression correlates with lung adenocarcinoma aggressiveness and poor clinical outcome in breast cancer	Benzerdjeb et al. (2016), Azimi et al. (2019)
SOCE	Therapeutic target	Liver cancer Glioblastoma	Inhibition with carboxyamidotriazole has antitumor activity in vitro and in vivo Multicenter Phase IB Trial of carboxyamidotriazole plus temozolomide in recurrent glioblastoma and anaplastic gliomas with promising activity and good brain penetration	Luzzi et al. (1998), Enfissi et al. (2004) Omuro et al. (2018)
TRPM2-AS	Prognostic marker	Prostate cancer	Overexpression is associated with poor clinical outcome	Orfanelli et al. (2014)
TRPM2	Therapeutic target	Solid tumors	Inhibition of the channel increases cytotoxicity in cell lines and reduces tumor growth in xenografts of neuroblastoma cells, as well as increasing chemotherapy sensitivity Downregulation of the channel reduces	Chen et al. (2014), Koh et al. (2015), Bao et al. (2016), Almasi et al. (2018)

(continued)

Table 1 (continued)

Ion channel/ transporter	Potential clinical use	Cancer type	Preclinical and clinical findings	References
			migration and invasion in gastric cancer cells and tumor growth in vivo	
TRPC5	Therapeutic target	Breast cancer Colorectal cancer	Inhibition reduces chemoresistance	Ma et al. (2012), Wang et al. (2015b)
TRPV6	Therapeutic target Diagnostic marker	Prostate cancer Ovarian cancer	Inhibitors SOR-C13 and SORC-27 used as a drug delivery system to enhance tumor detection and therapy Phase I study: SOR-C13 in patients with advanced tumors was well-tolerated and showed promising antitumor response	Bowen et al. (2013), Fu et al. (2017b), Xue et al. (2018)
CLIC1	Therapeutic target Prognosis marker	Head and neck cancer Ovarian cancer Lung cancer	Overexpression correlates with poor prognosis in oral squamous cell carcinoma and with intraperitoneal metastasis in epithelial ovarian cancers Inhibition increases ROS levels in lung cancer	Ye et al. (2015), Xu et al. (2018a), Lee et al. (2019)
SLC7A11	Therapeutic target	Glioma Pancreatic cancer Lung cancer	Blockage by sulphasalazine inhibits the entrance of intracellular cysteine necessary for GSH synthesis SLC7A11 inhibitors with temozolomide enhance cell death	Guan et al. (2009), Lo et al. (2010), Takeuchi et al. (2014), Sehm et al. (2016)
Kv10.1	Therapeutic target Tumor marker	Breast cancer Hepatocellular carcinoma Brain cancer	scFv62-TRAIL antibody in combination with chemotherapy drugs reduces tumor growth Astemizole reduces cell proliferation in vitro and tumor growth in vivo The clinical response to antidepressants is	Downie et al. (2008), Garcia-Quiroz et al. (2014), de Guadalupe Chavez-Lopez et al. (2015), Martinez et al. (2015), Hartung and Pardo (2016)

(continued)

Table 1 (continued)

Ion channel/ transporter	Potential clinical use	Cancer type	Preclinical and clinical findings	References
			associated with channel abundance in brain cancer patients	
Kv11.1	Therapeutic target Tumor and prognostic marker Prognostic marker	Breast cancer Pancreatic cancer Esophageal cancer Colon cancer Breast cancer	Anti-Kv 11.1 antibody-conjugated PEG-TiO2 nanoparticles display high specificity for Kv11.1 Channel activator NS1643 inhibits tumor growth in vivo Overexpression in resected esophageal squamous cell carcinomas is correlated with poor prognosis Protein expression of hERG1 and HIF-2α benefits patients for treatment with bevacizumab Channel expression is associated with prognosis	Sette et al. (2013), Fukushiro-Lopes et al. (2018), Ding et al. (2008), Iorio et al. (2018), Iorio et al. (2020)
Kv1.3	Therapeutic target	Lung cancer	Margatoxin significantly inhibited proliferation *in vitro* and *in vivo*	Jang et al. (2011)
mitoKv1.3	Therapeutic target	Melanoma Pancreatic cancer Leukemia	Inhibition by PCARBTP and PAPTP induces ROS and reduces tumor size in vivo preserving healthy cells Inhibition by clofazimine induces cell death in B-CLL cells	Leanza et al. (2013), (2017)
nfP2X$_7$	Therapeutic target	Skin cancer	Phase 1 clinical trial using an ointment with anti-nfP2X$_7$ antibodies (BIL010t), cause a reduction 65% of lesions in patients	Gilbert et al. (2017)
gBK	Therapeutic target	Glioma Lung cancer	Induces cytotoxic T lymphocyte response; tumor antigen for immunotherapy	Ge et al. (2012), Hoa et al. (2014)

(continued)

Table 1 (continued)

Ion channel/ transporter	Potential clinical use	Cancer type	Preclinical and clinical findings	References
nNav1.5	Therapeutic target Tumor marker	Breast cancer	Tumor antigen for immunotarget; biomarker for metastatic breast cancer	Chioni et al. (2005), Yamaci et al. (2017)
VDAC	Therapeutic target	Tumor cells expressing $HRAS^{V12}$ Fibrosarcoma Liver cancer	Erastin induces the anti-Warburg effect, mitochondria dysfunction, and induction of ROS by blockage of VDAC; it also reduces the synthesis of GSH by blockage of SLC7A11	Yagoda et al. (2007), Dixon et al. (2012), Maldonado et al. (2013)
mPTP	Therapeutic target	Colorectal cancer	Opening of the channel by JNK-dependent mPTP pathway induced by icaritin promotes cellular necrosis	Zhou et al. (2016)
UCP	Therapeutic target	Solid tumors	Inhibition by genipin sensitizes cancer cells to chemotherapy and reduces tumor growth in vivo	Shanmugam et al. (2018)

Beatty and Gladney 2015; Sullivan et al. 2016). In fact, the use of antioxidants in cancer is controversial, since they may either prevent tumor growth and genomic instability or favor tumor progression and migration (Bjelakovic et al. 2007; Klein et al. 2011; Porporato et al. 2014; Sayin et al. 2014; Le Gal et al. 2015; Harris et al. 2015; Prasad et al. 2017). For instance, in mouse models of B-RAF- and K-RAS-induced lung cancer, treatment with the antioxidants N-acetylcysteine (NAC) and vitamin E increased tumor cell proliferation and reduced survival by reducing ROS levels which leads to the reduction of p53 expression (Sayin et al. 2014). In melanoma, oxidative stress decreases metastasis in vivo; it is melanoma metastatic tumor cells overproduce glutathione and NADPH antioxidants, resisting the damage caused by oxidative stress and promoting metastasis (Piskounova et al. 2015). On the other hand, moderate to high levels of ROS/RNS in cancer cells promote initiation, proliferation, survival, and angiogenesis (Roderick and Cook 2008; Trachootham et al. 2009; Gorrini et al. 2013; Harris et al. 2015; Sullivan et al. 2016; Chio and Tuveson 2017).

Chemotherapy drugs like doxorubicin cause cell death by increasing the production of ROS/RNS, so, pro-oxidant drugs are also currently studied as anticancer options (Kong et al. 2000; Gorrini et al. 2013; Noh et al. 2015; Vilema-Enríquez et al. 2016). Actually, the strategy of delivering and augmenting the concentration of H_2O_2 in tumors has been proposed for lung cancer (Vilema-Enríquez et al. 2016).

Although various therapeutic approaches targeting the redox status in cancer cells have been proposed, clinical results remain elusive (Tong et al. 2015). Ion channels/transporters can be oxidized by direct interaction with ROS/RNS, particularly by H_2O_2, via their sulfhydryl groups and cysteine residues, or indirectly by altering signaling pathways that are involved in their regulation, expression, or function (Ramírez et al. 2016). Oxidative stress can increase $[Ca^{2+}]_i$ inducing protein phosphorylation and gene transcription, contributing to cancer cell survival (Roderick and Cook 2008). Depending on the duration, intensity, and type of oxidant, oxidative stress may cause either influx of Ca^{2+} into the cytosol via different channels and transporters in the plasma membrane or efflux of Ca^{2+} from the endoplasmic reticulum (ER), which in turn can cause Ca^{2+} overload that may lead to disruption of the mitochondrial metabolism and cell death (Ermak and Davies 2002). Thus, further research is needed to take advantage of the potential anticancer effects of oxidative stress and redox status. In this regard, the modulation of ion channels and transporters by ROS/RNS may bring a new therapeutic opportunity. Next, some potential candidates for this approach are discussed.

2.1 The SERCA-ATPase Pump and the Plasma Membrane Ca^{2+} ATPase

SERCA inhibitors have been proposed as an anticancer therapy since its blockage generates ER stress that leads to the activation of apoptotic pathways (Denmeade and Isaacs 2005). In breast cancer, the antioxidant and anti-inflammatory compound resveratrol induces the expression of SERCA3 decreasing cell viability (Izquierdo-Torres et al. 2017). Curcumin (a SERCA inhibitor) causes apoptosis by inducing ER stress in ovarian and thyroid cancer cells (Seo et al. 2016; Zhang et al. 2018). The blockage of SERCA with thapsigargin induces sustained elevation of $[Ca^{2+}]_i$ also leading to apoptosis in cancer cells (Denmeade and Isaacs 2005). Since the SERCA pump is widely expressed, the more specific thapsigargin-based prodrug mipsagargin has been used in a hepatocarcinoma (HCC) phase II clinical trial; the prodrug altered the tumor vasculature reducing tumor blood flow in HCC sites (Mahalingam et al. 2019).

The plasma membrane Ca^{2+} ATPase (PMCA) is responsible for pumping Ca^{+2} to the extracellular space and maintain $[Ca^{2+}]_i$ homeostasis. The platinum (II) complex [Pt(*O,O′*-acac)(γ-acac)(DMS)] decreased PMCA activity and induced higher levels of ROS by activating NADPH oxidase and mitochondrial ROS production in the chemotherapeutic-resistant breast cancer cell line MCF-7 (Muscella et al. 2011). Silencing of PMCA2 and PMCA4 combined with a Bcl-2 inhibitor (ABT-263) mediated cell death in MDA-MB-231 breast cancer cells (Curry et al. 2012, 2016).

2.2 ORAI Channels

Store-operated Ca^{2+} entry (SOCE) is the main mechanism for the entrance of Ca^{2+} in the cells; it is mediated by the STIMs Ca^{2+}sensors in the ER and the ORAI channels in the plasma membrane, both interact to restore the depletion of Ca^{2+} from the ER (Xie et al. 2016). SOCE inhibitors have antitumor activity in vitro, and some compounds have been studied in clinical trials (Chen et al. 2019). In fact, SOCE is necessary to induce cytotoxicity of cisplatin in non-small cell lung cancer cells, and depletion of STIM1 reduces the oxidative stress promoted by cisplatin (Gualdani et al. 2019). ORAIs and STIMs have been correlated with proliferation, apoptosis resistance, migration, and metastasis of many tumors (Fiorio Pla et al. 2016). ROS target ORAI channels, modulating $[Ca^{2+}]_i$, and H_2O_2 blocks Orai1 and Orai2, but not Orai3 because it lacks a cysteine residue at position 195 (Bogeski et al. 2010). Immune and cancer cells have a different Orai1/Orai3 isoform ratio in the cell membrane; this may alter Ca^{2+} signaling in oxidative stress because Orai1 can be blocked by H_2O_2 (Frisch et al. 2019). Orai3 is overexpressed and correlated with chemotherapy resistance in breast cancer cells (Hasna et al. 2018), besides Orai1 interacts with Kv10.1 channels and the secretory pathway Ca^{2+} ATPase (SPCA2) mediating a store-independent calcium entry (SICE) necessary to promote cell survival; interestingly the three proteins are overexpressed in aggressive tumor tissues (Peretti et al. 2019). Furthermore, Orai1 and Orai3 can interact with TRPC6 causing translocation of Orai channels to the plasma membrane; reduction of TRPC6 expression significantly inhibited SOCE in MCF-7 and MDA-MB-231 breast cancer cells (Jardin et al. 2018). Treatment with the phenolic compound (−)-oleocanthal downregulates TRPC6 channel expression reducing cell viability and migration of MCF-7 and MDA-MB-231 cells (Diez-Bello et al. 2019).

2.3 Members of the TRP Channel Family

One of the most studied families of ion channels in oxidative stress is the TRP family, among them TRPC5, TRPV1, and TRPA1 channels are directly activated by ROS and/or RNS by modification on their cysteine residues (Takahashi and Mori 2011); TRPM2 and TRPM7 may be activated via ROS-signaling pathways (Simon et al. 2013), although TRPM2 can also be directly activated by H_2O_2 in some cell types including microglia and pancreatic ß cells (Kühn et al. 2005).

In most nonmalignant cells, TRPM2 channels participate in a variety of cellular processes including insulin release, inflammatory response, and cell migration; and they are considered as redox sensors that induce Ca^{2+} influx leading to cell death by intracellular Ca^{2+} overload (Lange et al. 2009; Sumoza-Toledo et al. 2011; Faouzi and Penner 2014). TRPM2 channels have also been found in the nucleus, but its role is unclear (Zeng et al. 2010; Hopkins et al. 2015; Zhao et al. 2016). H_2O_2 can mediate TRPM2 activation via mitochondrial ADPR release, which can bind directly to the NUDT9-H domain of the channel (Hara et al. 2002). However, in some cancers, activation of TRPM2 by moderate levels of ROS has been considered as a protective mechanism for the ongoing growth and survival (Chen et al. 2013; Blake et al. 2017). In vitro and in vivo studies demonstrate that TRPM2 supports cancer cell survival; for instance, in neuroblastoma cells the activation and expression of the full-length TRPM2 (TRPM2-L) channel protects cell viability by modulating the expression of the hypoxia-inducible factor (HIF)-1/2α; activation of Src, Pyk2, and CREB; and increasing the levels of forkhead box transcription factor 3a (FOXO3a) and superoxide dismutase 2 (Chen et al. 2013, 2014; Hirschler-Laszkiewicz et al. 2018). In xenografts of neuroblastoma cells, tumor growth was decreased by expressing the dominant-negative isoform TRPM2-S that inhibits the functional TRPM2-L (Chen et al. 2014; Bao et al. 2016). In gastric cancer cells, expression of TRPM2 is necessary to induce migration and invasion through the PTEN/Akt signaling pathway (Almasi et al. 2019a), and PTEN downregulation is correlated with advanced stages of gastric cancer (Zhu et al. 2013). Silencing TRPM2 in lung cancer cells (A549 and H1299) increases ROS/RNS levels, induces G2/M arrest, activates JNK signaling pathway, and in SCID mice xenografts reduces cell migration and tumor growth (Almasi et al. 2019b). In prostate cancer, melanoma, and lung cancer, overexpression of the long noncoding TRPM2-AS (an antisense transcript for TRPM2 channel) has been correlated with increased proliferation and poor prognosis in patients (Orfanelli et al. 2008, 2014; Huang et al. 2017). Interestingly, inhibition of TRPM2 increases ROS, causes mitochondria dysfunction, impairs autophagy, and promotes sensitivity to chemotherapy in some cancer cells (Chen et al. 2014; Koh et al. 2015; Bao et al. 2016; Almasi et al. 2018). Thus, combining chemotherapeutic agents with TRPM2 inhibitors is a promising therapeutic approach, although possible side effects need further analysis because these channels participate in important physiological processes including protection against cardiac ischemia-reperfusion (Miller et al. 2014), activation of the immune response (Yamamoto et al. 2008), and insulin secretion from pancreatic B cells (Togashi et al. 2006).

In the case of TRPC5 channels, overexpression generates Ca^{2+} signals that activate NFATC3 (nuclear factor of activated T cells 3) which upregulates the synthesis of P-glycoprotein inducing chemotherapeutic drug efflux in adriamycin-resistant breast cancer cells (Ma et al. 2012). Extracellular vesicles released from breast cancer cells increase ROS which in turn activates autophagy and stimulates the release of growth-promoting factors in human mammary epithelial cells (HMECs) (Dutta et al. 2014). Interestingly, extracellular vesicles containing TRPC5 have been found in peripheral blood of breast cancer patients that underwent chemotherapy, suggesting a manner to transfer TRPC5 channels to other cells (Ma et al. 2014).

TRPA1 channels are activated by ROS by targeting cysteine residues in the intracellular site and are upregulated by NRF2, a transcription factor involved in protection against oxidative stress (Mukhopadhyay et al. 2011; Schaefer et al. 2013; Takahashi et al. 2018). Activation of TRPA1 generates Ca^{2+} influx stimulating proliferation pathways like RAS-ERK, PI3K/AKT, and mTOR, as well as triggering anti-apoptotic pathways. In xenograft tumor models, TRPA1 induces resistance to carboplatin (which induces ROS), and the inhibition of TRPA1 reduces tumor growth and increases chemotherapy sensitivity (Takahashi et al. 2018). In another context, TRPA1 is expressed in C-fiber nerves, and activation of the channel by chemotherapy drugs induces peripheral neuropathy; short-term treatments with antagonists have been suggested as a strategy for preventing peripheral neuropathy induced by chemotherapy (Trevisan et al. 2013). Furthermore, mice treated with doxorubicin and HC-030031 (a TRPA1 inhibitor) generated protection against doxorubicin cardiac injury (Wang et al. 2018b).

TRPV1 is also modulated by oxidizing agents potentiating its activity in neuronal tissues (Susankova et al. 2006; Özdemir et al. 2016). Combinations of antioxidants with chemotherapeutics (for instance, melatonin with doxorubicin or selenium with cisplatinum) in MCF-7 breast cancer cells promoted ROS production and apoptosis; this mechanism was due in part by inhibiting TRPV1 (Koşar et al. 2016; Sakallı et al. 2017). In contrast, a combination of the antioxidant alpha-lipoic acid (ALA) and cisplatinum increased TRPV1 activation resulting in increased ROS production, depolarization of the mitochondrial membrane, and apoptosis (Nur et al. 2017).

2.4 *Chloride Intracellular Channel Protein 1 (CLIC1)*

CLIC1 is considered a sensor and effector of oxidative stress; it is expressed in the nucleus and cytosol, but upon oxidation, a disulfide bond in cysteine residues of the CLIC1 monomer is formed, and it translocates to the plasma membrane as an active chloride channel (Littler et al. 2004). CLIC1 is overexpressed in various tumors including gastric, colon, and lung cancers, contributing in cell cycle progression, proliferation, migration, and invasion (Chen et al. 2007; Petrova et al. 2008; Averaimo et al. 2010; Wang et al. 2011). In the highly metastatic colon cancer LOVO cells and the SGC-7901 human gastric cancer cell line treated in hypoxic and reoxygenating conditions, CLIC1 channel expression is increased; inhibition of CLIC1 decreases ROS production and p-p38 MAPK/p-ERK levels, as well as reduces MMP-2 and MMP-9 protein levels which inhibits cell migration and invasion (Wang et al. 2014a; Zhao et al. 2015). CLIC1 silencing promotes apoptosis and decreases proliferation in human gallbladder cancer (He et al. 2018). Metformin inhibits CLIC1 reducing glioblastoma stem cell proliferation and invasiveness, compared to normal mesenchymal stem cells (Gritti et al. 2014). CLIC1 is a promising pharmacological target in stress-related diseases, including cancer, where CLIC1 increases tumorigenic and metastatic potential (Peretti et al. 2015).

2.5 *Amino Acid Transporter SLC7A11*

Metabolic reprogramming occurs in cancer cells to acquire the necessary nutrients to sustain their biosynthetic and bioenergetic processes, which also increases oxidative stress. The cystine/glutamate antiporter solute carrier family 7 member 11 (SLC7A11, also called xCT) imports a cysteine molecule coupled with the efflux of one glutamate molecule (Koppula et al. 2017). SLC7A11 regulates intracellular redox balance by maintaining intracellular levels of glutathione and inhibiting ferroptosis, protecting the cells from oxidative stress-induced cell death (Lewerenz et al. 2013; Zheng et al. 2019). SLC7A11 promotes cancer growth and drug resistance (Lewerenz et al. 2013), and in response to oxidative stress, the proto-oncogene *K-Ras* stimulates SLC7A11 transcription upregulating glutathione levels in the tumor cells (Lim et al. 2019). Inhibitors of this transporter have antitumor effects by altering the entrance of cysteine necessary for glutathione (GSH) synthesis (Robe et al. 2009; Takeuchi et al. 2014; Shitara et al. 2017). Sulfasalazine, a nonselective blocker of SLC7A11, has been studied as an anticancer drug alone or in combination with other anticancer therapies in animal models and clinical trials, but more selective inhibitors are needed to reduce high adverse effects in humans (Guan et al. 2009; Lo et al. 2010; Takeuchi et al. 2014; Peretti et al. 2015; Sehm et al. 2016; Shitara et al. 2017).

Thus, the diverse association between oxidative stress and ion channel and transporters represents a very important opportunity for cancer therapy. However, the specific channel and transporter inhibitors, as well as the particular anticancer drugs concomitantly used, should be carefully considered. High levels of ROS are also produced in the mitochondria, and ion channels and transporters of this organelle have been also associated with cancer and proposed as targets for therapy.

3 Mitochondrial Ion Channels and Transporters in Novel Potential Therapies for Cancer

The mitochondria play many important cellular functions including ATP and ROS production, apoptosis, as well as Ca^{2+} homeostasis (Sharma et al. 2019). Dysfunction of this organelle has been correlated with several diseases including cancer, where the involvement of several ion channels and transporters has been studied (Bachmann et al. 2018; Leanza et al. 2018).

3.1 *Voltage-Dependent Anion Channels*

The voltage-dependent anion channel (VDAC) transports several ions (K^+, Na^+, and Ca^{2+}), organic anions, ATP, ADP, Pi, and some metabolites depending on the state of the channel across the outer mitochondrial membrane (OMM) (Camara et al. 2017). These channels interact with members of the Bcl-2 family and with hexokinase, regulating apoptosis and with IP3R for the passage of Ca^{2+} from the endoplasmic reticulum (Mazure 2017; Leanza et al. 2018; Sharma et al. 2019). VDACs are overexpressed in different types of cancers where their expression is related to abnormal proliferation (Shoshan-Barmatz and Ben-Hail 2012). The interaction of hexokinase with VDAC favors cellular glycolysis which is of great relevance for cancer cells; methyl jasmonate (MJ) is an inhibitor of hexokinase-2 that prevents the interaction of hexokinase with VDAC on the mitochondrial membrane and has anticancer effects. The research into new analogs of MJ should help to find new agents against different types of cancer (Sucu et al. 2019). Furthermore, erastin leads to VDAC opening and induces mitochondria dysfunction, increases ROS, inhibits GSH synthesis, decreases glycolysis, and also induces non-apoptotic cell death by ferroptosis in some types of cancers (Yagoda et al. 2007; Dixon et al. 2012; Maldonado et al. 2013).

3.2 Mitochondrial Permeability Transition Pore

Mitochondrial permeability transition pore (mPTP) is a nonspecific channel located on the inner membrane of the mitochondria (IMM). Its prolonged activation depolarizes the mitochondrial membrane and generates ROS, leading to cell death. Thus, drugs that induce mPTP activation in tumor cells have gained great interest (Zoratti and Szabò 1995; Bernardi et al. 2015). Icaritin is an active natural ingredient of the Chinese plant *Epimedium* that decreases the mitochondrial membrane potential by opening mPTP, leading to necrosis and decreasing proliferation in colorectal cancer (CRC) cells. In accordance, mPTP blockers such as sanglifehrin A, cyclosporin A, and bongkrekic acid, as well as siRNA targeting mPTP decreased the cytotoxic effect of icaritin on CRC cells (Zhou et al. 2016). Similarly, the gold (III)-dithiocarbamate AUL12 contributes to mPTP opening and tumor cell death and shows very low systemic toxicity in vivo (Rasola and Bernardi 2014). Interestingly, various compounds that target the mitochondrial machinery are currently being studied in clinical trials (Suh et al. 2013).

3.3 Mitochondrial Calcium Uniporter

High levels of mitochondrial Ca^{2+} lead to the activation of the mitochondrial Ca^{2+} uniporter (MCU) triggering apoptosis (Mammucari et al. 2017). MCU also participates in the proliferation, invasion, and redox signaling in some types of cancers (Vultur et al. 2018). For instance, in triple-negative breast cancer cells, MCU silencing reduces the production of mitochondrial ROS and HIF1-α, impairing cell motility (Tosatto et al. 2016). Actually, MCU overexpression has been linked to lymph node migration, poor prognosis, and breast tumor size (Tang et al. 2015; Yu et al. 2017). Moreover, hepatocellular carcinoma progression and metastasis are associated with overexpression of the MCU-regulator 1 (MCUR1) protein (Jin et al. 2019a), and the anticancer properties of minocycline and doxycycline have been suggested to be related to their inhibitory effect on MCU (Cui et al. 2019). Finally, the thiourea derivative KB-R7943 inhibits MCU reducing Ca^{2+} release in HeLa cervical cancer cells (Santo-Domingo et al. 2007).

3.4 Uncoupling Protein 2

The uncoupling protein (UCP) is a proton (H^+) transporter located on the IMM (Berry et al. 2018). It has been suggested that UCP2 participates in the regulation of cell survival by reducing ROS and mitigating oxidative stress (Cannon et al. 2006; Baffy 2010). UCP2 is upregulated in different tumors, including hepatocellular carcinoma, colorectal, pancreatic, and thyroid cancer (Baffy 2010). UCP2 protects the cells from oxidative stress and prevents the apoptotic effects of different drugs (Derdak et al. 2008). The UCP2 inhibitor genipin reduces cell proliferation, enhances the response to chemotherapy, reverses chemotherapy resistance in some cancer cell lines, and reduces tumor growth in vivo (Mailloux et al. 2010; Dalla Pozza et al. 2012; Pons et al. 2015; Shanmugam et al. 2018). On the contrary, UCP2 expression in melanoma is associated with T-cell tumor infiltration, higher antitumor response, and prolonged survival (Cheng et al. 2019). Also, induced overexpression of UCP2 in melanoma cells generates an immunostimulatory microenvironment by producing chemokines and cytokines, enhancing $CD8^+$ T-cell infiltration in the tumor microenvironment, and suppressing tumor progression. Furthermore, the expression of UCP2 sensitizes melanoma cells against anti-programmed cell death 1 (PD-1) treatment (Cheng et al. 2019). Immune checkpoint-block therapy (like anti-PD-1) is a novel way to fight cancer, and targeting UCP2 expression may convert this immune therapy more efficient for some cancers.

Therefore, targeting different mitochondrial ion channels and transporters should be considered to design novel anticancer therapies (Leanza et al. 2018). Regarding the immune system, ion channels and transporters are becoming an attractive field in onco-immunology.

4 Ion Channels and Transporters in Cancer Immunotherapy

The participation of the immune system is fundamental for the recognition and elimination of tumor cells. Different immune cells infiltrate the tumor microenvironment activating the immune response, for instance, by $CD4^+$T or $CD8^+$T cells.

These cells bind directly to the MHC class I molecules presented by the tumor cells inducing the release of cytokines and cytotoxic granules, killing tumor cells (Ostroumov et al. 2018). However, cancer cells evade the immune response by several mechanisms that include defective antigen presentation, repression of T-cell activation, and production of immune-suppressive cytokines (Vinay et al. 2015; Liubomirski et al. 2019). In accordance, several types of immunotherapies are used in clinical practice including immune checkpoint inhibitors, immune system modulators, monoclonal antibodies, vaccines, and CAR T-cell therapy (Khalil et al. 2016).

4.1 Ion Channels and Leucocytes at a Glance

Some ion channels are implicated in the activation, differentiation, proliferation, chemotaxis, and migration of leucocytes (Feske et al. 2015). Lymphocyte function depends on ion channel-mediated Ca^{2+} signaling induced by antigen recognition. Briefly, activation of lymphocytes by binding of the antigen to the TRC (T cell) or BCR (B cells) receptor activates PLCγ1 in T cells and PLCγ2 in B cells increasing the formation of IP3. Then, the IP3 receptor (IP3R) is activated releasing Ca^{2+} from the endoplasmic reticulum. The depletion of Ca^{2+} from the ER activates either STIM1 or STIM2 subunits to oligomerize with IP3R in the ER and interacting with the ORAI channels in the plasma membrane forming functional CRAC channels. These channels allow the entrance of Ca^{2+} from the extracellular space into the lymphocyte. To maintain the electrical driving force for Ca^{2+} influx, activation of $K_{Ca}3.1$ channels by Ca^{2+} and activation of $K_V1.3$ channels by membrane depolarization are required. Following calcium influx, the calcineurin-NFAT pathway is activated increasing the transcription of genes associated with proliferation, cytokine production, and cytotoxicity (Panyi et al. 2014; Feske et al. 2015; Chiang et al. 2017). The blockage of these potassium channels has been proposed as a therapeutic strategy for immunosuppression in a variety of conditions including chronic inflammation, autoimmune diseases, and immunologic-derived cancers (Lam and Wulff 2011). Kv1.3 is also expressed in the IMM where it participates in apoptosis by interacting with Bax and inhibiting channel activity with the subsequent elevation of ROS and release of cytochrome C (Szabó et al. 2008). Two new inhibitors of mitoKv1.3 (PCARBTP and PAPTP) induce ROS production, promote cell death in chemoresistant cells, and reduce tumor growth in melanoma and pancreatic adenocarcinoma in vivo while preserving immune cells and healthy tissues. The authors propose that the selectivity to cancer cells may be partially due to the higher expression of mitoKv1.3 in cancer cells, which hyperpolarizes the IMM and alters the redox status (Leanza et al. 2017).

4.2 Cancer Immunotherapy Targeting Ion Channels

Several approaches targeting ion channels in cells from the immune system have been used. B cells from patients with chronic lymphocytic leukemia (B-CLL) have altered redox state and overexpress Kv1.3 channels in the plasma membrane and mitochondria compared with B cells from healthy subjects. Clofazimine induced cell death by blocking Kv1.3 channels in the mitochondria and activating the intrinsic apoptotic pathway in B-CLL cells. Furthermore, healthy B/T cells or B-CLL treated with the antioxidant enzymes catalase and superoxide dismutase was resistant to apoptosis induced by clofazimine, indicating a synergic action between inhibition of Kv1.3 and ROS production (Leanza et al. 2013). Kv1.3 channels with incomplete inactivation are overexpressed in Daudi B cells. Treatment of Daudi B cells with the antihuman CD20 antibody rituximab (used in patients with non-Hodgkin's lymphoma) downregulates Kv1.3 channels by activation of the FcγRIIB receptor, contributing to the induction of apoptosis (Wang et al. 2012). Also, in primary malignant T cells isolated from patients with Sézary syndrome, blockage of Kv1.3 inhibited activation and cell proliferation (Hu et al. 2019). Likewise, $K_{Ca}3.1$ is overexpressed in several cancers promoting cell proliferation, metastasis, and therapy resistance (Mohr et al. 2019). Treatment of CLL cells with clotrimazole or TRAM-34 ($K_{Ca}3.1$ channel blockers) decreases Ki67 expression and cell viability (Grössinger et al. 2014). Natural killer (NK) cells also express Kv1.3 and $K_{Ca}3.1$ in the plasma membrane. TRAM-34 increased the proliferation and degranulation levels of adherent NK cells in the presence of the leukemia cell line K562, and mice bearing K562 tumors treated with adherent NK cells and TRAM-34 formed smaller tumors (Koshy et al. 2013). Recently, it was described that radiation to the glioblastoma cell line GL-15 and primary cell cultures from tumors of patients with glioblastoma induced migration, and invasion mediated by $K_{Ca}3.1$channels. The blockage of $K_{Ca}3.1$ channels with TRAM-34 abolished the invasive phenotype of these cells (D'Alessandro et al. 2019). Besides, in the tumor microenvironment, high amounts of adenosine (ADO) are released from tumor cells in hypoxic conditions and regulatory T cells, as well as high amounts of ATP secreted from immune, stromal, apoptotic, and necrotic cells. ATP can be converted to ADO by the ectonucleotidases CD39 and CD75. In solid tumors, the excessive accumulation of ADO generates immunosuppression and failure of effector T cells to eliminate cancer cells, which is associated with tumor growth, metastasis, poor prognosis, and resistance to therapy (Allard et al. 2016). The function of $K_{Ca}3.1$ is inhibited by ADO in human T cells via A_{2A} receptors, reducing T-cell migration and cytokine release (Chimote et al. 2013). In addition, ADO inhibits chemotaxis of $CD8^+$ T cells from head and neck squamous cell carcinoma (HNSCC) patients via its A_{2A} receptor, reducing $K_{Ca}3.1$ channel activity and their ability to infiltrate the solid tumor. Enhancing $K_{Ca}3.1$ channel activity with the agonist 1-EBIO recovers the chemotaxis ability of $CD8^+$ T cells of HNSCC even in the presence of ADO (Chimote et al. 2018).

Adoptive T-cell transfer (ACT) therapy may be also used to target specific ion channels and transporters in cancer. ACT therapy options include tumor-infiltrating lymphocytes (TILs), T-cell receptor (TCR), and chimeric antigen receptor (CAR) therapies (June et al. 2018). $K_{Ca}1.1$ potassium channels (encoded by the KCNMA1 gene) have been associated with glioma, breast, prostate, and cervical cancer and express multiple splice variants (Liu et al. 2002; Bloch et al. 2007; Khaitan et al. 2009; Ge et al. 2012; Ramírez et al. 2018). Alternative splicing leads to the production of multiple mRNAs from a single gene, thus, encoding a diversity of proteins (Liu and Cheng 2013; Wang et al. 2015a). Some pathways deregulated in cancer frequently promote aberrant splicing, which in turn contributes to many aspects of tumor biology, including metabolism, apoptosis, cell cycle control, invasion, metastasis, and angiogenesis (David and Manley 2010; Wang et al. 2015a). Alternative splicing in ion channels modify their pharmacological profile, surface expression, intracellular localization, or electrophysiological properties; actually, in some instances, the splice variants lack conductive properties acting as dominant-negative subunits (Ramos Gomes et al. 2015). The gBK splice variant of KCNMA1 channels is strongly expressed in glioma cell lines and tumor tissues (Liu et al. 2002). This variant has two epitopes for T cells, namely, gBK1 and gBK2, that bind to the human leukocyte antigen HLA-A*0201 on the surface of dendritic cells (DCs). DC cells previously pulsed with gBK1 or gBK2 peptides induce cytotoxic T lymphocyte (CTL) response and cell death in glioma, gastric, lung, and breast cancer cell lines (Ge et al. 2012). Similar results were obtained with small cell lung cancer (SCLC) cell lines, where gBK-specific CTL-killing inhibits growth and stimulates IFN-γ, proposing gBK as a target for immunotherapy and vaccination in some types of cancer (Hoa et al. 2014). Since tumor cells expressed higher levels of gBK than noncancerous cells, targeting this splice variant may be a more selective therapy.

Another immunotherapy alternative is using antibodies against specific ion channels or transporters involved in cancer. The development of specific antibodies for cancer therapy has been studied for Kv10.1, Kv11.1, nfP2X_7, $\alpha2\delta1$ subunit (isoform 5 of voltage-gated Ca^{2+} channels), and MRP1 (Binyamin et al. 2004; Sette et al. 2013; Zhao et al. 2013; Hartung and Pardo 2016; Gilbert et al. 2017) among other proteins. Kv10.1 is abnormally expressed in approximately 70% of all types of cancers (Hemmerlein et al. 2006). The scFv62-TRAIL antibody targeting the pore of Kv10.1 (scFc62) and linked to the tumor necrosis factor-related apoptosis-inducing ligand (TRAIL) sensitized MDA-MB435S breast cancer cells to antineoplastic drugs commonly used in the clinic including paclitaxel and doxorubicin. The combination of the antibody with doxorubicin showed also significant inhibitory effects in vivo experiments, and in prostate cancer cells, the antibody induced apoptosis only in those expressing Kv10.1 channels (Hartung et al. 2011; Hartung and Pardo 2016). Thus, this antibody-based approach provides a very selective anticancer therapy. A monoclonal antibody against Kv11.1 conjugated with TiO2 nanoparticles (Kv11.1-Mab-PEG-TiO2 NPs) was designed and tested in the pancreatic ductal adenocarcinoma cell lines MIAPaCa-2 and Panc-1. Although that treatment with the Kv11.1-Mab-PEG-TiO2 NPs did not change cell viability, other options can be considered to generate cytotoxicity like testing the photocatalytic properties of TiO_2 which induce

ROS production (Sette et al. 2013), or other chemotherapeutic agents linked to Kv11.1 channels. The ion channel and transporter splice variants can also be used for cancer immunotherapy, and $P2X_7$ is an ATP-gated Ca^{2+} channel overexpressed in various cancers, promoting cell proliferation and invasiveness; a phase I clinical trial studying the topical administration of an antibody against the non-pore functional $P2X_7$ ($nfP2X_7$) variant reported reduced lesions in basal cell carcinoma, a very common skin cancer (Gilbert et al. 2017). In this way, the development of specific antibodies against malignant splice variants is emerging as a possible therapeutic approach to treat cancer. Despite that further investigation in cancer patients is needed; ion channels and transporters represent a promising alternative in cancer immunotherapy.

5 Splice Variants and Noncanonical Functions of Ion Channels in Cancer Therapy

5.1 Splice Variants

Targeting channel isoforms that are tumor-specific can provide more selectivity for drug development. In this direction, the pyrimido-indole compound CD-160130 is more effective in blocking the Kv11.1 isoform B ($IC_{50} = 1.8 \pm 0.26$ mM) compared to Kv11.1 isoform A ($IC_{50} = 13.4 \pm 3.0$ mM). Interestingly, leukemia cells mainly express Kv11.1 isoform B. Accordingly, CD-160130 induced apoptosis in vitro and prolonged survival in an acute myeloid leukemia mouse model at a dose of 10 mg/kg; it is worth mentioning that CD-160130 did not induce significant QT prolongation in mice and guinea pigs (Gasparoli et al. 2015).

Overexpression of different G proteins activated inwardly rectifying K^+ channel 1 (GIRK1) splice variants exerts opposite actions in breast cancer cells. While GIRK1a and GIRK1c overexpression reinforces parameters associated with malignancy; overexpression of GIRK1d has the contrary effect. A segment comprising aminoacids 235–402 present in GIRK1a and GIRK1c but not in GIRK1d seems to be the responsible component for the carcinogenic effect of these channels (Rezania et al. 2016). Overexpression of GIRK1 in the primary tumor is associated with lymph node metastasis and poor prognosis (Stringer et al. 2001). In addition, in breast cancer cells, the overexpression of GIRK1 affects wound healing, invasion, cellular velocities/motilities, and angiogenesis suggesting a pathophysiological role in breast cancer (Wagner et al. 2010; Rezania et al. 2016). Alternative transcripts have been also identified for Kv10.1 channels, namely, Kv10.1a and Kv10.1b. Two shorter splice variants, E65 and E70, isolated from the human brain and cancer cell lines lack the transmembrane segments. These variants produce cytoplasmic proteins without conducting properties but reduce the current of the full-length channels when co-expressed. E65 triggers the activation of cyclin-dependent kinases in *Xenopus laevis* oocytes, suggesting a role in cell cycle control (Gomes et al. 2015;

Ouadid-Ahidouch et al. 2016). TRPC channel splice variants play an important role in human ovarian cancer development. The nonselective TRPC channel blockers 2APB and SKF-96365 significantly inhibited the cell proliferation, while the increase of TRPC channel activity promoted the cell proliferation (Zeng et al. 2013). Some voltage-gated sodium (Nav) channels are expressed in the colon, small intestine, stomach, prostate, bladder, and breast, but the higher expression is found in the brain, as well as in skeletal and cardiac muscle. Interestingly, the neonatal splice variant of the Nav α-subunit subtype Nav1.5 (nNav1.5) displays a restricted expression pattern among tissues but is upregulated in human breast cancer. The high-level expression of this splice variant is associated with the estrogen receptor (ER) status. Thus, the nNav1.5 splice variant may be exploited both as a novel biomarker and a potential specific target for some common types of breast cancer (Yamaci et al. 2017).

5.2 Noncanonical Functions

Several splice variants may form non-conducting ion channels strongly suggesting that noncanonical functions of ion channels are also involved in carcinogenesis. For instance, Downie and colleagues developed a mutant Kv10.1 channel eliminating ion permeation and studied its oncogenic potential. This mutant fails to completely abolish xenograft tumor formation by transfected cells, strongly suggesting that the oncogenic mechanism of Kv10.1 comprises other molecular mechanisms independently of its primary function as an ion channel (Downie et al. 2008). Noncanonical functions of Cav channels are also associated with cancer. Proteolytic cleavage of the C-terminus of L-type Ca^{2+} channels α1C and α1D subunits (Cav1.2 and Cav1.3, respectively) produces a fragment that is translocated to the nucleus regulating the transcription of genes involved in tumor progression (Buchanan and McCloskey 2016). C-terminus cleavage of Cav1.2 channel generates the transcription factor Ca^{2+} channel-associated transcriptional regulator (CCAT), which regulates the expression of connexin CX31.1 and NR3, but also provides negative feedback regulating Cav1.2 channel expression (Gomez-Ospina et al. 2013). Alternatively, CCAT may result from the alternative splicing of the Cav1.2 gene (Barbado et al. 2009). The overexpression of this fragment affects also the expression of other ion channels including TRPV4 and $K_{Ca}2.3$, potentially leading to a cancer phenotype (Buchanan and McCloskey 2016). Electromagnetic field therapy, like tumor treating fields (TTFields), delivers a low-intensity, intermediate frequency, alternating electric field through noninvasive transducer arrays to tumor regions. This FDA-approved treatment for glioblastoma multiforme therapy disrupts mitosis and cytokinesis, stimulates calcium entry mediated by Cav1.2, and arrests the cells in the S and G1 phase of the cell cycle in glioblastoma cell lines (Neuhaus et al. 2019). Thus, the participation of ion channel splice variants in cancer and the diverse molecular mechanisms associating channels with cancer including noncanonical functions offer additional drug design and therapeutic opportunities to fight cancer. One of these opportunities arises by using current drugs prescribed for conditions different from cancer but affecting ion channels and transporters involved in cancer.

6 Repurposing Existing Drugs Targeting Ion Channels and Transporters for Cancer Therapy

Drug repurposing is an attractive strategy to reduce the cost and developing times of new antineoplastic agents. The safety, pharmacokinetics, and pharmacodynamic profile of currently used drugs are well-known; thus, these drugs may be quickly translated into phase II and III clinical studies (Oprea et al. 2011; Gupta et al. 2013; Pantziarka et al. 2014). In silico chemical genomic approaches have been used to predict drug repositioning candidates for cancer therapy based on large-scale drug-induced transcriptional signatures (Lee et al. 2016). Because several drugs used for different indications target ion channels and transporters involved in cancer, repurposing of these drugs is a very attractive and low-cost alternative to fight cancer.

6.1 Antihistamines

Histamine is involved in cell proliferation and tumor growth; thus, several antihistamines have been strongly suggested for repurposing as antineoplastic agents (Faustino-Rocha et al. 2017). Astemizole is a long-acting, non-sedating second-generation antihistamine indicated in the treatment of allergies. This drug is an antagonist of H_1-histamine receptors which are present in the gastrointestinal tract, uterus, blood vessels, and bronchial muscle, among other tissues (Garcia-Quiroz and Camacho 2011). Astemizole also targets several molecules involved in cancer development including ABC transporters (P-glycoprotein) and the potassium channels Kv10.1 and Kv11.1 (Pardo et al. 1999; Ishikawa et al. 2000; Garcia-Ferreiro et al. 2004; Camacho 2006; Garcia-Quiroz and Camacho 2011). This antihistamine has antiproliferative effects in cancer cell lines from breast such as MCF-7, SUM-229PE, T-47D, and BT-474, as well as in invasive ductal breast cancer primary cultures (Ouadid-Ahidouch et al. 2001; Roy et al. 2008; Garcia-Quiroz et al. 2012, 2019). It also inhibits proliferation and increases apoptosis in several cell lines from cervical, liver, prostate, and lung cancer (Chavez-Lopez et al. 2014, 2015, 2017; Bernal-Ramos et al. 2017;), as well as in cells from leukemia (Ishikawa et al. 2000) and in keratinocytes transfected with a human papillomavirus oncogene (Diaz et al. 2009). Moreover, astemizole inhibits the Kv10.1 mRNA expression both in vitro and in vivo in breast cancer and hepatocellular carcinoma, decreasing tumor development (Garcia-Quiroz et al. 2012, 2014; Chavez-Lopez et al. 2015). The antitumor activity of astemizole has been observed in several studies in animal tumor models. The oral administration of astemizole (50 mg/kg/day) reduced the growth rate of xenografts tumors induced by implantation of Kv10.1-transfected cells or MDA-MB435S breast cancer cells (Downie et al. 2008). In a rat model, astemizole was capable to prevent hepatocellular carcinoma (HCC) development induced by the carcinogen diethylnitrosamine (Chavez-Lopez et al. 2015). The daily administration of astemizole (50 mg/kg) in drinking water inhibited tumor growth in an in vivo preclinical model using athymic mice xenografted with two different human breast cancer cell lines: T-47D and a ductal infiltrating carcinoma breast cancer-derived primary cell culture (MBCDF) (Garcia-Quiroz et al. 2014). The dose of 50 mg/kg of astemizole was sufficient to inhibit tumor growth in mice without producing noticeable adverse effects (loss of body weight, diarrhea, or alterations in physical activity) (Downie et al. 2008; Garcia-Quiroz et al. 2014). In contrast, 30 mg/kg of astemizole induced ventricular contractions in dogs and torsade de pointes in one animal (Izumi-Nakaseko et al. 2016). Astemizole was withdrawn from the US market in 1999 due to its pro-arrhythmic potential; it soon became evident that most cases of toxicity involved either overdosing, drug interaction, or subjects with predisposed cardiac disease (Paakkari 2002). At the defined daily dose of prescribed astemizole (10 mg/day), the spontaneous cardiac adverse drug reaction reported in a lapse of 10 years were 110 cases per million of doses sold (Garcia-Quiroz and Camacho 2011). These side effects are mainly attributed to the blockade of the Kv11.1 cardiac potassium channels (IC_{50} of 48.4 ± 3.8 nM) (Suessbrich et al. 1996; Zhou et al. 1999). It is important to mention that not all Kv11.1 channel blockers produce torsade de points, for instance, verapamil and sertindole (D'Amico et al. 2013; Gentile et al. 2016), but several anticancer drugs have a pharmacological effect on Kv11.1 (Gentile et al. 2016). Therefore, the use of Kv11.1 blockers that do not induce cardiac side effect has been suggested for cancer treatment. One of the alternative proposed approaches is using drugs that bind to a specific state of the channel, like *R*-roscovitine that interacts with the channel in its open state, which is longer in tumors than in cardiac cells (D'Amico et al. 2013).

A very interesting property of astemizole is that its concomitant use with other antineoplastic agents has synergistic effects. Astemizole potentiates the growth-inhibitory activity of doxorubicin in doxorubicin-resistant human leukemia cells K562/DXR by inhibiting the P-glycoprotein (Ishikawa et al. 2000). The antihistamine also synergizes the calcitriol antiproliferative effects by downregulating CYP24A1 (which inactivates calcitriol), upregulating the vitamin D receptor (VDR), and targeting Kv10.1 (Garcia-Quiroz et al. 2012). The co-administration of astemizole and calcitriol to mice xenografted with human breast cancer cells inhibited tumor growth more efficiently than each drug alone (Garcia-Quiroz et al. 2014). Likewise, in a HCC model, astemizole increased VDR expression both in vitro and in vivo, enhanced vitamin D-induced decrease in cell viability and proliferation, increased apoptosis, decreased cell migration and invasion in vitro, as well as reduced the amount and mass of tumors (Xu et al. 2018b). Furthermore, in lung cancer cells, astemizole potentiated the inhibitory effect of vinorelbine on the colony formation of NCI-H1299 and cisplatin on the colony formation of NCI-H661 cells (Ellegaard et al. 2016). The combined effect of astemizole with the epidermal growth factor receptor type 1 (EGFR) inhibitor gefitinib further repressed the proliferation, survival, and Kv10.1 expression and increased the apoptosis more than the monotherapy in the lung cancer cell lines A549 and NCI-H1974 (Chavez-Lopez et al. 2017). In the same manner, astemizole and gefitinib synergistically inhibited the proliferation of breast cancer cells expressing the targets Kv10.1 and EGFR (Garcia-Quiroz et al. 2019). In addition, astemizole acts synergistically with radiation to increase the death of prostate cancer cells through a mechanism involving autophagy (Oprea et al. 2011).

Terfenadine, a second-generation H_1 receptor antagonist targets other molecules involved in cancer such as Kv11.1 (Suessbrich et al. 1996). This antihistamine induces apoptosis and inhibits tumor growth in murine models (Blaya et al. 2010). In human refractory prostate cancer cells, terfenadine upregulates and activates Bak and the cleavage of Mcl-1, leading to the loss of mitochondrial membrane potential and activation of caspase cascade resulting in DNA damage response and apoptosis (Wang et al. 2014b). Breast cancer cells resistant to HER-2/neu targeted therapy express high levels of H_1 receptors and are more sensitive to terfenadine. This drug leads to Sub-G0 cell accumulation, suppresses proliferation, promotes cell motility, and triggers the activation of extracellular signal-regulated kinase (ERK), initiating the mitochondrial apoptotic pathway in basal breast cancer. Moreover, in vivo experiments showed that terfenadine (10 mg/kg) therapy reduced the tumor growth of basal and trastuzumab-resistant breast cancer cells (Fernández-Nogueira et al. 2018). The combined treatment of terfenadine with epirubicin synergistically inhibits the growth and metastatic process of chemotherapy-resistant non-small cell lung cancer (NSCLC) cells both in vitro and in vivo (An et al. 2017), and ketoconazole potentiates terfenadine-induced apoptosis in human HepG2 cells through inhibition of p450 3A4 activity (Wang et al. 2002). Terfenadine was withdrawn from the market due to the induction of prolonged QT interval in cases of overdose, inappropriate co-medications or in subjects with predisposed cardiac disease. The FDA recommended terfenadine to be replaced by its active and nontoxic metabolite fexofenadine (Berul and Morad 1995; Paakkari 2002). Another antihistamine with important antineoplastic effects is loratadine, which is associated with significantly reduced all-cause mortality among patients with non-localized non-small cell lung cancer (NSCLC) or any non-localized cancer. Astemizole showed a similar significant association with reduced mortality in patients with non-localized cancer, and ebastine shows a similar tendency. Interestingly, submicromolar concentrations of these antihistamines sensitized NSCLC cells to chemotherapy and reverted multidrug resistance in NSCLC, breast, and prostate cancer cells (Ellegaard et al. 2016). Similar results with antihistamines were observed in ovarian cancer patients (Verdoodt et al. 2019).

6.2 Imipramine

Imipramine is a tricyclic antidepressant indicated for symptom relief of depression and other conditions including panic and obsessive-compulsive disorders, bulimia, and nocturnal enuresis; it acts by blocking the sodium-dependent serotonin and norepinephrine transporters reducing reuptake and increasing their concentration in the synaptic cleft (Gillman 2007). In addition, imipramine inhibits the current through Kv10.1 channels in a voltage-dependent manner and reduces the proliferation of cancer cells (Gavrilova-Ruch et al. 2002; Garcia-Ferreiro et al. 2004; Gomez-Varela et al. 2006). This drug also promotes apoptosis in the ovarian cancer cells SK-OV-3 (Asher et al. 2011). In brain cancer patients, the effect of imipramine is associated with the channel abundance; thus, the antidepressant improves the survival rate better in patients with moderate Kv10.1 expression (Martinez et al. 2015). These findings suggest that personalized therapy with this tricyclic antidepressant based on the expression of Kv10.1 channels may be used for brain malignancies. Besides, in 2013, Jahchan and colleagues used bioinformatic tools to identify potential candidate drugs for the treatment of small cell lung cancer from FDA-approved drugs and identified imipramine as a potential candidate. Imipramine at 20 μM decreased survival in H82, H69, and H187 human small cell lung cancer (SCLC) cells and Kp1, Kp2, and Kp3 mouse SCLC cells. In vivo, imipramine (25 mg/kg) inhibited the growth of SCLC allografts (mouse SCLC cell line Kp1), xenografts (human SCLC cell line H187), and one primary patient-derived xenograft (human SCLC tumor NJH29). This drug was effective also in cisplatin-resistant SCLC cells, suggesting that imipramine may be used as second-line therapy for SCLC patients who become refractory to cisplatin/etoposide (Jahchan et al. 2013; Kale et al. 2015). Imipramine also has cardiovascular side effects including orthostatic hypotension, atrioventricular conduction delay, reduced heart rate variability in response to exercise, tachycardia, syncope, and arrhythmias particularly observed in patients with concurrent cardiovascular disease or at high doses of treatment. This may be explained because imipramine blocks several neuronal and cardiac K^+, Na^+, and Ca^{2+} channels whit IC_{50} values ranging from 1 to 30 μM; its IC_{50} in cloned Kv11.1 channels is 3.4 ± 0.4 μM, and the complete blockage is achieved with 30 μM (Teschemacher et al. 1999; Garcia-Ferreiro et al. 2004).

To increase the antineoplastic effects, imipramine has been co-administered with other compounds. The combination of imipramine with doxorubicin enhanced the anti-invasive effect, whereas a combination with ticlopidine suppressed ATG7, a member of the autophagy survival signaling, resulting in cell death (Abdelaleem et al. 2019). The combined treatment of imipramine and radiotherapy in prostate cancer did not enhance the radiosensitivity of DU145 cells; unexpectedly, the treatment of imipramine alone was more effective (Barlaz Us et al. 2019). Several studies have evaluated the effect of imipramine blue, which is an organic triphenylmethane dye synthesized from imipramine and 4,4′-diethylaminobenzophenone. This compound was suggested because gentian violet (another triphenylmethane dye) also exhibits anticancer properties. This imipramine analog inhibits the invasion of glioma cells both in vitro and in vivo and enhances the efficacy of doxorubicin (Munson et al. 2012). In addition, imipramine blue inhibits breast cancer growth, progression, and metastasis (Rajamanickam et al. 2016); moreover, it has antineoplastic effects on head and neck cancer (Yang et al. 2016), Burkitt lymphoma (Klingenberg et al. 2014), as well as on acute (Metts et al. 2017) and chronic myeloid leukemia (Laidlaw et al. 2016).

6.3 Calcitriol

The endogenous synthesis of calcitriol begins in the skin by the action of ultraviolet radiation from sunlight but takes place mainly in the kidney and has been reported in other tissues such as skin, prostate, intestine, pancreatic islets, lymph nodes, brain, colon, and the mammary gland, where local calcitriol synthesis takes place (Deeb et al. 2007; Glowka et al. 2019). The coupling of calcitriol with the VDR allows dimerization with the retinoid receptor X (RXR); this heterodimer translocates to the nucleus and binds to VDR response elements (VDREs) in the promoter of target genes inducing gene expression.

Calcitriol acting via VDR promotes cytodifferentiation and apoptosis, modulates oncogene expression, and inhibits cell proliferation and migration, reducing or preventing cancer progression. Another potential antiproliferative mechanism of this secosteroid is its ability to downregulate Kv10.1 expression in cell lines (SUM-229PE and MCF-7) and primary cultures from breast cancer (Garcia-Becerra et al. 2010; Garcia-Quiroz et al. 2012), as well as in cervical (SiHa, HeLa) and prostate (PC-3) cancer cells and in syncytiotrophoblasts from normal human placenta (Avila et al. 2010). Kv10.1 repression by calcitriol in cervical cancer cells occurs at the transcriptional level and involves a functional nVDRE (negative-VDRE) in the Kv10.1 promoter (Cazares-Ordonez et al. 2015). Calcitriol also decreases Kv10.1 expression and tumor growth in vivo of the xenografted breast cancer cell lines T-47D and HCC-1806 and the MBCDF breast cancer primary culture (Garcia-Quiroz et al. 2014, 2016). The antineoplastic effect of calcitriol has also been observed in melanoma, pancreatic, prostate, and colorectal cancer, as well as in hepatocellular carcinoma. In fact, a large number of epidemiological studies have demonstrated an association between low circulating levels of the calcitriol precursor calcidiol, with higher risk to develop colorectal and breast cancer and hepatocellular carcinoma (Diaz et al. 2015).

In addition, the antineoplastic effects of calcitriol are potentiated in breast cancer in vitro and in vivo, by combining it with other antineoplastic agents including the natural compounds curcumin and resveratrol (García-Quiroz et al. 2019). Besides, the combination of calcitriol with the receptor tyrosine kinase inhibitors gefitinib, lapatinib, and neratinib is more effective to inhibit the growth of breast cancer cell lines in comparison with each compound alone (Segovia-Mendoza et al. 2015, 2017). Furthermore, the combinations of calcitriol or its analogs with chemotherapeutic agents such as antimetabolites, platinum compounds, or taxanes improve the antineoplastic effects in different types of cancer (Abu El Maaty and Wölfl 2017). Thus, calcitriol is an endogenous natural anticancer factor targeting ion channels and promising antineoplastic agent.

6.4 Clarithromycin

Clarithromycin is a macrolide antibiotic drug having a broad spectrum of antimicrobial activity for gram-positive and gram-negative organisms, atypical pathogens, and some anaerobes (Peters and Clissold 1992). Interestingly, in colorectal cancer, this macrolide modulates the PI3K/Akt pathway by targeting Kv11.1, modulating autophagic flux, and triggering apoptosis. This drug preferentially binds to Kv11.1 channels in their closed state and inhibits the formation of a macromolecular complex between the channel and the p85 subunit of PI3K, impairing this signaling pathway (Petroni et al. 2020). Additionally, clarithromycin targets the P-glycoprotein, which is overexpressed in different kinds of tumors and confers resistance to chemotherapy (Vermeer et al. 2016). This drug also enhances the cytotoxic effect of 5-fluorouracil both in vitro and in vivo (Petroni et al. 2020).

6.5 Fluoxetine

Fluoxetine is a selective serotonin reuptake inhibitor, initially intended for the treatment of depression; however, nowadays it is also prescribed to treat other conditions like obsessive-compulsive disorders (Wong et al. 1995). Interestingly, fluoxetine is also a non-torsadogenic Kv11.1 inhibitor successfully used in glioblastoma therapy without obvious cardiotoxicity and the added benefit of treating depression (Pointer et al. 2017). Kv11.1 channel blockers reduced glioblastoma cell proliferation and improved survival in patients who received one or more Kv11.1 blockers but only if their tumors exhibited high Kv11.1 expression levels (Pointer et al. 2017), which represents another example of the potential use of channel expression levels for personalized therapy.

6.6 Glibenclamide

Glibenclamide is a second-generation sulphonylurea, used for the treatment of non-insulin-dependent diabetes mellitus; this drug binds to the sulphonylurea receptor (SUR1) expressed in pancreatic B cells and blocks K_{ATP} channels, leading to insulin release (Payen et al. 2001). K_{ATP} channels are composed of at least two types of subunits, an inwardly rectifying K^+ channel (Kir6.x) and a regulatory subunit SUR. SUR1 belongs to the ATP-binding cassette (ABC) protein superfamily.

Glibenclamide inhibits the activity of various ABC transporters and multidrug resistance proteins (MRPs). In the human lung cancer cells GLC4/Sb30 that overexpress MRP1 and are resistant to the anticancer drugs doxorubicin and vincristine, glibenclamide (0.39–100 μM) inhibited MRP1 activity in a dose-dependent manner reverting drug resistance (Payen et al. 2001). This drug (0.5–200 μM) also decreased cell viability and induced apoptosis in the gastric cancer cell line MGC-803 by activating mitochondrial death pathways related to ROS generation, activation of JNK, and inhibition of Akt (Qian et al. 2008). Whereas in the breast cancer cell line MDA-MB-231, the sulphonylurea (10–50 μM) inhibited cell growth and induced G0/G1 arrest (Núñez et al. 2013). Glibenclamide (150 μM) also decreased the proliferation of several cervical cancer cell lines; the higher the expression of Kir6.2 subunit in the cervical cancer cells, the higher the inhibitory effect of the drug. The overexpression of the Kir6.2 subunit was also observed in cervical tumor tissues; therefore, glibenclamide is a potential therapy for this type of cancer (Vazquez-Sanchez et al. 2018). Interestingly, the combined treatment of glibenclamide with $CoCl_2$ decreased the expression of metalloproteinase-9 (MMP-9) and inhibited the growth in highly metastatic breast cancer cells (Rong et al. 2013). The antitumor effect of glibenclamide has been also observed in preclinical studies in melanoma (Suzuki et al. 2012), bladder carcinoma (Wondergem et al. 1998), prostate (Abdul and Hoosein 2002), and liver cancer (Malhi et al. 2000). The antineoplastic effects of glibenclamide may be explained by its ability to block K_{ATP} channels, ABC transporters, and MRPs and decrease the expression of MMP-9 (Payen et al. 2001; Rong et al. 2013).

6.7 *Verapamil*

Verapamil is an L-type Ca^{2+} channel blocker classified as a class IV antiarrhythmic agent that also blocks Kv11.1 currents (Zhang et al. 1999). This drug also exhibits anticancer effects attributed to its combined inhibitory activity against potassium and Ca^{2+} channels (Kale et al. 2015). Verapamil has antiproliferative effects on the breast cancer cells HT-39 both in vitro ($IC_{50} = 10$ μM) and in vivo (3.5 mg/day) (Taylor and Simpson 1992), as well as in prostate cancer (Rybalchenko et al. 2001), melanoma (Huber et al. 1989), and neuroblastoma (Schmidt et al. 1988) and in a nude mouse model of meningiomas (Jensen and Wurster 2001). Interestingly, verapamil overcomes the vincristine resistance both in vitro and in vivo in P388 leukemia cells (Yusa and Tsuruo 1989), doxorubicin-resistant myeloma (Durie and Dalton 1988), and vinblastine-resistant pediatric tumors (Cairo et al. 1989). In a prospective study in 99 patients with anthracycline-resistant metastatic breast carcinoma, verapamil given in conjunction with chemotherapy increased survival (Belpomme et al. 2000). In a randomized trial of 72 patients with advanced non-small cell lung cancer (NSCLC), verapamil plus chemotherapy (vindesine/ifosfamide) improved patient outcome (Millward et al. 1993). The reversal mechanism of MDR by verapamil is because the antiarrhythmic drug interacts with specific binding sites on the P-glycoprotein (Yusa and Tsuruo 1989); however, the clinical use of this agent has been hampered because of the unacceptable toxicity and side effects at the doses required to modulate the P-glycoprotein (Arora et al. 2005). Thus, the synthesis of new analogs of verapamil deserves further investigation.

6.8 *Nifedipine and Mibefradil*

Nifedipine is a potent L-type Ca^{2+} channel blocker indicated as an antihypertensive drug from several years ago and has an acceptable safety profile. In vitro studies showed that nifedipine reduces the mitogenic effect of endothelin-1 by blocking Ca^{2+} channels in lung cancer cells (Kale et al. 2015). In endometrial carcinoma cells, nifedipine induced autophagy through Beclin1 and the mTor pathway (Bao et al. 2012). In addition, the Ca^{2+} channel blockers nifedipine, mibefradil, and tetrandrine modulated the androgen receptor-mediated gene expression and induced cytotoxicity in LNCaP, LAPAC-4, and C4-2 androgen receptor-positive prostate cancer cells (Loughlin 2014). The antitumor effect of cisplatin was enhanced by nifedipine in cisplatin-sensitive human glioblastoma U-87MG cells and cisplatin-resistant U87-MG-CR cells both in vitro and in vivo (Kondo et al. 1995), as well as in lung carcinoma cells (Onoda et al. 1988). However, the potential use of nifedipine as antineoplastic is controversial because it cannot be used in hypotensive cancer patients. Nevertheless, the alternate dosing systems like the continuous release system developed by Bayer may help to control the blood Ca^{2+} levels and avoid rapid hypotension (Kale et al. 2015). Mibefradil, a T-type channel blocker was approved as an antihypertensive drug by the FDA in 1997 but voluntarily withdrawn from the market by Roche Laboratories in 1998 after reports of dangerous and even fatal interactions with at least other 25 drugs, including antibiotics, antihistamines, and anticancer drugs (SoRelle 1998). This drug has important antineoplastic effects in glioblastoma (Keir et al. 2013), breast cancer, and retinoblastoma (Bertolesi et al. 2002). Holdhoff and colleagues designed a phase I study to determine the safety and the maximum tolerated dose of mibefradil when given sequentially with temozolomide in recurrent high-grade gliomas. The study enrolled 27 patients; mibefradil followed by temozolomide was well tolerated; and the lack of toxicity and response in some patients warrants further investigation (Holdhoff et al. 2017). Besides, mibefradil regulates the gating of Kv10.1 channels inducing an apparent inactivation, probably by binding to the voltage sensor domain (Gómez-Lagunas et al. 2017), which adds a new potential mechanism of the anticancer effects of this drug.

6.9 Celecoxib

Celecoxib has been used as an anti-inflammatory, analgesic, and antipyretic drug, but it also has antineoplastic properties. The mechanism of action of celecoxib as an antineoplastic agent has been not sufficiently investigated (Toloczko-Iwaniuk et al. 2019). This drug decreases the proliferation of rat pheochromocytoma PC12 cells in a dose-dependent manner by blocking Ca_V-mediated currents (Zhang et al. 2007). The clinical efficacy and safety of celecoxib have been evaluated in combination with chemotherapy in metastatic or postoperative recurrent gastric patients, which offers more clinical benefits (Guo et al. 2019). A case report described that HCC practically disappeared in a patient after 8 months of treatment with celecoxib and pentoxifylline (Jimenez-Luevano et al. 2018). The combination of celecoxib with antineoplastic agents as capecitabine could be a good option for patients with thymic carcinoma (Wood et al. 2018). In addition, the combination of the anti-inflammatory drug with erlotinib may be efficacious for patients with advanced non-small cell lung carcinoma and wild-type EGFR (Jin et al. 2019b). Preclinical and clinical studies have demonstrated promising results of the role of celecoxib in the treatment and prevention of some cancers such as colon, breast, prostate, and head and neck (Toloczko-Iwaniuk et al. 2019). Whether calcium channels are involved in all these effects remains elusive.

6.10 Bromocriptine

Bromocriptine is an ergot and dopamine D_2 receptor agonist used to treat Parkinson's disease, acromegaly, hyperprolactinemia, galactorrhea, and diabetes mellitus. The drug is active also against prolactinomas and growth hormone-producing adenomas. This drug reduces tumor mass in 80–90% of patients with microadenomas and in 70% of patients with macroadenomas (Seo et al. 2018). Prolactin constitutes a growth factor for breast cancer cells, is associated with poor prognosis, and reduced efficacy of antitumor therapies in metastatic breast carcinoma. A clinical study evaluated the effect of taxotere versus taxotere plus bromocriptine in metastatic breast cancer patients pretreated with anthracyclines. The results suggested that the inhibition of prolactin secretion by antiprolactinemic drugs such as bromocriptine might enhance the efficacy of chemotherapy for metastatic breast cancer (Lissoni et al. 2002). More recently, bromocriptine (0.001–100 μM) was proved to inhibit drug-resistant tumor cells in a hormone-independent manner. The combination of bromocriptine with either doxorubicin or paclitaxel resulted in a synergic effect in the MDR P-glycoprotein overexpressing CEM/ADR5000 leukemic cells (Seo et al. 2018).

Thus, several approved drugs originally prescribed for other indications may be repurposed for cancer therapy because of their antineoplastic properties acting on ion channels or transporters. This approach should accelerate the development of clinical trials, especially for poor prognosis cancers. Toxins targeting ion channels and transporters represent an additional alternative to fight cancer.

7 Therapeutic Potential of Animal Venoms Against Channels and Transporters in Cancer

Over ten million of active peptides and proteins are estimated to be present in animal venoms; in many cases small amounts of the venom are sufficient to kill either preys or predators and microbial invaders (Wulff et al. 2019). Several venom components (salts, nucleotides, biogenic amines, enzymes such as phospholipase, hyaluronidase, L-amino acid oxidase, metalloproteinase, serine protease, mucoproteins, peptides, and proteins) possess antineoplastic effects via regulating the expression or activity of ion channels (Ding et al. 2014; Chen et al. 2018; Wulff et al. 2019).

7.1 Scorpion and Spider Venom Peptides as Antineoplastic Agents

Chlorotoxin is one of the most abundant peptides from the *Leiurus quinquestriatus hebraeus* deathstalker scorpion venom and exhibits great specificity for gliomas and tumors of neuroectodermal origin blocking glioma-specific chloride ion channels with high affinity and MMP-2, decreasing cells invasion (Lyons et al. 2002; McFerrin and Sontheimer 2006). Chloride channels are either absent or in low abundance in healthy tissues and in tumors of non-glial origin; however, their expression increases as gliomas progress and are crucial in tumor cell invasion and migration; chlorotoxin potently blocks these channels (Dardevet et al. 2015). Another target of chlorotoxin on the surface of glioma cells is MMP-2, which is upregulated in gliomas and related cancers but is not expressed in the normal brain. Chlorotoxin binds to MMP-2, inhibits its catalytic activity in a dose-dependent manner, and reduces its surface expression by inducing its internalization (Deshane et al. 2003). This toxin also targets other cancer cells including those from melanoma, small cell lung carcinoma, neuroblastoma, medulloblastoma, Ewing's sarcoma, and pheochromocytoma (Dardevet et al. 2015). A chlorotoxin:CY5.5 bioconjugate that emits near-IR fluorescent signal was developed as a contrast agent with the potential to improve intraoperative detection and resection of malignancies. This bioconjugate demonstrated preferential accumulation in a wide variety of tumors, including prostate and intestinal cancer, and sarcoma (Veiseh et al. 2007). In addition, a chlorotoxin-modified doxorubicin-loaded liposome delivery system for targeting gliomas was developed to improve chemotherapeutic efficacy. The liposomes enhanced the cellular uptake by the murine (C6) and human glioma (U87MG and 251MG) cell lines and mice brain microvascular endothelial cells

(BMECs), which increased drug cytotoxicity. The encapsulated doxorubicin enhanced the targeting efficiency to subcutaneous and intracranial gliomas improving the antitumor efficacy and lowering blood toxicity (Xiang et al. 2011). Furthermore, a platinum (IV) complex was conjugated to chlorotoxin in order to deliver cisplatin to cancer cells (Graf et al. 2012). In vivo assays showed that the iodine 125- and 131-labeled chlorotoxin specifically bind to brain tumor cells, making this peptide a promising candidate in radiotherapy of the postsurgical brain tumors (Srairi-Abid et al. 2019). Several chlorotoxin-based clinical trials for cancer therapy and diagnoses have been developed (Xiang et al. 2011; Pennington et al. 2018). The synthetic version of chlorotoxin reached phase III clinical trials under the name of TM-601 (Srairi-Abid et al. 2019). The FDA approved the iodine-131 radioconjugate of synthetic chlorotoxin (131I-TM-601) for glioma therapy and diagnostics (Dardevet et al. 2015).

KAaH1 is a peptide from *Androctonus australis* scorpion blocking Kv1.1 and Kv1.3 channels with an IC_{50} of 5 and 50 nM, respectively (Srairi-Abid et al. 2005). Both channels are expressed in U87 (glioblastoma), MDA-MB-231 (breast cancer), and LS174 (colon adenocarcinoma) cell lines. KAaH1 inhibited the migration of the three cell lines. KAaH2 is slightly active only on Kv1.1 channels and inhibits U87 cell proliferation probably via either the EGFR signaling pathway or other K^+ channels (Aissaoui et al. 2018). Kv10.1 and Kv11.1 channels could be potential candidates for the KAaH2 effect since EGFR forms a multimeric complex with the Kv11.1 channel (Aissaoui et al. 2018), and EGFR regulates Kv10.1 currents (Wu et al. 2012). Margatoxin is a toxin from *Centruroides margaritatus* scorpion venom that inhibits human lung adenocarcinoma A549 cell proliferation by selective inhibition of Kv1.3 channels. This toxin increases the expression of $P21^{Waf1/Cip1}$ and decreases that of Cdk4 and cyclin D3; in vivo, the toxin reduced the tumor volume when injected into the tumor (Jang et al. 2011). Moreover, margatoxin also inhibits the proliferation of the weakly metastatic rat prostate cancer cell line AT2, but not that of the strongly metastatic prostate cancer cell line Mat-LyLu (Fraser et al. 2000). Although different studies were focused mainly on the implication of Kv1.3 in regulating cell proliferation, several reports suggested that the mechanism could be also extrapolated for other K^+ channels subtypes (Srairi-Abid et al. 2019). Iberiotoxin is a peptide from *Mesobuthus tumulus* scorpion venom that inhibits Ca^{2+}-activated K^+ channels and decreases the human malignant glioma cell number in a dose- and time-dependent manner. The toxin arrested glioma cells in the S phase of the cell cycle, which eventually led to cell death. The expression of these Ca^{2+}-activated K^+ channels is upregulated in human glioma, and the expression levels increase with the grade of the tumor (Weaver et al. 2004). The antiproliferative effects of iberiotoxin have been also observed in 1321N1 astrocytomas (Basrai et al. 2002) and PC-3 prostate cancer cell lines (Bloch et al. 2007). Charybdotoxin is another scorpion toxin inhibiting Ca^{2+}-activated K^+ channels that slow the migration of melanoma cells in a dose-dependent manner (Schwab et al. 1999). The scorpion peptide κ-hefutoxin 1 inhibits the Kv10.1 channel in a dose-dependent manner (IC_{50}~26 μM) (Moreels et al. 2017b). Other toxins targeting Kv11.1 channels are ErgoTx from the *Centruroides noxious* scorpion venom, BeKm-1 from the scorpion

Buthus eupeus, and BmTx3 from the scorpion *Butus martensi*. The analgesic-antitumor peptide (AGAP) from scorpion venom is an inhibitor of voltage-gated sodium channels and decreases proliferation and migration of glioma cells (Zhao et al. 2011; Chen et al. 2018).

Psalmotoxin is a specific acid-sensing ion channel (ASIC1) blocker from the *Psalmopoeus cambridgei* tarantula venom. This toxin inhibits cation currents in malignant astroglioma and glioblastoma multiforme cells, arresting the cell cycle in the G0/G1 phase and upregulating p21 and p27 protein expression due to a reduction of the phosphorylation of ERK1/2 (Wu et al. 2019). Aa1a and Ap1a spider venom peptides are gating modifiers of the Kv10.1 channel. Ap1a peptide is more selective (>30-fold) for Kv10.1 than for Kv11.1 (Ma et al. 2018). Several peptides from the Chilean tarantula *Grammostla rosea* venom blocks Kv11.1 channels currents transiently expressed in CHO cells in a reversible manner (Wanke and Restano-Cassulini 2007).

7.2 Blarina brevicauda *Saliva Peptides, Snake Venoms, and Anemone Toxins with Antineoplastic Effects*

Soricidin is a novel paralytic peptide found in the saliva of the northern short-tailed shrew *Blarina brevicauda* that modulates TRPV6 channels, which are highly expressed in ovarian, breast, prostate, colon, and thyroid cancers, as well as certain leukemias and lymphomas. Actually, the overexpression of these channels is associated with tumor development and progression, and its inhibition decreases cancer cell proliferation and promotes apoptosis (Pennington et al. 2018; Wulff et al. 2019). SOR-C13 and SORC-27 are two shorter peptides derived from the C-terminus of soricidin and bind to TRPV6 channels with high affinity targeting tumor sites in mice bearing ovarian or prostate tumors; thus, these peptides may be used as either drug carriers or diagnostic agents in TRPV6-enriched tumors (Bowen et al. 2013). In an ovarian cancer xenograft mouse model, daily i.p. injection of SORC-13 and SORC-27 inhibited tumor growth (Xue et al. 2018). A phase I study of SORC-13 in patients with advanced tumors of epithelial origin showed stable disease in 54.5% of the patients (ranging from 2.8 to 12.5 months) without drug-related serious adverse events. The best response in this study was a 27% reduction in a pancreatic tumor with a 55% reduction in the levels of the tumor marker CA19-9 (Fu et al. 2017a, b).

Snake venom components inhibit cell proliferation and promote cell death; the mechanisms of action include increasing Ca^{2+} influx, cytochrome C release, and modified protein expression. These venom toxins also prevent metastasis, promote toxicity and free-radical generation, inhibit nucleic acid synthesis, decrease the expression and activity of matrix metalloproteinase, and inhibit integrins preventing migration and invasion of cancer cells, as well as angiogenesis (Chen et al. 2018). APETx1 and APETx4 toxins from the sea anemone *Anthopleura elegantissima* block Kv11.1 ($IC_{50} = 34$ nM) and Kv10.1 ($IC_{50} = 1.1$ μM) channels. APETx1 modifies the voltage dependence of Kv11.1, while APETx4 seems to keep Kv10.1 channels in a closed state (Diochot et al. 2003; Moreels et al. 2017a). APETx4 induces a concentration-dependent cytotoxic and proapoptotic effect in various cancer and noncancer cell lines (Moreels et al. 2017a). The large diversity of venom components and their multiple antineoplastic effects makes them excellent candidates to establish new therapeutic and specific strategies with fewer side effects in cancer treatment. Finally, we will discuss how ion channels and transporter-based nanomedicine can be used to fight cancer.

8 Ion Channel and Transporter-Based Nanomedicine in Cancer Therapy

The medical applications of nanotechnology led to the emergence of nanomedicine with the idea of inserting nanorobots into patients to treat several diseases, including cancer (Freitas 2005). Then some nanomaterials including liposomes, nanofibers, polymeric micelles, magnetic, inorganic, and polymeric nanoparticles have been used to decrease the side effects of chemotherapy by increasing its specificity, direct targeting, easy absorption, and sustained release and reducing drug degradation (Xie et al. 2019; Yang et al. 2019). Additional advantages of drug carrier systems over traditional chemotherapy include better solubility of hydrophobic drugs, higher stability, and improved blood half-life of the therapeutic agent. Nevertheless, drug delivery is complicated due to multiple physical barriers that limit diffusion and tumor penetration of the released drugs (Sun et al. 2008b). However, nanoparticles do not necessarily have to penetrate into tumor cells because the antineoplastic efficacy could be improved if the delivered drugs target cell membrane proteins involved in cancer like ion channels and transporters. For instance, nanoprobes composed of polyethylene glycol (PEG)-coated iron oxide nanoparticles were functionalized with chlorotoxin and the fluorescent molecule Cy5.5 targeting glioma tumors with high affinity, high resolution, and good therapeutic effect (Veiseh et al. 2005, 2009; Sun et al. 2008a). Chlorotoxin was also conjugated in nanoparticles with the chemotherapeutic agent methotrexate, demonstrating preferential accumulation and increased cytotoxicity in tumor cells, as well as prolonged nanoparticle retention within tumors (Sun et al. 2008b). A cancer cell-specific magnetic nanovector functionalized with siRNA and chlorotoxin was developed with the aims of efficient siRNA delivery and noninvasive monitoring by magnetic resonance imaging (MRI). The nanovector demonstrated both increased siRNA internalization by tumor cells and intracellular trafficking toward enhanced knockdown of targeted gene expression. In addition, this nanovector enhanced MRI contrast in vitro, potentially enabling monitoring of the treatment in vivo. The elevated specificity and potency of this nanovector system make it a potential gene therapy approach for malignant tumors (Veiseh et al. 2010).

Drug resistance has been also addressed by nanomedicine. Functionalized nanoparticles deliver and concentrate drugs at the plasma membrane where ABC

Table 2 Examples of potential ion channel and transporter-based anticancer therapies

Therapeutic approach	Mechanism involved	Examples of channels or transporters	References
Induction of oxidative stress	To enhance ROS/RNS production in cancer cells leading to selective cell death versus normal cells	SERCA TRPM2 SLC7A11 VDAC	Denmeade et al. (2012), Maldonado et al. (2013), Sehm et al. (2016), Almasi et al. (2019b)
Inhibition/activation of mitochondrial channels and proteins	Alteration of the mitochondria permeability, activating apoptotic pathways, and/or inducing ROS production	VDAC mPTP MCU UCP2	Zhou et al. (2016), Cui et al. (2019), Sucu et al. (2019)
Immunotherapy	Design of specific antibodies and enhancement of the immune response	gBK Kv11.1 Kv10.1 $nfP2X_7$	Ge et al. (2012), Sette et al. (2013), Hoa et al. (2014), Hartung and Pardo (2016), Gilbert et al. (2017)
Targeting splice variants of channels	Aberrant expression of splice variants in tumor cells provide better selectivity	GIRK1 nNav1.5 $nfP2X_7$ gBK	Wagner et al. (2010); Ge et al. (2012), Hoa et al. (2014), Rezania et al. (2016), Gilbert et al. (2017), Yamaci et al. (2017)
Targeting noncanonical functions of channels	To avoid translocation of channel fragments, related to the transcription of cancer genes	Cav1.2	Xiao et al. (2010), Buchanan McCloskey (2016), Chen et al. (2018)
Repurposing of drugs targeting ion channels and transporters	Increase the anticancer molecular mechanisms of the treatment and decrease the administered doses and side effects	K^+ and Ca^{2+} channels	Kale et al. (2015)
Use of animal venoms	Higher tumor selectivity	TRPV6	Bowen et al. (2013), Xue et al. (2018)
Nanomedicine	More selective drug carrier systems and modifications of functional biomolecules	Cl^- channels MDR1	Veiseh et al. (2005, 2009), Meng et al. (2010), Wang et al. (2018a)
Photothermal therapy	Induced cell death by NIR irradiation based on the photothermal performance of rGO-P and the specific interactions between potassium channels and calmodulin	Kv10.1	Chai et al. (2018)

transporters are located and are saturated with either the antineoplastic drug, ABC transporter blockers, or inhibitors (Xue and Liang 2012). Mesoporous silica nanoparticles were used to deliver doxorubicin and P-glycoprotein siRNA to the drug-resistant cancer cell line KB-V1 (derived from HeLa cervical cancer cells). The dual delivery of doxorubicin and siRNA in KB-V1cells increased the intracellular and intranuclear drug concentration to levels exceeding free doxorubicin or the drug being delivered by mesoporous in the absence of siRNA co-delivery (Meng et al.

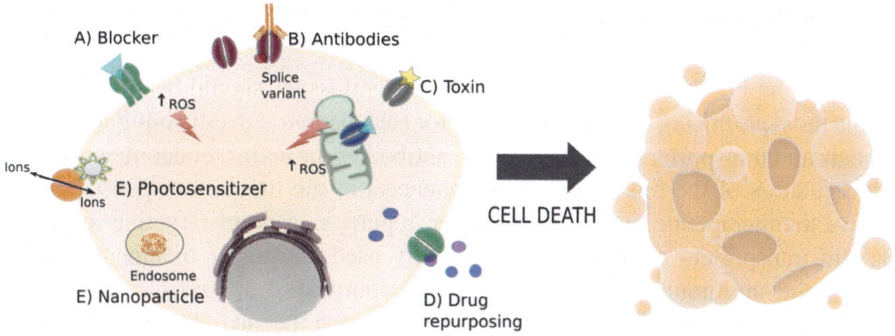

Fig. 1 Novel therapeutic opportunities targeting ion channels or transporters in cancer. (**a**) Inducing higher ROS production in cancer cells by blocking ion channels or transporters may convert cancer cells more sensitive to chemotherapy. (**b**) Immunotherapy against channels overexpressed in cancer cells offers a more selective approach. (**c**) Toxins targeting ion channels and transporters may also help to improve selectivity and the antitumor response. (**d**) Identifying and investigating new uses of approved drugs hold potential and many advantages for anticancer therapy. (**e**) New technologies like nanoparticles and phototherapy targeting ion channels and transporters are potential more selective anticancer therapies

2010). Wang and colleagues observed similar results; the co-delivery of doxorubicin and MDR1-siRNA by mesoporous silica nanoparticles-polymerpolyethylenimine improved oral squamous carcinoma treatment in vitro and in vivo (Wang et al. 2018a). In breast cancer, the dual delivery system of doxorubicin and siRNA against P-glycoprotein resulted in synergistic inhibition of tumor growth (Meng et al. 2013). Similar results were observed using doxorubicin/P-glycoprotein siRNA-loaded nanomicelles (Suo et al. 2016) and RGD peptide (arginine-glycine-aspartic acid)-modified liposomes containing also doxorubicin and P-glycoprotein siRNA (Xue and Liang 2012).

Photothermal therapy (PTT) converts near-infrared light (NIR) stimulation into local mild heat that can stimulate a photosensitizer to control biological processes in a remote and noninvasive manner. Photothermal transducers of graphene oxide linked to a calmodulin-binding peptide (rGO-P) have been designed to activate or inactivate potassium channels based on their binding to Ca^{2+}-calmodulin (Chai et al. 2018). Kv10.1 channel activity is inhibited by $[Ca^{2+}]i$ through the binding of Ca^{2+}-calmodulin (Lörinczi et al. 2016). HEK-293 cells transfected with Kv10.1, treated with rGO-P, and stimulated with NIR irradiation displayed an open state of Kv10.1 channels because the channel cannot be closed via binding of Ca^{2+}-calmodulin, suggesting a new strategy to regulate ion channels involved in cancer like Kv10.1 (Chai et al. 2018). The application of nanomaterials to PTT is emerging as a new strategy for cancer therapy and showing encouraging results in vivo (Zou et al. 2016).

A number of studies suggest that targeting ion channels and transporters is a feasible approach to achieve therapeutic efficacy and tumor selectivity in cancer treatment. Table 2 summarizes some of these findings, and Fig. 1 shows the diversity of the potential anticancer therapeutic options based on ion channels and transporters.

9 Conclusions

Cancer is a multifactorial disease; therefore, several issues should be considered to find the most promising anticancer therapy for each patient. Specific inhibitors of ion channels and transporters including drugs, antibodies, or toxins, concomitantly used with regulators of the redox status and enhancers of the immune response, should improve the therapeutic outcome. Moreover, repurposing drugs targeting ion channels and transporters combined with currently used anticancer treatments should enhance the antitumor response. This drug repurposing approach adds several anticancer mechanisms to the treatment, may decrease the side effects by lowering the administrated doses, and should decrease treatment costs, as well as accelerating the development of clinical trials. The specificity of many toxins targeting ion channels and transporters should allow the directed and specific treatment of tumors. Exploiting ion channel and transporter-based approaches to fight cancer and the use of nanotechnology and nanomedicine should improve the current therapies for the benefit of cancer patients.

Acknowledgments *Conflict of Interest:* The authors declare no conflict of interests. *Contributions:* AR, JG-Q, and LAA-E wrote the first draft of the manuscript. YS-P and JC revised the first draft and contributed to manuscript writing of the following versions. All authors read and approved the final version of the manuscript.

References

Abdelaleem M et al (2019) Prospects for repurposing CNS drugs for cancer treatment. Oncol Rev. https://doi.org/10.4081/oncol.2019.411

Abdul M, Hoosein N (2002) Expression and activity of potassium ion channels in human prostate cancer. Cancer Lett. https://doi.org/10.1016/s0304-3835(02)00348-8

Abu El Maaty MA, Wölfl S (2017) Effects of 1, 25(OH)2D3 on cancer cells and potential applications in combination with established and putative anti-cancer agents'. Nutrients. https://doi.org/10.3390/nu9010087

Aissaoui D et al (2018) Functional role of Kv1.1 and Kv1.3 channels in the neoplastic progression steps of three cancer cell lines, elucidated by scorpion peptides. Int J Biol Macromol. https://doi.org/10.1016/j.ijbiomac.2018.01.144

Allard B et al (2016) Immunosuppressive activities of adenosine in cancer. Curr Opin Pharmacol. https://doi.org/10.1016/j.coph.2016.04.001

Almasi S et al (2018) TRPM2 channel–mediated regulation of autophagy maintains mitochondrial function and promotes gastric cancer cell survival via the JNK-signaling pathway. J Biol Chem. https://doi.org/10.1074/jbc.M117.817635

Almasi S, Sterea AM et al (2019a) TRPM2 ion channel promotes gastric cancer migration, invasion and tumor growth through the AKT signaling pathway. Sci Rep. https://doi.org/10.1038/s41598-019-40330-1

Almasi S, Long CY et al (2019b) TRPM2 silencing causes G2/M arrest and apoptosis in lung cancer cells via increasing intracellular ROS and RNS levels and activating the JNK pathway. Cell Physiol Biochem. https://doi.org/10.33594/000000052

An L et al (2017) Terfenadine combined with epirubicin impedes the chemo-resistant human non-small cell lung cancer both in vitro and in vivo through EMT and Notch reversal. Pharmacol Res. https://doi.org/10.1016/j.phrs.2017.07.021

Arbabian A et al (2013) Modulation of endoplasmic reticulum calcium pump expression during lung cancer cell differentiation. FEBS J. https://doi.org/10.1111/febs.12064

Arcangeli A, Becchetti A (2010) New trends in cancer therapy: targeting ion channels and transporters. Pharmaceuticals. https://doi.org/10.3390/ph3041202

Arora A et al (2005) Modulation of P-glycoprotein-mediated multidrug resistance in K562 leukemic cells by indole-3-carbinol. Toxicol Appl Pharmacol. https://doi.org/10.1016/j.taap.2004.06.017

Asher V et al (2010) Eag and HERG potassium channels as novel therapeutic targets in cancer. World J Surg Oncol. https://doi.org/10.1186/1477-7819-8-113

Asher V et al (2011) The role of Eag and HERG channels in cell proliferation and apoptotic cell death in SK-OV-3 ovarian cancer cell line. Cancer Cell Int. https://doi.org/10.1186/1475-2867-11-6

Averaimo S et al (2010) Chloride intracellular channel 1 (CLIC1): sensor and effector during oxidative stress. FEBS Lett. https://doi.org/10.1016/j.febslet.2010.02.073

Avila E et al (2010) Calcitriol down-regulates human ether a go-go 1 potassium channel expression in cervical cancer cells. Anticancer Res. https://doi.org/10.1016/j.yexcr.2009.11.008

Azimi I, Roberts-Thomson SJ, Monteith GR (2014) Calcium influx pathways in breast cancer: opportunities for pharmacological intervention. Br J Pharmacol. https://doi.org/10.1111/bph.12486

Azimi I et al (2019) ORAI1 and ORAI3 in breast cancer molecular subtypes and the identification of ORAI3 as a hypoxia sensitive gene and a regulator of hypoxia responses. Cancer. https://doi.org/10.3390/cancers11020208

Bachmann M et al (2018) Targeting mitochondrial ion channels to fight cancer. Int J Mol Sci. https://doi.org/10.3390/ijms19072060

Baffy G (2010) Uncoupling protein-2 and cancer. Mitochondrion. https://doi.org/10.1016/j.mito.2009.12.143

Bao XX et al (2012) Nifedipine induced autophagy through Beclin1 and mTOR pathway in endometrial carcinoma cells. Chin Med J (Engl). https://doi.org/10.3760/cma.j.issn.0366-6999.2012.17.028

Bao L et al (2016) Depletion of the human ion channel TRPM2 in neuroblastoma demonstrates its key role in cell survival through modulation of mitochondrial reactive oxygen species and bioenergetics. J Biol Chem. https://doi.org/10.1074/jbc.M116.747147

Barbado M et al (2009) Gene regulation by voltage-dependent calcium channels. Biochim Biophys Acta. https://doi.org/10.1016/j.bbamcr.2009.02.004

Barlaz Us S et al (2019) Effect of imipramine on radiosensitivity of prostate cancer: an in vitro study. Cancer Invest. https://doi.org/10.1080/07357907.2019.1662434

Basrai D et al (2002) BK channel blockers inhibit potassium-induced proliferation of human astrocytoma cells. Neuroreport. https://doi.org/10.1097/00001756-200203250-00008

Beatty GL, Gladney WL (2015) Immune escape mechanisms as a guide for cancer immunotherapy. Clin Cancer Res. https://doi.org/10.1158/1078-0432.CCR-14-1860

Belpomme D et al (2000) Verapamil increases the survival of patients with anthracycline-resistant metastatic breast carcinoma. Ann Oncol. https://doi.org/10.1023/a:1026556119020

Benzerdjeb, N. et al. (2016) 'Orai3 is a predictive marker of metastasis and survival in resectable lung adenocarcinoma', *Oncotarget*. doi: https://doi.org/10.18632/oncotarget.13149

Bernal-Ramos G et al (2017) Astemizole inhibits cell proliferation in human prostate tumorigenic cells expressing ether a-go-go-1 potassium channels. Cell Mol Biol (Noisy-le-Grand). https://doi.org/10.14715/cmb/2017.63.12.4

Bernardi P et al (2015) The mitochondrial permeability transition pore: channel formation by F-ATP synthase, integration in signal transduction, and role in pathophysiology. Physiol Rev. https://doi.org/10.1152/physrev.00001.2015

Berry BJ et al (2018) Use the protonmotive force: mitochondrial uncoupling and reactive oxygen species. J Mol Biol. https://doi.org/10.1016/j.jmb.2018.03.025
Bertolesi GE et al (2002) The Ca($^{2+}$) channel antagonists mibefradil and pimozide inhibit cell growth via different cytotoxic mechanisms. Mol Pharmacol. https://doi.org/10.1124/mol.62.2.210
Berul CI, Morad M (1995) Regulation of potassium channels by nonsedating antihistamines. Circulation. https://doi.org/10.1161/01.CIR.91.8.2220
Binyamin L et al (2004) Targeting an extracellular epitope of the human multidrug resistance protein 1 (MRP1) in malignant cells with a novel recombinant single chain Fv antibody. Int J Cancer. https://doi.org/10.1002/ijc.20177
Birben E et al (2012) Oxidative stress and antioxidant defense. World Allergy Organ J. https://doi.org/10.1097/WOX.0b013e3182439613
Bjelakovic G et al (2007) Mortality in randomized trials of antioxidant supplements for primary and secondary prevention: systematic review and meta-analysis. JAMA. https://doi.org/10.1001/jama.297.8.842
Blake SD et al (2017) Transient receptor potential, Melastatin-2 (TRPM2) blockade: perspectives on potential novel clinical utility in cancer. Transl Cancer Res. https://doi.org/10.21037/tcr.2017.03.11
Blaya B et al (2010) Histamine and histamine receptor antagonists in cancer biology. Inflamm Allergy Drug Targets. https://doi.org/10.2174/187152810792231869
Bloch M et al (2007) KCNMA1 gene amplification promotes tumor cell proliferation in human prostate cancer. Oncogene. https://doi.org/10.1038/sj.onc.1210036
Bogeski I et al (2010) Differential redox regulation of ORAI ion channels: a mechanism to tune cellular calcium signaling. Sci Signal. https://doi.org/10.1126/scisignal.2000672
Bootman MD et al (2001) Calcium signalling – an overview. Semin Cell Dev Biol. https://doi.org/10.1006/scdb.2000.0211
Bowen CV et al (2013) In vivo detection of human TRPV6-rich tumors with anti-cancer peptides derived from soricidin. PLoS One. https://doi.org/10.1371/journal.pone.0058866
Buchanan PJ, McCloskey KD (2016) Ca_V channels and cancer: canonical functions indicate benefits of repurposed drugs as cancer therapeutics. Eur Biophys J. https://doi.org/10.1007/s00249-016-1144-z
Cairo MS et al (1989) Clinical trial of continuous infusion verapamil, bolus vinblastine, and continuous infusion VP-16 in drug-resistant pediatric tumors. Cancer Res 49(4):1063–1066
Camacho J (2006) Ether a go-go potassium channels and cancer. Cancer Lett. https://doi.org/10.1016/j.canlet.2005.02.016
Camara AKS et al (2017) Mitochondrial VDAC1: a key gatekeeper as potential therapeutic target. Front Physiol. https://doi.org/10.3389/fphys.2017.00460
Cannon B et al (2006) Uncoupling proteins: a role in protection against reactive oxygen species-or not? Biochim Biophys Acta Bioenerg. https://doi.org/10.1016/j.bbabio.2006.05.016
Cazares-Ordonez V et al (2015) A cis-acting element in the promoter of human ether a go-go 1 potassium channel gene mediates repression by calcitriol in human cervical cancer cells. Biochem Cell Biol. https://doi.org/10.1139/bcb-2014-0073
Chai R et al (2018) Remote-controlling potassium channels in living cells through photothermal inactivation of calmodulin. Adv Healthc Mater. https://doi.org/10.1002/adhm.201800674
Chavez-Lopez MG et al (2014) Antiproliferative and proapoptotic effects of astemizole on cervical cancer cells. Int J Gynecol Cancer. https://doi.org/10.1097/IGC.0000000000000151
Chavez-Lopez MG et al (2015) Astemizole-based anticancer therapy for hepatocellular carcinoma (HCC), and Eag1 channels as potential early-stage markers of HCC. Tumour Biol. https://doi.org/10.1007/s13277-015-3299-0
Chavez-Lopez MG et al (2017) The combination astemizole-gefitinib as a potential therapy for human lung cancer. Onco Targets Ther. https://doi.org/10.2147/OTT.S144506
Chen CD et al (2007) Overexpression of CLIC1 in human gastric carcinoma and its clinicopathological significance. Proteomics. https://doi.org/10.1002/pmic.200600663

Chen SJ et al (2013) Role of TRPM2 in cell proliferation and susceptibility to oxidative stress. Am J Physiol Cell Physiol. https://doi.org/10.1152/ajpcell.00069.2012

Chen SJ et al (2014) A splice variant of the human ion channel TRPM2 modulates neuroblastoma tumor growth through hypoxia-inducible factor (HIF)-1/2α. J Biol Chem. https://doi.org/10.1074/jbc.M114.620922

Chen N et al (2018) Animal protein toxins: origins and therapeutic applications. Biophys Rep. https://doi.org/10.1007/s41048-018-0067-x

Chen YF et al (2019) Store-operated Ca2+ entry in tumor progression: From molecular mechanisms to clinical implications. Cancer. https://doi.org/10.3390/cancers11070899

Cheng WC et al (2019) Uncoupling protein 2 reprograms the tumor microenvironment to support the anti-tumor immune cycle. Nat Immunol. https://doi.org/10.1038/s41590-018-0290-0

Chiang EY et al (2017) Potassium channels Kv1.3 and KCa3.1 cooperatively and compensatorily regulate antigen-specific memory T cell functions. Nat Commun. https://doi.org/10.1038/ncomms14644

Chimote AA et al (2013) Selective inhibition of KCa3.1 channels mediates adenosine regulation of the motility of human T cells. J Immunol. https://doi.org/10.4049/jimmunol.1300702

Chimote AA et al (2018) A defect in KCa3.1 channel activity limits the ability of CD8+ T cells from cancer patients to infiltrate an adenosine-rich microenvironment. Sci Signal. https://doi.org/10.1126/scisignal.aaq1616

Chio IIC, Tuveson DA (2017) ROS in cancer: the burning question. Trends Mol Med. https://doi.org/10.1016/j.molmed.2017.03.004

Chioni AM et al (2005) A novel polyclonal antibody specific for the Nav1.5 voltage-gated Na+ channel "neonatal" splice form. J Neurosci Methods. https://doi.org/10.1016/j.jneumeth.2005.03.010

Chung FY et al (2006) Sarco/endoplasmic reticulum calcium-ATPase 2 expression as a tumor marker in colorectal cancer. Am J Surg Pathol. https://doi.org/10.1097/00000478-200608000-00006

Clapham DE (2003) TRP channels as cellular sensors. Nature. https://doi.org/10.1038/nature02196

Comes N et al (2013) The voltage-dependent K(+) channels Kv1.3 and Kv1.5 in human cancer. Front Physiol. https://doi.org/10.3389/fphys.2013.00283

Crociani O et al (2014) hERG1 channels regulate VEGF-A secretion in human gastric cancer: clinicopathological correlations and therapeutical implications. Clin Cancer Res. https://doi.org/10.1158/1078-0432.CCR-13-2633

Cui C et al (2019) Progress in understanding mitochondrial calcium uniporter complex-mediated calcium signalling: a potential target for cancer treatment. Br J Pharmacol. https://doi.org/10.1111/bph.14632

Curry MC et al (2012) Distinct regulation of cytoplasmic calcium signals and cell death pathways by different plasma membrane calcium ATPase isoforms in MDA-MB-231 breast cancer cells. J Biol Chem. https://doi.org/10.1074/jbc.M112.364737

Curry M, Roberts-Thomson SJ, Monteith GR (2016) PMCA2 silencing potentiates MDA-MB-231 breast cancer cell death initiated with the Bcl-2 inhibitor ABT-263. Biochem Biophys Res Commun. https://doi.org/10.1016/j.bbrc.2016.09.030

D'Alessandro G et al (2019) Radiation increases functional $K_{Ca}3.1$ expression and invasiveness in glioblastoma. Cancer. https://doi.org/10.3390/cancers11030279

D'Amico M, Gasparoli L, Arcangeli A (2013) Potassium channels: novel emerging biomarkers and targets for therapy in cancer. Recent Pat Anticancer Drug Discov 8(1):53–65

Dalla Pozza E et al (2012) Role of mitochondrial uncoupling protein 2 in cancer cell resistance to gemcitabine. Biochim Biophys Acta Mol Cell Res. https://doi.org/10.1016/j.bbamcr.2012.06.007

Dang D, Rao R (2016) Calcium-ATPases: gene disorders and dysregulation in cancer. Biochim Biophys Acta Mol Cell Res. https://doi.org/10.1016/j.bbamcr.2015.11.016

Dardevet L et al (2015) Chlorotoxin: a helpful natural scorpion peptide to diagnose glioma and fight tumor invasion. Toxins (Basel). https://doi.org/10.3390/toxins7041079

David CJ, Manley JL (2010) Alternative pre-mRNA splicing regulation in cancer: pathways and programs unhinged. Genes Dev. https://doi.org/10.1101/gad.1973010

Debes JD et al (2004) Inverse association between prostate cancer and the use of calcium channel blockers. Cancer Epidemiol Biomarkers. https://doi.org/10.1158/1055-9965.epi-03-0093

Deeb KK, Trump DL, Johnson CS (2007) Vitamin D signalling pathways in cancer: potential for anticancer therapeutics. Nat Rev Cancer. https://doi.org/10.1038/nrc2196

Deliot N, Constantin B (2015) Plasma membrane calcium channels in cancer: alterations and consequences for cell proliferation and migration. Biochim Biophys Acta. https://doi.org/10.1016/j.bbamem.2015.06.009

Denicola GM et al (2011) Oncogene-induced Nrf2 transcription promotes ROS detoxification and tumorigenesis. Nature. https://doi.org/10.1038/nature10189

Denmeade SR, Isaacs JT (2005) The SERCA pump as a therapeutic target: making a "smart bomb" for prostate cancer. Cancer Biol Ther. https://doi.org/10.4161/cbt.4.1.1505

Denmeade SR et al (2012) Engineering a prostate-specific membrane antigen-activated tumor endothelial cell prodrug for cancer therapy. Sci Transl Med. https://doi.org/10.1126/scitranslmed.3003886

Derdak Z et al (2008) The mitochondrial uncoupling protein-2 promotes chemoresistance in cancer cells. Cancer Res. https://doi.org/10.1158/0008-5472.CAN-08-0053

Deshane J, Garner CC, Sontheimer H (2003) Chlorotoxin inhibits glioma cell invasion via matrix metalloproteinase-2. J Biol Chem. https://doi.org/10.1074/jbc.M205662200

Diaz L et al (2009) Estrogens and human papilloma virus oncogenes regulate human ether-a-go-go-1 potassium channel expression. Cancer Res. https://doi.org/10.1158/0008-5472.CAN-08-2036

Diaz L et al (2015) Mechanistic effects of calcitriol in cancer biology. Nutrients. https://doi.org/10.3390/nu7065020

Diez-Bello R et al (2019) (−)-Oleocanthal inhibits proliferation and migration by modulating Ca^{2+} entry through TRPC6 in breast cancer cells. Biochim Biophys Acta Mol Cell Res. https://doi.org/10.1016/j.bbamcr.2018.10.010

Ding XW et al (2008) Overexpression of hERG1 in resected esophageal squamous cell carcinomas: a marker for poor prognosis. J Surg Oncol. https://doi.org/10.1002/jso.20891

Ding J et al (2014) Scorpion venoms as a potential source of novel cancer therapeutic compounds. Exp Biol Med (Maywood). https://doi.org/10.1177/1535370213513991

Diochot S et al (2003) APETx1, a new toxin from the sea anemone *Anthopleura elegantissima*, blocks voltage-gated human ether-a-go-go-related gene potassium channels. https://doi.org/10.1124/mol.64.1.59

Dixon SJ et al (2012) Ferroptosis: an iron-dependent form of nonapoptotic cell death. Cell. https://doi.org/10.1016/j.cell.2012.03.042

Downie BR et al (2008) Eag1 expression interferes with hypoxia homeostasis and induces angiogenesis in tumors. J Biol Chem. https://doi.org/10.1074/jbc.M801830200

Dubois C et al (2014) Remodeling of channel-forming ORAI proteins determines an oncogenic switch in prostate cancer. Cancer Cell. https://doi.org/10.1016/j.ccr.2014.04.025

Durie BG, Dalton WS (1988) Reversal of drug-resistance in multiple myeloma with verapamil. Br J Haematol. https://doi.org/10.1111/j.1365-2141.1988.tb06190.x

Dutta S et al (2014) Interactions between exosomes from breast cancer cells and primary mammary epithelial cells leads to generation of reactive oxygen species which induce DNA damage response, stabilization of p53 and autophagy in epithelial cells. PLoS One. https://doi.org/10.1371/journal.pone.0097580

Ellegaard AM et al (2016) Repurposing cationic amphiphilic antihistamines for cancer treatment. EBioMedicine. https://doi.org/10.1016/j.ebiom.2016.06.013

Enfissi A et al (2004) The blocking of capacitative calcium entry by 2-aminoethyl diphenylborate (2-APB) and carboxyamidotriazole (CAI) inhibits proliferation in Hep G2 and Huh-7 human hepatoma cells. Cell Calcium. https://doi.org/10.1016/j.ceca.2004.04.004

Ermak G, Davies KJA (2002) Calcium and oxidative stress: from cell signaling to cell death. Mol Immunol. https://doi.org/10.1016/S0161-5890(01)00108-0

Faouzi M, Penner R (2014) TRPM2. Handb Exp Pharmacol. https://doi.org/10.1007/978-3-642-54215-2_16

Farias LM et al (2004) Ether a go-go potassium channels as human cervical cancer markers. Cancer Res. https://doi.org/10.1158/0008-5472.CAN-04-1204

Faustino-Rocha AI et al (2017) Antihistamines as promising drugs in cancer therapy. Life Sci. https://doi.org/10.1016/j.lfs.2016.12.008

Fernández-Nogueira P et al (2018) Histamine receptor 1 inhibition enhances antitumor therapeutic responses through extracellular signal-regulated kinase (ERK) activation in breast cancer. Cancer Lett. https://doi.org/10.1016/j.canlet.2018.03.014

Feske S, Wulff H, Skolnik EY (2015) Ion channels in innate and adaptive immunity. Annu Rev Immunol. https://doi.org/10.1146/annurev-immunol-032414-112212

Fiorio Pla A, Kondratska K, Prevarskaya N (2016) STIM and ORAI proteins: crucial roles in hallmarks of cancer. Am J Physiol Cell Physiol. https://doi.org/10.1152/ajpcell.00364.2015

Fitzpatrick AL et al (2001) Hypertension, heart rate, use of antihypertensives, and incident prostate cancer. Ann Epidemiol. https://doi.org/10.1016/s1047-2797(01)00246-0

Fraser SP, Grimes JA, Djamgoz MBA (2000) Effects of voltage-gated ion channel modulators on rat prostatic cancer cell proliferation: comparison of strongly and weakly metastatic cell lines. Prostate. https://doi.org/10.1002/1097-0045(20000615)44:1<61::aid-pros9>3.0.co;2-3

Freitas RA Jr (2005) What is nanomedicine? Nanomedicine. https://doi.org/10.1016/j.nano.2004.11.003

Frisch J et al (2019) STIM-Orai channels and reactive oxygen species in the tumor microenvironment. Cancer. https://doi.org/10.3390/cancers11040457

Fu S et al (2017a) Erratum to: first-in-human phase I study of SOR-C13, a TRPV6 calcium channel inhibitor, in patients with advanced solid tumors. Invest New Drugs. https://doi.org/10.1007/s10637-017-0455-y

Fu S et al (2017b) First-in-human phase I study of SOR-C13, a TRPV6 calcium channel inhibitor, in patients with advanced solid tumors. Invest New Drugs. https://doi.org/10.1007/s10637-017-0438-z

Fukushiro-Lopes DF et al (2018) Preclinical study of a Kv11.1 potassium channel activator as antineoplastic approach for breast cancer. Oncotarget. https://doi.org/10.18632/oncotarget.22925

Gao T et al (2000) Role of the C terminus of the alpha 1C (CaV1.2) subunit in membrane targeting of cardiac L-type calcium channels. J Biol Chem. https://doi.org/10.1074/jbc.M003465200

Gao T et al (2001) C-terminal fragments of the alpha 1C (CaV1.2) subunit associate with and regulate L-type calcium channels containing C-terminal-truncated alpha 1C subunits. J Biol Chem. https://doi.org/10.1074/jbc.M008000200

Garcia-Becerra R et al (2010) Calcitriol inhibits ether-a go-go potassium channel expression and cell proliferation in human breast cancer cells. Exp Cell Res. https://doi.org/10.1016/j.yexcr.2009.11.008

Garcia-Ferreiro RE et al (2004) Mechanism of block of hEag1 K+ channels by imipramine and astemizole. J Gen Physiol. https://doi.org/10.1085/jgp.200409041

Garcia Quiroz J, Camacho J (2011) Astemizole: an old anti histamine as a new promising anti cancer drug. Anticancer Agents Med Chem. https://doi.org/10.2174/187152011795347513

Garcia-Quiroz J et al (2012) Astemizole synergizes calcitriol antiproliferative activity by inhibiting CYP24A1 and upregulating VDR: a novel approach for breast cancer therapy. PLoS One. https://doi.org/10.1371/journal.pone.0045063

Garcia-Quiroz J et al (2014) In vivo dual targeting of the oncogenic ether-a-go-go-1 potassium channel by calcitriol and astemizole results in enhanced antineoplastic effects in breast tumors. BMC Cancer. https://doi.org/10.1186/1471-2407-14-745

Garcia-Quiroz J et al (2016) Calcitriol stimulates gene expression of cathelicidin antimicrobial peptide in breast cancer cells with different phenotype. J Biomed Sci. https://doi.org/10.1186/s12929-016-0298-4

Garcia-Quiroz J et al (2019) Astemizole, an inhibitor of ether-a-go-go-1 potassium channel, increases the activity of the tyrosine kinase inhibitor gefitinib in breast cancer cells. Rev Invest Clin. https://doi.org/10.24875/RIC.18002840
García-Quiroz J et al (2019) Synergistic antitumorigenic activity of calcitriol with curcumin or resveratrol is mediated by angiogenesis inhibition in triple negative breast cancer xenografts. Cancer. https://doi.org/10.3390/cancers11111739
Gasparoli L et al (2015) New pyrimido-indole compound CD-160130 preferentially inhibits the KV11.1B isoform and produces antileukemic effects without cardiotoxicity. Mol Pharmacol. https://doi.org/10.1124/mol.114.094920
Gavrilova-Ruch O et al (2002) Effects of imipramine on ion channels and proliferation of IGR1 melanoma cells. J Membr Biol. https://doi.org/10.1007/s00232-001-0181-3
Ge L et al (2012) Glioma big potassium channel expression in human cancers and possible T cell epitopes for their immunotherapy. J Immunol. https://doi.org/10.4049/jimmunol.1102965
Gélébart P et al (2002) Expression of endomembrane calcium pumps in colon and gastric cancer cells. Induction of SERCA3 expression during differentiation. J Biol Chem. https://doi.org/10.1074/jbc.M201747200
Gentile S et al (2016) hERG1 potassium channel in cancer cells: a tool to reprogram immortality. Eur Biophys J. https://doi.org/10.1007/s00249-016-1169-3
Gessner G, Heinemann SH (2003) Inhibition of hEAG1 and hERG1 potassium channels by clofilium and its tertiary analogue LY97241. Br J Pharmacol 138(1):161–171. https://doi.org/10.1038/sj.bjp.0705025
Gilbert SM et al (2017) A phase I clinical trial demonstrates that nfP2X7-targeted antibodies provide a novel, safe and tolerable topical therapy for basal cell carcinoma. Br J Dermatol. https://doi.org/10.1111/bjd.15364
Gillman PK (2007) Tricyclic antidepressant pharmacology and therapeutic drug interactions updated. Br J Pharmacol. https://doi.org/10.1038/sj.bjp.0707253
Glowka E, Stasiak J, Lulek J (2019) Drug delivery systems for vitamin D supplementation and therapy. Pharmaceutics. https://doi.org/10.3390/pharmaceutics11070347
Gomes FR et al (2015) Alternatively spliced isoforms of K_V10.1 potassium channels modulate channel properties and can activate cyclin-dependent kinase in Xenopus oocytes. J Biol Chem. https://doi.org/10.1074/jbc.M115.668749
Gómez-Lagunas F et al (2017) Gating modulation of the tumor-related Kv10.1 channel by mibefradil. J Cell Physiol. https://doi.org/10.1002/jcp.25448
Gomez-Ospina N et al (2013) A promoter in the coding region of the calcium channel gene CACNA1C generates the transcription factor CCAT. PLoS One. https://doi.org/10.1371/journal.pone.0060526
Gomez-Varela D et al (2006) Different relevance of inactivation and F468 residue in the mechanisms of hEag1 channel blockage by astemizole, imipramine and dofetilide. FEBS Lett. https://doi.org/10.1016/j.febslet.2006.08.030
Gomez-Varela D et al (2007) Monoclonal antibody blockade of the human Eag1 potassium channel function exerts antitumor activity. Cancer. https://doi.org/10.1158/0008-5472.CAN-07-0107
Gorrini C, Harris IS, Mak TW (2013) Modulation of oxidative stress as an anticancer strategy. Nat Rev Drug Discov. https://doi.org/10.1038/nrd4002
Graf N et al (2012) Platinum(IV)-chlorotoxin (CTX) conjugates for targeting cancer cells. J Inorg Biochem. https://doi.org/10.1016/j.jinorgbio.2012.02.012
Gritti M et al (2014) Metformin repositioning as antitumoral agent: Selective antiproliferative effects in human glioblastoma stem cells, via inhibition of CLIC1-mediated ion current. Oncotarget. https://doi.org/10.18632/oncotarget.2617
Grössinger EM et al (2014) Targeting proliferation of chronic lymphocytic leukemia (CLL) cells through KCa3.1 blockade. Leukemia. https://doi.org/10.1038/leu.2014.37
Gualdani R et al (2019) Store-operated calcium entry contributes to cisplatin-induced cell death in non-small cell lung carcinoma. Cancer. https://doi.org/10.3390/cancers11030430

Guan J et al (2009) The xc- cystine/glutamate antiporter as a potential therapeutic target for small-cell lung cancer: use of sulfasalazine. Cancer Chemother Pharmacol. https://doi.org/10.1007/s00280-008-0894-4

Guo Q et al (2019) A comprehensive evaluation of clinical efficacy and safety of celecoxib in combination with chemotherapy in metastatic or postoperative recurrent gastric cancer patients: a preliminary, three-center, clinical trial study. Medicine (Baltimore). https://doi.org/10.1097/MD.0000000000016234

Gupta SC et al (2013) Cancer drug discovery by repurposing: teaching new tricks to old dogs. Trends Pharmacol Sci. https://doi.org/10.1016/j.tips.2013.06.005

Hara Y et al (2002) LTRPC2 Ca2+-permeable channel activated by changes in redox status confers susceptibility to cell death. Mol Cell. https://doi.org/10.1016/S1097-2765(01)00438-5

Harr MW, Distelhorst CW (2010) Apoptosis and autophagy: decoding calcium signals that mediate life or death. Cold Spring Harb Perspect Biol. https://doi.org/10.1101/cshperspect.a005579

Harris IS et al (2015) Glutathione and thioredoxin antioxidant pathways synergize to drive cancer initiation and progression. Cancer Cell. https://doi.org/10.1016/j.ccell.2014.11.019

Hartung F, Pardo LA (2016) Guiding TRAIL to cancer cells through Kv10.1 potassium channel overcomes resistance to doxorubicin. Eur Biophys J. https://doi.org/10.1007/s00249-016-1149-7

Hartung F, Stühmer W, Pardo LA (2011) Tumor cell-selective apoptosis induction through targeting of K_V10.1 via bifunctional TRAIL antibody. Mol Cancer. https://doi.org/10.1186/1476-4598-10-109

Hasna J et al (2018) Orai3 calcium channel and resistance to chemotherapy in breast cancer cells: the p53 connection. Cell Death Differ. https://doi.org/10.1038/s41418-017-0007-1

He YM et al (2018) Effect of CLIC1 gene silencing on proliferation, migration, invasion and apoptosis of human gallbladder cancer cells. J Cell Mol Med. https://doi.org/10.1111/jcmm.13499

He S et al (2020) HERG channel and cancer: a mechanistic review of carcinogenic processes and therapeutic potential. Biochim Biophys Acta Rev Cancer. https://doi.org/10.1016/j.bbcan.2020.188355

Hemmerlein B et al (2006) Overexpression of Eag1 potassium channels in clinical tumours. Mol Cancer. https://doi.org/10.1186/1476-4598-5-41

Hirschler-Laszkiewicz I et al (2018) The human ion channel TRPM2 modulates neuroblastoma cell survival and mitochondrial function through Pyk2, CREB, and MCU activation. Am J Physiol Cell Physiol. https://doi.org/10.1152/ajpcell.00098.2018

Hoa NT et al (2014) Small cell lung cancer cells express the late stage gBK tumor antigen: a possible immunotarget for the terminal disease. Am J Transl Res. https://doi.org/10.1158/1538-7445.am2014-2895

Hoelder S, Clarke PA, Workman P (2012) Discovery of small molecule cancer drugs: successes, challenges and opportunities. Mol Oncol. https://doi.org/10.1016/j.molonc.2012.02.004

Holdhoff M et al (2017) Timed sequential therapy of the selective T-type calcium channel blocker mibefradil and temozolomide in patients with recurrent high-grade gliomas. Neuro Oncol. https://doi.org/10.1093/neuonc/nox020

Hopkins MM et al (2015) Inhibition of the transient receptor potential melastatin-2 channel causes increased DNA damage and decreased proliferation in breast adenocarcinoma cells. Int J Oncol. https://doi.org/10.3892/ijo.2015.2919

Hu T et al (2019) Expression and function of Kv1.3 channel in malignant T cells in Sézary syndrome. Oncotarget. https://doi.org/10.18632/oncotarget.27122

Huang X, Jan LY (2014) Targeting potassium channels in cancer. J Cell Biol. https://doi.org/10.1083/jcb.201404136

Huang L et al (2009) ATP-sensitive potassium channels control glioma cells proliferation by regulating ERK activity. Carcinogenesis. https://doi.org/10.1093/carcin/bgp034

Huang C et al (2017) Downregulation of a novel long noncoding RNA TRPM2-AS promotes apoptosis in non–small cell lung cancer. Tumor Biol. https://doi.org/10.1177/1010428317691191

Huber KR et al (1989) Effect of verapamil on cell cycle transit and c-myc gene expression in normal and malignant murine cells. Br J Cancer. https://doi.org/10.1038/bjc.1989.150

Iorio J et al (2018) hERG1 channel expression associates with molecular subtypes and prognosis in breast cancer. Cancer Cell Int. https://doi.org/10.1186/s12935-018-0592-1

Iorio J et al (2020) hERG1 and HIF-2α behave as biomarkers of positive response to bevacizumab in metastatic colorectal cancer patients. Transl Oncol. https://doi.org/10.1016/j.tranon.2020.01.001

Ishikawa M et al (2000) Reversal of acquired resistance to doxorubicin in K562 human leukemia cells by astemizole. Biol Pharm Bull. https://doi.org/10.1248/bpb.23.112

Izquierdo-Torres E et al (2017) ATP2A3 gene as an important player for resveratrol anticancer activity in breast cancer cells. Mol Carcinog. https://doi.org/10.1002/mc.22625

Izumi-Nakaseko H et al (2016) Possibility as an anti-cancer drug of astemizole: evaluation of arrhythmogenicity by the chronic atrioventricular block canine model. J Pharmacol Sci. https://doi.org/10.1016/j.jphs.2016.04.024

Jahchan NS et al (2013) A drug repositioning approach identifies tricyclic antidepressants as inhibitors of small cell lung cancer and other neuroendocrine tumors. Cancer Discov. https://doi.org/10.1158/2159-8290.CD-13-0183

Jang SH et al (2011) Anti-proliferative effect of Kv1.3 blockers in A549 human lung adenocarcinoma in vitro and in vivo. Eur J Pharmacol. https://doi.org/10.1016/j.ejphar.2010.10.066

Jardin I et al (2018) Trpc6 channels are required for proliferation, migration and invasion of breast cancer cell lines by modulation of orai1 and orai3 surface exposure. Cancer. https://doi.org/10.3390/cancers10090331

Jehle J et al (2011) Novel roles for hERG K(+) channels in cell proliferation and apoptosis. Cell Death Dis. https://doi.org/10.1038/cddis.2011.77

Jensen RL, Wurster RD (2001) Calcium channel antagonists inhibit growth of subcutaneous xenograft meningiomas in nude mice. Surg Neurol. https://doi.org/10.1016/s0090-3019(01)00444-x

Jimenez-Luevano MA et al (2018) Treatment of hepatocarcinoma with celecoxib and pentoxifylline: a case report. Rev Med Inst Mex Seguro Soc

Jin M et al (2019a) MCUR1 facilitates epithelial-mesenchymal transition and metastasis via the mitochondrial calcium dependent ROS/Nrf2/Notch pathway in hepatocellular carcinoma. J Exp Clin Cancer Res. https://doi.org/10.1186/s13046-019-1135-x

Jin YH et al (2019b) Efficacy of erlotinib and celecoxib for patients with advanced non-small cell lung cancer: a retrospective study. Medicine. https://doi.org/10.1097/MD.0000000000014785

June CH et al (2018) CAR T cell immunotherapy for human cancer. Science. https://doi.org/10.1126/science.aar6711

Kale VP, Amin SG, Pandey MK (2015) Targeting ion channels for cancer therapy by repurposing the approved drugs. Biochim Biophys Acta. https://doi.org/10.1016/j.bbamem.2015.03.034

Keir ST et al (2013) Mibefradil, a novel therapy for glioblastoma multiforme: cell cycle synchronization and interlaced therapy in a murine model. J Neurooncol. https://doi.org/10.1007/s11060-012-0995-0

Khaitan D et al (2009) Role of KCNMA1 gene in breast cancer invasion and metastasis to brain. BMC Cancer. https://doi.org/10.1186/1471-2407-9-258

Khalil DN et al (2016) The future of cancer treatment: immunomodulation, CARs and combination immunotherapy. Nat Rev Clin Oncol. https://doi.org/10.1038/nrclinonc.2016.25

Kischel P et al (2019) Ion channels: new actors playing in chemotherapeutic resistance. Cancer. https://doi.org/10.3390/cancers11030376

Klein EA et al (2011) Vitamin E and the risk of prostate cancer: the selenium and vitamin E cancer prevention trial (SELECT). JAMA. https://doi.org/10.1001/jama.2011.1437

Klingenberg M et al (2014) The NADPH oxidase inhibitor imipramine-blue in the treatment of Burkitt lymphoma. Mol Cancer Ther. https://doi.org/10.1158/1535-7163.MCT-13-0688

Koh DW et al (2015) Enhanced cytotoxicity in triple-negative and estrogen receptor-positive breast adenocarcinoma cells due to inhibition of the transient receptor potential melastatin-2 channel. Oncol Rep. https://doi.org/10.3892/or.2015.4131

Kondo S et al (1995) Combination therapy with cisplatin and nifedipine induces apoptosis in cisplatin-sensitive and cisplatin-resistant human glioblastoma cells. Br J Cancer. https://doi.org/10.1038/bjc.1995.57

Kong Q, Beel JA, Lillehei KO (2000) A threshold concept for cancer therapy. Med Hypotheses. https://doi.org/10.1054/mehy.1999.0982

Koppula P et al (2017) The glutamate/cystine antiporter SLC7A11/xCT enhances cancer cell dependency on glucose by exporting glutamate. J Biol Chem. https://doi.org/10.1074/jbc.M117.798405

Korošec B et al (2006) Alterations in the ATP2A2 gene in correlation with colon and lung cancer. Cancer Genet Cytogenet. https://doi.org/10.1016/j.cancergencyto.2006.06.016

Koşar PA et al (2016) Synergic effects of doxorubicin and melatonin on apoptosis and mitochondrial oxidative stress in MCF-7 breast cancer cells: involvement of TRPV1 channels. J Membr Biol. https://doi.org/10.1007/s00232-015-9855-0

Koshy S et al (2013) Blocking $K_{Ca}3.1$ channels increases tumor cell killing by a subpopulation of human natural killer lymphocytes. PLoS One. https://doi.org/10.1371/journal.pone.0076740

Kühn FJP, Heiner I, Lückhoff A (2005) TRPM2: a calcium influx pathway regulated by oxidative stress and the novel second messenger ADP-ribose. Pflugers Arch Eur J Physiol. https://doi.org/10.1007/s00424-005-1446-y

Laidlaw KM et al (2016) Cooperation of imipramine blue and tyrosine kinase blockade demonstrates activity against chronic myeloid leukemia. Oncotarget. https://doi.org/10.18632/oncotarget.10541

Lam J, Wulff H (2011) The lymphocyte potassium channels Kv1.3 and KCa3.1 as targets for immunosuppression. Drug Dev Res. https://doi.org/10.1002/ddr.20467

Lange I et al (2009) TRPM2 functions as a lysosomal Ca^{2+}-release channel in cells. Sci Signal. https://doi.org/10.1126/scisignal.2000278

Lansu K, Gentile S (2013) Potassium channel activation inhibits proliferation of breast cancer cells by activating a senescence program. Cell Death Dis. https://doi.org/10.1038/cddis.2013.174

Lastraioli E et al (2015a) hERG1 potassium channels: novel biomarkers in human solid cancers. Biomed Res Int. https://doi.org/10.1155/2015/896432

Lastraioli E et al (2015b) hERG1 channels drive tumour malignancy and may serve as prognostic factor in pancreatic ductal adenocarcinoma. Br J Cancer. https://doi.org/10.1038/bjc.2015.28

Lastraioli E et al (2019) The hERG1 potassium channel behaves as prognostic factor in gastric dysplasia endoscopic samples. Onco Targets Ther. https://doi.org/10.2147/OTT.S226257

Le Gal K et al (2015) Antioxidants can increase melanoma metastasis in mice. Sci Transl Med. https://doi.org/10.1126/scitranslmed.aad3740

Leanza L et al (2013) Clofazimine, Psora-4 and PAP-1, inhibitors of the potassium channel Kv1.3, as a new and selective therapeutic strategy in chronic lymphocytic leukemia. Leukemia. https://doi.org/10.1038/leu.2013.56

Leanza L et al (2017) Direct pharmacological targeting of a mitochondrial ion channel selectively kills tumor cells *in vivo*. Cancer Cell. https://doi.org/10.1016/j.ccell.2017.03.003

Leanza L et al (2018) Pharmacological modulation of mitochondrial ion channels. Br J Pharmacol. https://doi.org/10.1111/bph.14544

Lee H, Kang S, Kim W (2016) Drug repositioning for cancer therapy based on large-scale drug-induced transcriptional signatures. PLoS One. https://doi.org/10.1371/journal.pone.0150460

Lee JR et al (2019) The inhibition of chloride intracellular channel 1 enhances Ca2+ and reactive oxygen species signaling in A549 human lung cancer cells. Exp Mol Med. https://doi.org/10.1038/s12276-019-0279-2

Lewerenz J et al (2013) The cystine/glutamate antiporter system xc- in health and disease: from molecular mechanisms to novel therapeutic opportunities. Antioxid Redox Signal. https://doi.org/10.1089/ars.2011.4391

Lim JKM et al (2019) Cystine/glutamate antiporter xCT (SLC7A11) facilitates oncogenic RAS transformation by preserving intracellular redox balance. Proc Natl Acad Sci U S A. https://doi.org/10.1073/pnas.1821323116

Lis A et al (2007) CRACM1, CRACM2, and CRACM3 are store-operated Ca2+ channels with distinct functional properties. Curr Biol. https://doi.org/10.1016/j.cub.2007.03.065

Lissoni P et al (2002) A clinical study of taxotere versus taxotere plus the antiprolactinemic agent bromocriptine in metastatic breast cancer pretreated with anthracyclines. Anticancer Res 22 (2B):1131–1134

Litan A, Langhans SA (2015) Cancer as a channelopathy: ion channels and pumps in tumor development and progression. Front Cell Neurosci. https://doi.org/10.3389/fncel.2015.00086

Littler DR et al (2004) The intracellular chloride ion channel protein CLIC1 undergoes a redox-controlled structural transition. J Biol Chem. https://doi.org/10.1074/jbc.M308444200

Liu S, Cheng C (2013) Alternative RNA splicing and cancer. Wiley Interdiscip Rev RNA. https://doi.org/10.1002/wrna.1178

Liu X et al (2002) Cloning and characterization of glioma BK, a novel BK channel isoform highly expressed in human glioma cells. J Neurosci. https://doi.org/10.1523/jneurosci.22-05-01840.2002

Liu GX et al (2015) Expression of eag1 channel associated with the aggressive clinicopathological features and subtype of breast cancer. Int J Clin Exp Pathol 8(11):15093–15099

Liubomirski Y et al (2019) Tumor-stroma-inflammation networks promote pro-metastatic chemokines and aggressiveness characteristics in triple-negative breast cancer. Front Immunol. https://doi.org/10.3389/fimmu.2019.00757

Lo M et al (2010) Potential use of the anti-inflammatory drug, sulfasalazine, for targeted therapy of pancreatic cancer. Curr Oncol. https://doi.org/10.3747/co.v17i3.485

Lörinczi E et al (2016) Calmodulin regulates human ether à Go-Go 1 (hEAG1) potassium channels through interactions of the Eag domain with the cyclic nucleotide binding homology domain. J Biol Chem. https://doi.org/10.1074/jbc.M116.733576

Loughlin KR (2014) Calcium channel blockers and prostate cancer. Urol Oncol. https://doi.org/10.1016/j.urolonc.2013.08.001

Luzzi KJ et al (1998) Inhibition of angiogenesis in liver metastases by carboxyamidotriazole (CAI). Angiogenesis. https://doi.org/10.1023/A:1009259521092

Lyons SA, O'Neal J, Sontheimer H (2002) Chlorotoxin, a scorpion-derived peptide, specifically binds to gliomas and tumors of neuroectodermal origin. Glia. https://doi.org/10.1002/glia.10083

Ma X et al (2012) Transient receptor potential channel TRPC5 is essential for P-glycoprotein induction in drug-resistant cancer cells. Proc Natl Acad Sci U S A. https://doi.org/10.1073/pnas.1202989109

Ma X et al (2014) Essential role for TrpC5-containing extracellular vesicles in breast cancer with chemotherapeutic resistance. Proc Natl Acad Sci U S A. https://doi.org/10.1073/pnas.1400272111

Ma L et al (2018) Novel venom-derived inhibitors of the human EAG channel, a putative antiepileptic drug target. Biochem Pharmacol. https://doi.org/10.1016/j.bcp.2018.08.038

Mahalingam D et al (2019) A phase II, multicenter, single-arm study of mipsagargin (G-202) as a second-line therapy following sorafenib for adult patients with progressive advanced hepatocellular carcinoma. Cancer. https://doi.org/10.3390/cancers11060833

Mailloux RJ, Adjeitey CNK, Harper ME (2010) Genipin-induced inhibition of uncoupling protein-2 sensitizes drug-resistant cancer cells to cytotoxic agents. PLoS One. https://doi.org/10.1371/journal.pone.0013289

Maldonado EN et al (2013) Voltage-dependent anion channels modulate mitochondrial metabolism in cancer cells: regulation by free tubulin and erastin. J Biol Chem. https://doi.org/10.1074/jbc.M112.433847

Malhi H et al (2000) KATP channels regulate mitogenically induced proliferation in primary rat hepatocytes and human liver cell lines. Implications for liver growth control and potential therapeutic targeting. J Biol Chem. https://doi.org/10.1074/jbc.M001576200

Mammucari C, Gherardi G, Rizzuto R (2017) Structure, activity regulation, and role of the mitochondrial calcium uniporter in health and disease. Front Oncol. https://doi.org/10.3389/fonc.2017.00139

Mancini M, Toker A (2009) NFAT proteins: emerging roles in cancer progression. Nat Rev Cancer 9(11):810–820. https://doi.org/10.1038/nrc2735

Manoli S et al (2019) The activity of Kv11.1 potassium channel modulates F-actin organization during cell migration of pancreatic ductal adenocarcinoma cells. Cancer 11(2). pii: E135. https://doi.org/10.3390/cancers11020135

Martinez R et al (2015) Analysis of the expression of Kv10.1 potassium channel in patients with brain metastases and glioblastoma multiforme: impact on survival. BMC Cancer. https://doi.org/10.1186/s12885-015-1848-y

Martinez-Delgado G, Felix R (2017) Emerging role of CaV1.2 channels in proliferation and migration in distinct cancer cell lines. Oncology. https://doi.org/10.1159/000464293

Mazure NM (2017) VDAC in cancer. Biochim Biophys Acta Bioenerg. https://doi.org/10.1016/j.bbabio.2017.03.002

McFerrin MB, Sontheimer H (2006) A role for ion channels in glioma cell invasion. Neuron Glia Biol. https://doi.org/10.1017/S17440925X06000044

Meng H et al (2010) Engineered design of mesoporous silica nanoparticles to deliver doxorubicin and P-glycoprotein siRNA to overcome drug resistance in a cancer cell line. ACS Nano. https://doi.org/10.1021/nn100690m

Meng H et al (2013) Codelivery of an optimal drug/siRNA combination using mesoporous silica nanoparticles to overcome drug resistance in breast cancer in vitro and in vivo. ACS Nano. https://doi.org/10.1021/nn3044066

Metts J et al (2017) Imipramine blue sensitively and selectively targets FLT3-ITD positive acute myeloid leukemia cells. Sci Rep. https://doi.org/10.1038/s41598-017-04796-1

Miller BA, Zhang W (2011) TRP channels as mediators of oxidative stress. Adv Exp Med Biol 704:531–544. https://doi.org/10.1007/978-94-007-0265-3_29

Miller BA et al (2014) TRPM2 channels protect against cardiac ischemia-reperfusion injury: role of mitochondria. J Biol Chem. https://doi.org/10.1074/jbc.M113.533851

Millward MJ et al (1993) Oral verapamil with chemotherapy for advanced non-small cell lung cancer: a randomised study. Br J Cancer. https://doi.org/10.1038/bjc.1993.189

Mohr CJ et al (2019) Cancer-associated intermediate conductance Ca^{2+}-activated K^+ channel K Ca 3.1. Cancer. https://doi.org/10.3390/cancers11010109

Monen SH, Schmidt PH, Wondergem R (1998) Membrane potassium channels and human bladder tumor cells. I. Electrical properties. J Membr Biol. https://doi.org/10.1007/s002329900331

Moreels L, Peigneur S, Galan DT et al (2017a) APETx4, a novel sea anemone toxin and a modulator of the cancer-relevant potassium channel K_V10.1. Mar Drugs. https://doi.org/10.3390/md15090287

Moreels L, Peigneur S, Yamaguchi Y et al (2017b) Expanding the pharmacological profile of kappa-hefutoxin 1 and analogues: a focus on the inhibitory effect on the oncogenic channel Kv10.1. Peptides. https://doi.org/10.1016/j.peptides.2016.08.008

Mukhopadhyay I et al (2011) Expression of functional TRPA1 receptor on human lung fibroblast and epithelial cells. J Recept Signal Transduct Res. https://doi.org/10.3109/10799893.2011.602413

Munson JM et al (2012) Anti-invasive adjuvant therapy with imipramine blue enhances chemotherapeutic efficacy against glioma. Sci Transl Med. https://doi.org/10.1126/scitranslmed.3003016

Muscella A et al (2011) The platinum (II) complex [Pt(O,O'-acac)(γ-acac)(DMS)] alters the intracellular calcium homeostasis in MCF-7 breast cancer cells. Biochem Pharmacol 81(1):91–103. https://doi.org/10.1016/j.bcp.2010.09.012

Neuhaus E et al (2019) Alternating electric fields (TTFields) activate Cav1.2 channels in human glioblastoma cells. Cancers (Basel) 11(1). https://doi.org/10.3390/cancers11010110

Nielsen N, Lindemann O, Schwab A (2014) TRP channels and STIM/ORAI proteins: sensors and effectors of cancer and stroma cell migration. Br J Pharmacol. https://doi.org/10.1111/bph.12721

Noh J et al (2015) Amplification of oxidative stress by a dual stimuli-responsive hybrid drug enhances cancer cell death. Nat Commun. https://doi.org/10.1038/ncomms7907

Núñez M et al (2013) Glibenclamide inhibits cell growth by inducing G0/G1 arrest in the human breast cancer cell line MDA-MB-231. BMC Pharmacol Toxicol. https://doi.org/10.1186/2050-6511-14-6

Nur G, Nazıroğlu M, Deveci HA (2017) Synergic prooxidant, apoptotic and TRPV1 channel activator effects of alpha-lipoic acid and cisplatin in MCF-7 breast cancer cells. J Recept Signal Transduct Res. https://doi.org/10.1080/10799893.2017.1369121

Omuro A et al (2018) Multicenter phase IB trial of carboxyamidotriazole orotate and temozolomide for recurrent and newly diagnosed glioblastoma and other anaplastic gliomas. J Clin Oncol 36 (17):1702–1709. https://doi.org/10.1200/JCO.2017.76.9992

Onoda JM et al (1988) Cisplatin and nifedipine: synergistic antitumor effects against an inherently cisplatin-resistant tumor. Cancer Lett. https://doi.org/10.1016/0304-3835(88)90260-1

Oprea TI et al (2011) Drug repurposing from an academic perspective. Drug Discov Today Ther Strateg 8(3–4):61–69. https://doi.org/10.1016/j.ddstr.2011.10.002

Orfanelli U et al (2008) Identification of novel sense and antisense transcription at the TRPM2 locus in cancer. Cell Res. https://doi.org/10.1038/cr.2008.296

Orfanelli U et al (2014) Antisense transcription at the TRPM2 locus as a novel prognostic marker and therapeutic target in prostate cancer. Oncogene. https://doi.org/10.1038/onc.2014.144

Ortiz CS et al (2011) Eag1 potassium channels as markers of cervical dysplasia. Oncol Rep. https://doi.org/10.3892/or.2011.1441

Ostroumov D et al (2018) CD4 and CD8 T lymphocyte interplay in controlling tumor growth. Cell Mol Life Sci. https://doi.org/10.1007/s00018-017-2686-7

Ouadid-Ahidouch H et al (2001) Changes in the K^+ current-density of MCF-7 cells during progression through the cell cycle: possible involvement of a h-ether.a-gogo K^+ channel. Receptors Channels

Ouadid-Ahidouch H, Ahidouch A, Pardo LA (2016) Kv10.1 K^+ channel: from physiology to cancer. Pflugers Arch 468(5):751–762. https://doi.org/10.1007/s00424-015-1784-3

Özdemir ÜS et al (2016) *Hypericum perforatum* attenuates spinal cord injury-induced oxidative stress and apoptosis in the dorsal root ganglion of rats: involvement of TRPM2 and TRPV1 channels. Mol Neurobiol 53(6):3540–3551. https://doi.org/10.1007/s12035-015-9292-1

Paakkari I (2002) Cardiotoxicity of new antihistamines and cisapride. Toxicol Lett 127 (1–3):279–284. https://doi.org/10.1016/s0378-4274(01)00510-0

Pantziarka P et al (2014) The repurposing drugs in oncology (ReDO) project. Ecancermedicalscience 8:442. https://doi.org/10.3332/ecancer.2014.442

Panyi G, Beeton C, Felipe A (2014) Ion channels and anti-cancer immunity. Philos Trans R Soc Lond B Biol Sci. https://doi.org/10.1098/rstb.2013.0106

Papp B, Brouland JP (2011) Altered endoplasmic reticulum calcium pump expression during breast tumorigenesis. Breast Cancer: Basic Clin Res. https://doi.org/10.4137/BCBCR.S7481

Pardo LA, Stühmer W (2008) Eag1: an emerging oncological target. Cancer Res. https://doi.org/10.1158/0008-5472.CAN-07-5710

Pardo LA et al (1999) Oncogenic potential of EAG $K(^+)$ channels. EMBO J. https://doi.org/10.1093/emboj/18.20.5540

Payen L et al (2001) The sulphonylurea glibenclamide inhibits multidrug resistance protein (MRP1) activity in human lung cancer cells. Br J Pharmacol. https://doi.org/10.1038/sj.bjp.0703863

Pennington MW, Czerwinski A, Norton RS (2018) Peptide therapeutics from venom: current status and potential. Bioorg Med Chem. https://doi.org/10.1016/j.bmc.2017.09.029

Peretti M et al (2015) Chloride channels in cancer: focus on chloride intracellular channel 1 and 4 (CLIC1 AND CLIC4) proteins in tumor development and as novel therapeutic targets. Biochim Biophys Acta. https://doi.org/10.1016/j.bbamem.2014.12.012

Peretti M et al (2019) Original association of ion transporters mediates the ECM-induced breast cancer cell survival: Kv10.1-Orai1-SPCA2 partnership. Sci Rep. https://doi.org/10.1038/s41598-018-37602-7

Perez-Neut M et al (2016) Activation of hERG3 channel stimulates autophagy and promotes cellular senescence in melanoma. Oncotarget 7(16):21991–22004. https://doi.org/10.18632/oncotarget.7831

Peters DH, Clissold SP (1992) Chlarythromicin. a review of its antimicrobial, pharmacokinetic properties and therapeutic potential. Drugs. https://doi.org/10.2165/00003495-199244010-00009

Petroni G et al (2020) Clarithromycin inhibits autophagy in colorectal cancer by regulating the hERG1 potassium channel interaction with PI3K. Cell Death Dis. https://doi.org/10.1038/s41419-020-2349-8

Petrova DT et al (2008) Expression of chloride intracellular channel protein 1 (CLIC1) and tumor protein D52 (TPD52) as potential biomarkers for colorectal cancer. Clin Biochem. https://doi.org/10.1016/j.clinbiochem.2008.07.012

Piskounova E et al (2015) Oxidative stress inhibits distant metastasis by human melanoma cells. Nature 527(7577):186–191. https://doi.org/10.1038/nature15726

Pointer KB et al (2017) Administration of non-torsadogenic human ether-a-go-go-related gene inhibitors is associated with better survival for high hERG-expressing glioblastoma patients. Clin Cancer Res 23(1):73–80. https://doi.org/10.1158/1078-0432.CCR-15-3169

Pons DG et al (2015) UCP2 inhibition sensitizes breast cancer cells to therapeutic agents by increasing oxidative stress. Free Radic Biol Med 86:67–77. https://doi.org/10.1016/j.freeradbiomed.2015.04.032

Poprac P et al (2017) Targeting free radicals in oxidative stress-related human diseases. Trends Pharmacol Sci 38(7):592–607. https://doi.org/10.1016/j.tips.2017.04.005

Porporato PE et al (2014) A mitochondrial switch promotes tumor metastasis. Cell Rep 8 (3):754–766. https://doi.org/10.1016/j.celrep.2014.06.043

Prasad S, Gupta SC, Tyagi AK (2017) Reactive oxygen species (ROS) and cancer: role of antioxidative nutraceuticals. Cancer Lett 387:95–105. https://doi.org/10.1016/j.canlet.2016.03.042

Qian X et al (2008) Glibenclamide exerts an antitumor activity through reactive oxygen species-c-jun NH2-terminal kinase pathway in human gastric cancer cell line MGC-803. Biochem Pharmacol. https://doi.org/10.1016/j.bcp.2008.09.009

Queiroz FM et al (2006) Ether a go-go potassium channel expression in soft tissue sarcoma patients. Mol Cancer. https://doi.org/10.1186/1476-4598-5-42

Rajamanickam S et al (2016) Inhibition of FoxM1-mediated DNA repair by imipramine blue suppresses breast cancer growth and metastasis. Clin Cancer Res 22(14):3524–3536. https://doi.org/10.1158/1078-0432.CCR-15-2535

Ramírez A et al (2016) Ion channels and oxidative stress as a potential link for the diagnosis or treatment of liver diseases. Oxid Med Cell Longev 2016:3928714. https://doi.org/10.1155/2016/3928714

Ramírez A et al (2018) Calcium-activated potassium channels as potential early markers of human cervical cancer. Oncol Lett. https://doi.org/10.3892/ol.2018.8187

Ramos Gomes F et al (2015) Alternatively spliced isoforms of K_V10.1 potassium channels modulate channel properties and can activate cyclin-dependent kinase in Xenopus oocytes. J Biol Chem. https://doi.org/10.1074/jbc.M115.668749

Rani V et al (2016) Oxidative stress and metabolic disorders: pathogenesis and therapeutic strategies. Life Sci. https://doi.org/10.1016/j.lfs.2016.02.002

Rasola A, Bernardi P (2014) The mitochondrial permeability transition pore and its adaptive responses in tumor cells. Cell Calcium 56(6):437–445. https://doi.org/10.1016/j.ceca.2014.10.003

Rezania S et al (2016) Overexpression of KCNJ3 gene splice variants affects vital parameters of the malignant breast cancer cell line MCF-7 in an opposing manner. BMC Cancer. https://doi.org/10.1186/s12885-016-2664-8

Robe PA et al (2009) Early termination of ISRCTN45828668, a phase 1/2 prospective, randomized study of sulfasalazine for the treatment of progressing malignant gliomas in adults. BMC Cancer. https://doi.org/10.1186/1471-2407-9-372

Roderick HL, Cook SJ (2008) Ca^{2+} signalling checkpoints in cancer: remodelling Ca 2+ for cancer cell proliferation and survival. Nat Rev Cancer. https://doi.org/10.1038/nrc2374

Rong Z et al (2013) Combined treatment of glibenclamide and CoCl2 decreases MMP9 expression and inhibits growth in highly metastatic breast cancer. J Exp Clin Cancer Res. https://doi.org/10.1186/1756-9966-32-32

Roy J et al (2008) Pharmacological separation of hEAG and hERG K+ channel function in the human mammary carcinoma cell line MCF-7. Oncol Rep. https://doi.org/10.3892/or.19.6.1511

Rybalchenko V et al (2001) Verapamil inhibits proliferation of LNCaP human prostate cancer cells influencing K+ channel gating. Mol Pharmacol. https://doi.org/10.1124/mol.59.6.1376

Sakallı ÇE et al (2017) Selenium potentiates the anticancer effect of cisplatin against oxidative stress and calcium ion signaling-induced intracellular toxicity in MCF-7 breast cancer cells: involvement of the TRPV1 channel. J Recept Signal Transduct Res. https://doi.org/10.3109/10799893.2016.1160931

Santo-Domingo J et al (2007) The plasma membrane Na $^{+/}$Ca $^{2+}$ exchange inhibitor KB-R7943 is also a potent inhibitor of the mitochondrial Ca^{2+} uniporter. Br J Pharmacol. https://doi.org/10.1038/sj.bjp.0707260

Sayin VI et al (2014) Cancer: antioxidants accelerate lung cancer progression in mice. Sci Transl Med. https://doi.org/10.1126/scitranslmed.3007653

Schaefer EAM et al (2013) Stimulation of the chemosensory TRPA1 cation channel by volatile toxic substances promotes cell survival of small cell lung cancer cells. Biochem Pharmacol. https://doi.org/10.1016/j.bcp.2012.11.019

Schmidt WF et al (1988) Antiproliferative effect of verapamil alone on brain tumor cells in vitro. Cancer Res 48(13):3617–3621

Schwab A et al (1999) $K(^{+})$ channel-dependent migration of fibroblasts and human melanoma cells. Cell Physiol Biochem. https://doi.org/10.1159/000016309

Segovia-Mendoza M et al (2015) Calcitriol and its analogues enhance the antiproliferative activity of gefitinib in breast cancer cells. J Steroid Biochem Mol Biol. https://doi.org/10.1016/j.jsbmb.2014.12.006

Segovia-Mendoza M et al (2017) The addition of calcitriol or its synthetic analog EB1089 to lapatinib and neratinib treatment inhibits cell growth and promotes apoptosis in breast cancer cells. Am J Cancer Res 7(7):1486–1500

Sehgal P et al (2017) Inhibition of the sarco/endoplasmic reticulum (ER) Ca^{2+}-ATPase by thapsigargin analogs induces cell death via ER Ca^{2+} depletion and the unfolded protein response. J Biol Chem. https://doi.org/10.1074/jbc.M117.796920

Sehm T et al (2016) Temozolomide toxicity operates in a xCT/SLC7a11 dependent manner and is fostered by ferroptosis. Oncotarget. https://doi.org/10.18632/oncotarget.11858

Seo J et al (2016) Curcumin induces apoptosis by inhibiting sarco/endoplasmic reticulum Ca^{2+} ATPase activity in ovarian cancer cells. Cancer Lett. https://doi.org/10.1016/j.canlet.2015.11.021

Seo EJ et al (2018) Repurposing of bromocriptine for cancer therapy. Front Pharmacol. https://doi.org/10.3389/fphar.2018.01030

Serrano-Novillo C et al (2019) Implication of voltage-gated potassium channels in neoplastic cell proliferation. Cancers (Basel). https://doi.org/10.3390/cancers11030287

Sette A et al (2013) Development of novel anti-Kv 11.1 antibody-conjugated PEG-TiO2 nanoparticles for targeting pancreatic ductal adenocarcinoma cells. J Nanopart Res. https://doi.org/10.1007/s11051-013-2111-6
Shankar DB et al (2005) The role of CREB as a proto-oncogene in hematopoiesis and in acute myeloid leukemia. Cancer Cell. https://doi.org/10.1016/j.ccr.2005.02.018
Shanmugam MK et al (2018) Potential role of genipin in cancer therapy. Pharmacol Res. https://doi.org/10.1016/j.phrs.2018.05.007
Sharma A et al (2019) Causal roles of mitochondrial dynamics in longevity and healthy aging. EMBO Rep. https://doi.org/10.15252/embr.201948395
Shitara K et al (2017) Dose-escalation study for the targeting of CD44v+ cancer stem cells by sulfasalazine in patients with advanced gastric cancer (EPOC1205). Gastric Cancer. https://doi.org/10.1007/s10120-016-0610-8
Shoshan-Barmatz V, Ben-Hail D (2012) VDAC, a multi-functional mitochondrial protein as a pharmacological target. Mitochondrion. https://doi.org/10.1016/j.mito.2011.04.001
Silva R et al (2015) Modulation of P-glycoprotein efflux pump: induction and activation as a therapeutic strategy. Pharmacol Ther. https://doi.org/10.1016/j.pharmthera.2014.11.013
Simon F, Varela D, Cabello-Verrugio C (2013) Oxidative stress-modulated TRPM ion channels in cell dysfunction and pathological conditions in humans. Cell Signal. https://doi.org/10.1016/j.cellsig.2013.03.023
SoRelle R (1998) Withdrawal of Posicor from market. Circulation. https://doi.org/10.1161/01.cir.98.9.831
Srairi-Abid N et al (2005) A new type of scorpion Na+-channel-toxin-like polypeptide active on K+ channels. Biochem J. https://doi.org/10.1042/BJ20041407
Srairi-Abid N et al (2019) Anti-tumoral effect of scorpion peptides: emerging new cellular targets and signaling pathways. Cell Calcium. https://doi.org/10.1016/j.ceca.2019.05.003
Stringer BK, Cooper AG, Shepard SB (2001) Overexpression of the G-protein inwardly rectifying potassium channel 1 (GIRK1) in primary breast carcinomas correlates with axillary lymph node metastasis. Cancer Res
Sucu BO et al (2019) Synthesis of novel methyl jasmonate derivatives and evaluation of their biological activity in various cancer cell lines. Bioorg Chem. https://doi.org/10.1016/j.bioorg.2019.103146
Suessbrich H et al (1996) Blockade of HERG channels expressed in Xenopus oocytes by the histamine receptor antagonists terfenadine and astemizole. FEBS Lett. https://doi.org/10.1016/0014-5793(96)00355-9
Suh DH et al (2013) Mitochondrial permeability transition pore as a selective target for anti-cancer therapy. Front Oncol. https://doi.org/10.3389/fonc.2013.00041
Sullivan LB, Gui DY, Van Der Heiden MG (2016) Altered metabolite levels in cancer: Implications for tumour biology and cancer therapy. Nat Rev Cancer. https://doi.org/10.1038/nrc.2016.85
Sumoza-Toledo A et al (2011) Dendritic cell maturation and chemotaxis is regulated by TRPM2-mediated lysosomal Ca 2+ release. FASEB J. https://doi.org/10.1096/fj.10-178483
Sun C, Veiseh O et al (2008a) In vivo MRI detection of gliomas by chlorotoxin-conjugated superparamagnetic nanoprobes. Small. https://doi.org/10.1002/smll.200700784
Sun C, Fang C et al (2008b) Tumor-targeted drug delivery and MRI contrast enhancement by chlorotoxin-conjugated iron oxide nanoparticles. Nanomedicine (Lond). https://doi.org/10.2217/17435889.3.4.495
Suo A et al (2016) Comb-like amphiphilic polypeptide-based copolymer nanomicelles for co-delivery of doxorubicin and P-gp siRNA into MCF-7 cells. Korean J Couns Psychother. https://doi.org/10.1016/j.msec.2016.02.007
Susankova K et al (2006) Reducing and oxidizing agents sensitize heat-activated vanilloid receptor (TRPV1) current. Mol Pharmacol. https://doi.org/10.1124/mol.106.023069
Suzuki Y et al (2012) Depolarization potentiates TRAIL-induced apoptosis in human melanoma cells: role for ATP-sensitive K+ channels and endoplasmic reticulum stress. Int J Oncol. https://doi.org/10.3892/ijo.2012.1483

Szabó I et al (2008) Mitochondrial potassium channel Kv1.3 mediates Bax-induced apoptosis in lymphocytes. Proc Natl Acad Sci U S A. https://doi.org/10.1073/pnas.0804236105
Takahashi N, Mori Y (2011) TRP channels as sensors and signal integrators of redox status changes. Front Pharmacol. https://doi.org/10.3389/fphar.2011.00058
Takahashi N et al (2018) Cancer cells co-opt the neuronal redox-sensing channel TRPA1 to promote oxidative-stress tolerance. Cancer Cell. https://doi.org/10.1016/j.ccell.2018.05.001
Takeuchi S et al (2014) Sulfasalazine and temozolomide with radiation therapy for newly diagnosed glioblastoma. Neurol India. https://doi.org/10.4103/0028-3886.128280
Tang S et al (2015) Mitochondrial Ca 2+ uniporter is critical for store-operated Ca 2+ entry-dependent breast cancer cell migration. Biochem Biophys Res Commun. https://doi.org/10.1016/j.bbrc.2015.01.092
Taylor JM, Simpson RU (1992) Inhibition of cancer cell growth by calcium channel antagonists in the athymic mouse. Cancer Res 52(9)
Taylor JT et al (2008) Calcium signaling and T-type calcium channels in cancer cell cycling. World J Gastroenterol. https://doi.org/10.3748/wjg.14.4984
Teisseyre A, Gasiorowska J, Michalak K (2015) Voltage-gated potassium channels Kv1.3--potentially new molecular target in cancer diagnostics and therapy. Adv Clin Exp Med. https://doi.org/10.17219/acem/22339
Teschemacher AG et al (1999) Inhibition of the current of heterologously expressed HERG potassium channels by imipramine and amitriptyline. Br J Pharmacol. https://doi.org/10.1038/sj.bjp.0702800
Tochhawng L et al (2013) Redox regulation of cancer cell migration and invasion. Mitochondrion. https://doi.org/10.1016/j.mito.2012.08.002
Togashi K et al (2006) TRPM2 activation by cyclic ADP-ribose at body temperature is involved in insulin secretion. EMBO J. https://doi.org/10.1038/sj.emboj.7601083
Toloczko-Iwaniuk N et al (2019) Celecoxib in cancer therapy and prevention – review. Curr Drug Targets. https://doi.org/10.2174/1389450119666180803121737
Tong L et al (2015) Reactive oxygen species in redox cancer therapy. Cancer Lett. https://doi.org/10.1016/j.canlet.2015.07.008
Tosatto A et al (2016) The mitochondrial calcium uniporter regulates breast cancer progression via HIF-1α. EMBO Mol Med. https://doi.org/10.15252/emmm.201606255
Trachootham D, Alexandre J, Huang P (2009) Targeting cancer cells by ROS-mediated mechanisms: a radical therapeutic approach? Nat Rev Drug Discov. https://doi.org/10.1038/nrd2803
Trevisan G et al (2013) Novel therapeutic strategy to prevent chemotherapy-induced persistent sensory neuropathy by TRPA1 blockade. Cancer Res. https://doi.org/10.1158/0008-5472.CAN-12-4370
Varghese E et al (2019) Anti-cancer agents in proliferation and cell death: the calcium connection. Int J Mol Sci. https://doi.org/10.3390/ijms20123017
Vazquez-Sanchez AY et al (2018) Expression of KATP channels in human cervical cancer: potential tools for diagnosis and therapy. Oncol Lett. https://doi.org/10.3892/ol.2018.8165
Veiseh O et al (2005) Optical and MRI multifunctional nanoprobe for targeting gliomas. Nano Lett. https://doi.org/10.1021/nl0502569
Veiseh M et al (2007) Tumor paint: a chlorotoxin: Cy5.5 bioconjugate for intraoperative visualization of cancer foci. Cancer Res. https://doi.org/10.1158/0008-5472.CAN-06-3948
Veiseh O et al (2009) Inhibition of tumor-cell invasion with chlorotoxin-bound superparamagnetic nanoparticles. Small. https://doi.org/10.1002/smll.200800646
Veiseh O et al (2010) Chlorotoxin bound magnetic nanovector tailored for cancer cell targeting, imaging, and siRNA delivery. Biomaterials. https://doi.org/10.1016/j.biomaterials.2010.07.016
Verdoodt F et al (2019) Antihistamines and ovarian cancer survival: nationwide cohort study and in vitro cell viability assay. J Natl Cancer Inst. pii: djz217. https://doi.org/10.1093/jnci/djz217
Vermeer LM et al (2016) Evaluation of ketoconazole and its alternative clinical CYP3A4/5 inhibitors as inhibitors of drug transporters: the in vitro effects of ketoconazole, ritonavir,

clarithromycin, and itraconazole on 13 clinically-relevant drug transporters. Drug Metab Dispos. https://doi.org/10.1124/dmd.115.067744

Vilema-Enríquez G et al (2016) Molecular and cellular effects of hydrogen peroxide on human lung cancer cells: potential therapeutic implications. Oxid Med Cell Longev. https://doi.org/10.1155/2016/1908164

Vinay DS et al (2015) Immune evasion in cancer: mechanistic basis and therapeutic strategies. Semin Cancer Biol. https://doi.org/10.1016/j.semcancer.2015.03.004

Vultur A et al (2018) The role of the mitochondrial calcium uniporter (MCU) complex in cancer. Pflugers Arch Eur J Physiol. https://doi.org/10.1007/s00424-018-2162-8

Wagner V et al (2010) Cloning and characterisation of GIRK1 variants resulting from alternative RNA editing of the KCNJ3 gene transcript in a human breast cancer cell line. J Cell Biochem. https://doi.org/10.1002/jcb.22564

Wang YJ et al (2002) Ketoconazole potentiates terfenadine-induced apoptosis in human Hep G2 cells through inhibition of cytochrome p450 3A4 activity. J Cell Biochem. https://doi.org/10.1002/jcb.10282

Wang W et al (2011) The expression and clinical significance of CLIC1 and HSP27 in lung adenocarcinoma. Tumour Biol. https://doi.org/10.1007/s13277-011-0223-0

Wang LH et al (2012) Rituximab inhibits Kv1.3 channels in human B lymphoma cells via activation of FcγRIIB receptors. Biochim Biophys Acta Mol Cell Res. https://doi.org/10.1016/j.bbamcr.2011.11.012

Wang P et al (2014a) Chloride intracellular channel 1 regulates colon cancer cell migration and invasion through ROS/ERK pathway. World J Gastroenterol. https://doi.org/10.3748/wjg.v20.i8.2071

Wang WT et al (2014b) Terfenadine induces anti-proliferative and apoptotic activities in human hormone-refractory prostate cancer through histamine receptor-independent Mcl-1 cleavage and Bak up-regulation. Naunyn Schmiedebergs Arch Pharmacol. https://doi.org/10.1007/s00210-013-0912-x

Wang CY et al (2015a) Meta-analysis of public microarray datasets reveals voltage-gated calcium gene signatures in clinical cancer patients. PLoS One. https://doi.org/10.1371/journal.pone.0125766

Wang T et al (2015b) Inhibition of transient receptor potential channel 5 reverses 5-fluorouracil resistance in human colorectal cancer cells. J Biol Chem. https://doi.org/10.1074/jbc.M114.590364

Wang D et al (2018a) Codelivery of doxorubicin and MDR1-siRNA by mesoporous silica nanoparticles-polymerpolyethylenimine to improve oral squamous carcinoma treatment. Int J Nanomedicine. https://doi.org/10.2147/IJN.S150610

Wang Z et al (2018b) Inhibition of TRPA1 attenuates doxorubicin-induced acute cardiotoxicity by suppressing oxidative stress, the inflammatory response, and endoplasmic reticulum stress. Oxid Med Cell Longev. https://doi.org/10.1155/2018/5179468

Wanke E, Restano-Cassulini R (2007) Toxins interacting with ether-a-go-go-related gene voltage-dependent potassium channels. Toxicon. https://doi.org/10.1016/j.toxicon.2006.09.025

Weaver AK, Liu X, Sontheimer H (2004) Role for calcium activated potassium channels (BK) in growth control of human malignant glioma cells. J Neurosci Res. https://doi.org/10.1002/jnr.20240

Wondergem R et al (1998) Membrane potassium channels and human bladder tumor cells: II. Growth properties. J Membr Biol. https://doi.org/10.1007/s002329900332

Wong DT, Bymaster FP, Engleman EA et al (1995) Prozac (fluoxetine, Lilly 110140), the first selective serotonin uptake inhibitor and an antidepressant drug: twenty years since its first publication. Life Sci. https://doi.org/10.1016/0024-3205(95)00209-o

Wood K et al (2018) Capecitabine and celecoxib as a promising therapy for thymic neoplasms. Am J Clin Oncol. https://doi.org/10.1097/COC.0000000000000400

Wu W et al (2012) Human ether-a-go-go gene potassium channels are regulated by EGFR tyrosine kinase. Biochim Biophys Acta. https://doi.org/10.1016/j.bbamcr.2011.10.010

Wu T et al (2019) Spider venom peptides as potential drug candidates due to their anticancer and antinociceptive activities. J Venom Anim Toxins Incl Trop Dis. https://doi.org/10.1590/1678-9199-JVATITD-14-63-18
Wulff H et al (2019) Antibodies and venom peptides: new modalities for ion channels. Nat Rev Drug Discov. https://doi.org/10.1038/s41573-019-0013-8
Xiang Y et al (2011) Chloride channel-mediated brain glioma targeting of chlorotoxin-modified doxorubicine-loaded liposomes. J Control Release. https://doi.org/10.1016/j.jconrel.2011.03.014
Xiao X et al (2010) Targeting CREB for cancer therapy: friend or foe. Curr Cancer Drug Targets
Xie J et al (2016) SOCE and cancer: recent progress and new perspectives. Int J Cancer. https://doi.org/10.1002/ijc.29840
Xie J et al (2019) Nanomaterial-based blood-brain-barrier (BBB) crossing strategies. Biomaterials. https://doi.org/10.1016/j.biomaterials.2019.119491
Xu Y et al (2018a) Expression of CLIC1 as a potential biomarker for oral squamous cell carcinoma: a preliminary study. Onco Targets Ther. https://doi.org/10.2147/OTT.S181936
Xu J et al (2018b) Astemizole promotes the anti-tumor effect of vitamin D through inhibiting miR-125a-5p-meidated regulation of VDR in HCC. Biomed Pharmacother. https://doi.org/10.1016/j.biopha.2018.08.153
Xue X, Liang XJ (2012) Overcoming drug efflux-based multidrug resistance in cancer with nanotechnology. Chin J Cancer. https://doi.org/10.5732/cjc.011.10326
Xue H et al (2018) Inhibition of transient receptor potential vanilloid 6 channel, elevated in human ovarian cancers, reduces tumour growth in a xenograft model. J Cancer 9(17):3196–3207. https://doi.org/10.7150/jca.20639
Yagoda N et al (2007) RAS-RAF-MEK-dependent oxidative cell death involving voltage-dependent anion channels. Nature. https://doi.org/10.1038/nature05859
Yamaci RF et al (2017) Neonatal Nav1.5 protein expression in normal adult human tissues and breast cancer. Pathol Res Pract. https://doi.org/10.1016/j.prp.2017.06.003
Yamamoto S et al (2008) TRPM2-mediated Ca2+ influx induces chemokine production in monocytes that aggravates inflammatory neutrophil infiltration. Nat Med. https://doi.org/10.1038/nm1758
Yang WH et al (2016) Imipramine blue halts head and neck cancer invasion through promoting F-box and leucine-rich repeat protein 14-mediated Twist1 degradation. Oncogene. https://doi.org/10.1038/onc.2015.291
Yang MY et al (2019) Carrier-free nanodrug: a novel strategy of cancer diagnosis and synergistic therapy. Int J Pharm. https://doi.org/10.1016/j.ijpharm.2019.118663
Ye Y et al (2015) CLIC1 a novel biomarker of intraperitoneal metastasis in serous epithelial ovarian cancer. Tumor Biol. https://doi.org/10.1007/s13277-015-3052-8
Yu C et al (2017) Mitochondrial calcium uniporter as a target of microRNA-340 and promoter of metastasis via enhancing the Warburg effect. Oncotarget. https://doi.org/10.18632/oncotarget.19747
Yusa K, Tsuruo T (1989) Reversal mechanism of multidrug resistance by verapamil: direct binding of verapamil to P-glycoprotein on specific sites and transport of verapamil outward across the plasma membrane of K562/ADM cells. Cancer Res 49(18):5002–5006
Zeng X et al (2010) Novel role for the transient receptor potential channel TRPM2 in prostate cancer cell proliferation. Prostate Cancer Prostatic Dis. https://doi.org/10.1038/pcan.2009.55
Zeng B et al (2013) TRPC channels and their splice variants are essential for promoting human ovarian cancer cell proliferation and tumorigenesis. Curr Cancer Drug Targets 13(1):103–116
Zhang S et al (1999) Mechanism of block and identification of the verapamil binding domain to HERG potassium channels. Circ Res. https://doi.org/10.1161/01.res.84.9.989
Zhang Y et al (2007) Effects of celecoxib on voltage-gated calcium channel currents in rat pheochromocytoma (PC12) cells. Pharmacol Res. https://doi.org/10.1016/j.phrs.2007.07.004
Zhang L et al (2018) Curcumin induces endoplasmic reticulum stress-associated apoptosis in human papillary thyroid carcinoma BCPAP cells via disruption of intracellular calcium

homeostasis. Medicine (United States) 97(24):e11095. https://doi.org/10.1097/MD.0000000000011095

Zhao Y et al (2011) Analgesic-antitumor peptide inhibits proliferation and migration of SHG-44 human malignant glioma cells. J Cell Biochem. https://doi.org/10.1002/jcb.23166

Zhao W et al (2013) 1B50-1, a mAb raised against recurrent tumor cells, targets liver tumor-initiating cells by binding to the calcium channel α2δ1 subunit. Cancer Cell. https://doi.org/10.1016/j.ccr.2013.02.025

Zhao W, Lu M, Zhang Q (2015) Chloride intracellular channel 1 regulates migration and invasion in gastric cancer by triggering the ROS-mediated p38 MAPK signaling pathway. Mol Med Rep. https://doi.org/10.3892/mmr.2015.4459

Zhao LY et al (2016) The overexpressed functional transient receptor potential channel TRPM2 in oral squamous cell carcinoma. Sci Rep. https://doi.org/10.1038/srep38471

Zheng Z et al (2019) The Xc− inhibitor sulfasalazine improves the anti-cancer effect of pharmacological vitamin C in prostate cancer cells via a glutathione-dependent mechanism. Cell Oncol. https://doi.org/10.1007/s13402-019-00474-8

Zhou Z et al (1999) Block of HERG potassium channels by the antihistamine astemizole and its metabolites desmethylastemizole and norastemizole. J Cardiovasc Electrophysiol. https://doi.org/10.1111/j.1540-8167.1999.tb00264.x

Zhou C et al (2016) Icaritin activates JNK-dependent mPTP necrosis pathway in colorectal cancer cells. Tumor Biol. https://doi.org/10.1007/s13277-015-4134-3

Zhu X et al (2013) Loss and reduced expression of PTEN correlate with advanced-stage gastric carcinoma. Exp Ther Med. https://doi.org/10.3892/etm.2012.749

Zoratti M, Szabò I (1995) The mitochondrial permeability transition. Biochim Biophys Acta. https://doi.org/10.1016/0304-4157(95)00003-a

Zou L et al (2016) Current approaches of photothermal therapy in treating cancer metastasis with nanotherapeutics. Theranostics. https://doi.org/10.7150/thno.14988

Rev Physiol Biochem Pharmacol (2022) 183: 103–134
https://doi.org/10.1007/112_2020_48

Published online: 8 September 2020

Ion Channels in Cancer: Orchestrators of Electrical Signaling and Cellular Crosstalk

Jerry J. Fan and Xi Huang

Contents

Abstract Ion channels are pore-forming transmembrane proteins that govern ion flux to regulate a myriad of biological processes in development, physiology, and disease. Across various types of cancer, ion channel expression and activity are often

J. J. Fan (✉) and X. Huang (✉)
Program in Developmental and Stem Cell Biology, The Hospital for Sick Children, Toronto, ON, Canada

Arthur and Sonia Labatt Brain Tumour Research Centre, The Hospital for Sick Children, Toronto, ON, Canada

Department of Molecular Genetics, University of Toronto, Toronto, ON, Canada
e-mail: jj.fan@mail.utoronto.ca; xi.huang@sickkids.ca

dysregulated. We review the contribution of ion channels to multiple stages of tumorigenesis based on data from in vivo model systems. As intertumoral and intratumoral heterogeneities are major obstacles in developing effective therapies, we provide perspectives on how ion channels in tumor cells and their microenvironment represent targetable vulnerabilities in the areas of tumor-stromal cell interactions, cancer neuroscience, and cancer mechanobiology.

Keywords Bioelectrical signaling · Cancer · Ion channels · Mechanobiology · Membrane potential · Metastasis · Tumor heterogeneity · Tumor initiation · Tumor microenvironment · Tumor progression

1 Introduction

1.1 Tumor Heterogeneity and the Multi-Step Tumorigenic Process Are Necessary Considerations for Personalized Cancer Medicine

Cancer is the second leading cause of death worldwide. Current standard-of-care is often non-specific and toxic to normal cells, causing side effects which reduce quality of life in survivors. Targeted therapeutic strategies have been developed through efforts in cancer genomics. Identification of oncogenic mutations, cellular mechanisms, and signaling networks have aided risk stratification and prediction of therapeutic response (The Cancer Genome Atlas Research Network 2008, 2011, 2012; The Cancer Genome Atlas Network 2012a, b). Tumorigenesis is a multi-step process which drives normal cells toward oncogenic transformation (initiation), stimulates tumor cells to survive and expand (progression), promotes dissemination and colonization of tumor cells at distant sites (metastasis), and confers therapeutic resistance (resistance and relapse). It is crucial to identify actionable targets in the exact step of tumorigenesis to develop effective therapeutic approaches.

Major obstacles remain in the fight against cancer. Malignant cells possess distinct phenotypic and functional characteristics between different tumors (intertumoral heterogeneity) and within individual tumors (intratumoral heterogeneity). Tumor heterogeneity is a vital consideration in validating therapeutic targets to eliminate tumor cell populations. Intertumoral heterogeneity hinders subtype-agnostic use of targeted therapies. Intratumoral heterogeneity creates therapy-resistant subclones that underlie disease relapse. Intratumoral heterogeneity can arise from several sources. First, it can arise from subclonal genetic mutations (different tumor cells and their progeny acquire different mutations). Second, functional heterogeneity exists in the form of quiescent cancer stem cells, fast-cycling tumor cells, post-mitotic tumor cells, and tumor stromal cells, each with unique molecular profiles and chemosensitivity. Third, tumor cells can possess distinct biophysical heterogeneity, such as electrical and mechanical properties. Importantly, all cells regulate ionic homeostasis to maintain proper resting membrane potentials, and many cells within tumors are capable of generating electrical activity in response to cell extrinsic or intrinsic stimuli. Thus, tumor cells can form electrical networks with input from their local microenvironment. Recent work on electrical and chemical synapses in cancer has revealed that cancer is an electrically active entity and that these electrical networks can be therapeutically targeted. As central regulators of cellular electrical properties, ion channels are implicated in all steps of tumorigenesis.

1.2 Why Target Ion Channels in Cancer?

As pore-forming transmembrane proteins, ion channels are regulated by chemical and physical stimuli to mediate ion flux along electrochemical gradients (Hille 2001). The voltage-gated ion channel superfamily is encoded by over 140 genes in the human genome, making it the third largest group of signaling molecules after G-protein-coupled receptors and kinases (Alexander et al. 2019). Broadly, ion channels can be classified as permeating anions or cations. Anion channels conduct chloride and other less abundant anions, while cation channels can be classified into potassium, sodium, and calcium-permeating families, in addition to the non-selective cation channels which include transient receptor potential (TRP) channels. Ion channels are present in all cells and regulate a myriad of biological processes ranging from rapid electrical signaling in excitatory cells to slower processes such as proliferation, volume regulation, migration, apoptosis, and hormone secretion. For example, ion channels control cell excitability through action potential propagation and regulate membrane potential during cell cycle progression and proliferation (Lang et al. 2005). Coordinated ion channel activity mediates cell migration (Schwab et al. 2012) through differential subcellular channel distribution and local ion flux at leading or trailing edge membranes (Schwab et al. 1995; Schneider et al. 2000; Huang et al. 2015). In addition, ion channels relay information from the extracellular environment (e.g., ion concentration, osmolarity, voltage, mechanics) and integrate external cues into signaling cascades within tumor cells. Ion channels promote tumor growth and survival, render cancer cells resistant to apoptotic and anti-proliferative signals, or play tumor-suppressive roles to prevent aberrant proliferation or oncogenic transformation.

Many ion channels have well-studied pharmacology and frequently localize at the cell surface, making them accessible drug targets. Small molecule ion channel modulators have been identified by high-throughput screening for proof of concept preclinical studies. Medicinal chemistry of existing chemical structures or rational design of small molecules can be guided by information on ion channel protein structures, domains that confer ion selectivity, and electrophysiological characteristics. Therefore, as the molecular targets of approximately 15% of US FDA-approved drugs (Overington et al. 2006), ion channels represent prime candidates for drug repurposing to treat cancer. Through use of classic methods (such as patch clamp) in conjunction with more recent approaches (such as automated patch clamp, genetically encoded voltage/ion indicators, optogenetics, and chemogenetics), ion channel function in tumorigenesis can now be dissected with unprecedented resolution.

1.3 Criteria for Selection of Studies in this Review

In this review we highlight ways in which the fields of ion channel and cancer biology intersect (Table 1). First, we review how ion channel expression and alterations have been implicated in multiple steps of tumorigenesis. Second, we highlight in vivo evidence of ion channel function in cancer. While many pioneering studies were performed using cancer cells lines, in vitro systems do not adequately model complex tumor cellular architecture, microenvironment, drug bioavailability, and organismal toxicity. Thus, we focus on in vivo findings primarily in the form of orthotopic xenograft tumor models or genetically engineered animal models. We discuss examples of pharmacological targeting of ion channels with emphasis on drug repurposing, medicinal chemistry, use of preclinical models, and consideration for side effects. Third, we discuss emergent areas of ion channel function in cancer. Ion channels may mediate tumor cell co-option of neuronal synapses to establish electrical networks in cancer. Cell non-autonomous interactions and propagation of electrical activity in cancer require ion channels and gap junctions. Furthermore, mechanosensitive ion channels can perceive and respond to the altered tissue mechanics in cancer to regulate malignant progression. We provide perspectives on these aspects as the field moves forward.

We refer readers to excellent complementary reviews of ion channels in cancer (Bates 2015; Prevarskaya et al. 2018) and their breakdown by ion type: potassium (Pardo and Stühmer 2014; Huang and Jan 2014), sodium (Djamgoz and Onkal 2012; Fraser et al. 2014; Roger et al. 2015), calcium (Yang et al. 2010; Monteith et al. 2017), TRP channels (Santoni and Farfariello 2011; Prevarskaya et al. 2011; Ouadid-Ahidouch et al. 2013), chloride (Cuddapah and Sontheimer 2011; Peretti et al. 2015), and bioelectrical signaling (Tuszynski et al. 2017; Payne et al. 2019). As we discuss ion channel classes, in each section we present them in the order of potassium, calcium, sodium, chloride, and non-selective channels.

2 Ion Channel Expression in Human Cancer

Molecular classification and gene expression analysis allow the distinction between cancers with favorable diagnosis, which can be managed with conservative treatment, from those associated with poor prognosis, which require more aggressive therapy. The expression of a single ion channel or group (gene signature) may offer value in stratifying risk and determining the treatment plan. Ion channel expression

Table 1 Ion channels that are implicated in cancer growth in vivo

Tumorigenic process	Ion channel	Nature of dysregulation	Cancer type	Model	Therapeutic relevance	Reference
Tumor initiation	CACNA1D	Mutation	Endocrine	Human sequencing	n.d.	(Scholl et al. 2013)
	KCNJ5	Mutation	Endocrine	Human sequencing	Macrolide antibiotics normalize mutant KCNJ5 currents in cancer cells	(Choi et al. 2011; Scholl et al. 2017)
	KCNQ1	Loss of function	Gastrointestinal	Mouse genetics	n.d.	(Starr et al. 2009; Than et al. 2014)
	CFTR	Loss of function	Gastrointestinal	Mouse genetics, Human patients	n.d.	(Starr et al. 2009; Than et al. 2016) (Neglia et al. 1995; Maisonneuve et al. 2013)
Tumor promoting	KCNA3	Expression	Leukemia, melanoma, pancreatic	Mouse xenograft	PAP-1 derivatives inhibit KCNA3 to elicit tumor-specific apoptosis	(Leanza et al. 2012, 2013, 2017)
	EAG1	Overexpression	Melanoma, pancreatic, prostate	Mouse xenograft	EAG1 antibody visualizes and targets tumor cells	(Pardo et al. 1999) (Hartung et al. 2011; Napp et al. 2016)
	EAG2/eag	Overexpression	Medulloblastoma	Mouse xenograft, Human patient, *Drosophila*	FDA-approved thioridazine inhibits EAG2 and displays anti-tumor efficacy in a medulloblastoma patient	(Huang et al. 2012, 2015)
	KCNT2	Overexpression	Medulloblastoma	Mouse xenograft	n.d.	(Huang et al. 2015)
	Paralytic	Expression	*Drosophila* brain neoplasia	*Drosophila*	n.d.	(Piggott et al. 2019)
	CLIC1/clic	Overexpression	Medulloblastoma	Mouse genetics, Mouse xenograft, *Drosophila*	n.d.	(Francisco et al. 2020)
	PIEZO1/piezo	Overexpression	Glioma	Mouse xenograft, *Drosophila*	n.d.	(Chen et al. 2018)
	TRPA1	Overexpression	Breast, lung	Mouse xenograft	TRPA1 inhibitor AM-0902 reduces tumor growth and confers chemosensitivity	(Takahashi et al. 2018)
	TRPML1	Overexpression	Melanoma, bladder, head and neck cancer	Mouse xenograft	n.d.	(Jung et al. 2019; Kasitinon et al. 2019)
	TRPM3	Overexpression	Clear cell renal cell carcinoma	Mouse xenograft	NSAID mefenamic acid inhibits TRPM3 and reduces *in vivo* tumor growth	(Mikhaylova et al. 2012; Hall et al. 2014)
Tumor suppressive	TRPV1	Overexpression	Glioma, gastrointestinal	Mouse genetics, Mouse xenograft	TRPV1 agonist arvanil prolongs survival of glioma-bearing mice	(Stock et al. 2012; de Jong et al. 2014)
Metastasis	EAG2	Overexpression	Medulloblastoma	Mouse xenograft, Human patient	FDA-approved thioridazine inhibits EAG2 and reduces metastatic burden in a medulloblastoma patient	(Huang et al. 2015)
	KCNN4	Overexpression	Glioma	Mouse xenograft	n.d.	(Turner et al. 2014)
	Orai1/STIM1	Expression	Breast	Mouse xenograft	SKF96365 inhibits store-operated Ca^{2+} entry and reduces breast-to-lung metastasis	(Yang et al. 2009)
	Orai1/KCNN3	Expression	Breast	Mouse xenograft	Ohmline disrupts KCNN3-Orai1 localization and impairs breast-to-bone metastasis	(Girault et al. 2011; Chantome et al. 2013)

The table lists ion channels implicated in specific stages of tumorigenesis and evidence on pharmacological targeting of ion channels. n.d. not determined is frequently dysregulated in cancer, occurring through various mechanisms, such as amplification, copy number alterations, mutations, and overexpression.

2.1 Dysregulated Expression

In breast and prostate cancer, decreased expression of potassium channel KCNA3 correlates with increased tumor grading (Abdul and Hoosein 2006; Brevet et al. 2009; Comes et al. 2013). Potassium channel EAG1 (KCNH1) is overexpressed in a wide array of cancers, including breast, prostate, colon, lung, liver, and soft tissue sarcoma (Hemmerlein et al. 2006; Mello de Queiroz et al. 2006). In addition to solid tumors, EAG1 expression is also elevated in myelodysplastic syndrome and leukemia (AML and CML), where high *EAG1* expression in AML predicts poor outcome (Agarwal et al. 2010). In normal tissues, EAG1 is mainly expressed in the brain with restricted expression in the periphery (Hemmerlein et al. 2006). Broad EAG1 overexpression across tumor types offers great therapeutic potential. Through conjugation to fluorescent or apoptosis-inducing ligands, EAG1-targeting antibodies can be used to visualize (Napp et al. 2016) or kill EAG1-expressing tumor cells (Hartung et al. 2011).

Potassium channel HERG1 (KCNH2) is highly expressed in colorectal cancers, while HERG1 is not detected in the normal colonic mucosa. In agreement with its expression being further elevated in metastatic disease, HERG1 regulates colon cancer cell invasiveness in vitro (Lastraioli et al. 2004). In the pediatric brain tumor medulloblastoma, potassium channels EAG2 (KCNH5) and KCNT2 are upregulated in several molecular subgroups (Huang et al. 2012, 2015).

In non-small cell lung cancer (Bonnet et al. 2007) and glioma (Preußat et al. 2003), potassium channel KCNA5 expression is inversely correlated with higher tumor grade. Low expression of KCNQ1 is associated with poor patient prognosis in gastrointestinal cancer (Than et al. 2014) and colon cancer (den Uil et al. 2016). Elevated expression of small conductance potassium channel KCNN4 is found in glioma and clear cell renal carcinoma. KCNN4 overexpression is associated with poor survival and increased metastatic potential, with *KCNN4* mRNA expression being further elevated in metastatic tumors compared to non-metastatic renal carcinomas (Turner et al. 2014; Rabjerg et al. 2015). In non-small cell lung cancer, promoter hypomethylation and increased *KCNN4* expression are associated with poor progression-free survival and overall survival (Bulk et al. 2015).

Store-operated calcium channel *ORAI1* is overexpressed in gastrointestinal stromal tumors and correlates with high-risk grading. Loss of ORAI1-mediated store-operated calcium suppresses tumor cell proliferation and induces apoptosis in vitro (Wang et al. 2017b). Mechanosensitive cation channel PIEZO1 is overexpressed in glioma, glioblastoma, breast cancer, and gastric cancer and is associated with poor prognosis (Li et al. 2015; Zhang et al. 2018; Chen et al. 2018). Transient receptor potential channel *TRPA1* is overexpressed in a variety of cancer types, including breast, kidney, lung, and malignant peripheral nerve sheath tumors. In breast and lung cancer, *TRPA1* overexpression promotes oxidative stress tolerance and chemoresistance and is associated with worse patient survival (Takahashi et al. 2018). TRP channel TRPM3 is overexpressed in clear cell renal carcinoma relative to normal kidney. In particular, TRMP3 expression is elevated in VHL-mutant tumors relative to VHL-wild type tumors (Hall et al. 2014). The vanilloid receptor TRPV1 is highly expressed in high-grade astrocytomas compared to non-tumor brain, and *TRPV1* expression positively correlates with tumor grading (Stock et al. 2012). Elevated expression of TRPML1 is associated with a worse prognosis in melanoma (Kasitinon et al. 2019) and in HRAS-driven bladder and head and neck cancers (Jung et al. 2019). Increased expression of TRPML2 is found in glioma of higher pathological grades (Morelli et al. 2016).

Chloride intracellular channel CLIC1 is overexpressed in multiple brain cancer types, including medulloblastoma, glioma, ependymoma, atypical teratoid rhabdoid, and primitive neuroectodermal tumors (Francisco et al. 2020). Additionally, CLIC1 is overexpressed in glioblastoma, pancreatic, lung, and gallbladder cancer, and elevated CLIC1 expression correlates with worse overall survival (Wang et al. 2011; Setti et al. 2013; Ding et al. 2015; Lu et al. 2015; Jia et al. 2016). CLIC1 overexpression in gastric cancer correlates with increased metastasis, invasion, and poor prognosis (Chen et al. 2007; Li et al. 2018). Calcium-activated chloride channel TMEM16A is upregulated in over 75% of pancreatic cancers, and high TMEM16A expression is associated with poor patient survival (Crottès et al. 2019).

In a cohort of low-grade to high-grade gliomas, an 18-ion channel gene signature predicts survival. Downregulation of 16 out of 18 ion channel genes is associated with high-grade gliomas and shorter survival (Wang et al. 2015). Among these, high *KCNB1* and *KCNJ10* expression correlate with favorable prognosis, while high *CLIC1* and *CLIC4* expression correlate with worse survival. In a separate cohort, *KCNB1* expression inversely correlates with glioma prognosis, and *KCNB1* overexpression induces autophagy and reduces tumor cell growth (Wang et al. 2017a). In a comparison between glioblastoma stem cells and normal neural cell types, KCNB1 is among four GSC-enriched ion channels associated with survival. shRNA-mediated *KCNB1* knockdown reduces GSC viability in vitro (Pollak et al. 2017). Analysis of different patient cohorts may underlie these inconsistent findings, and further analysis is required to determine whether KCNB1 has oncogenic or tumor suppressive functions. Alternatively KCNB1 may play opposing roles in GSCs compared to more differentiated tumor cell types, a hypothesis that has not been definitively tested. Given the wide array of overexpressed ion channels, gene regulatory pathways may be leveraged as a cancer-specific vulnerability. For example, oncogenic mutations in chromatin remodeling factors may promote ion channel gene transcription and oncogenic transcription factors may bind regulatory elements of ion channel genes. Functional validation is crucial in determining whether ion channel overexpression is causal or correlative in cancer.

2.2 Structural Variations and Copy Number Alterations

Structural variations can lead to oncogene amplification, tumor suppressor deletion, or ectopic fusion genes with hypermorphic, hypomorphic, or neomorphic properties. *KCNK9* is overexpressed or amplified in breast, lung, ovarian, and colorectal cancers (Mu et al. 2003; Kim et al. 2004; Innamaa et al. 2013). Functionally, KCNK9 overexpression confers resistance to hypoxia and serum deprivation, resulting in enhanced growth of human breast cancer cells in vitro (Mu et al. 2003). *TRPA1* is amplified in a subset of breast cancers and malignant peripheral nerve sheath tumors (Takahashi et al. 2018). The impact of copy number alterations on ion channels in cancer has not been fully characterized. Integration of copy number and gene expression data should reveal additional cases of amplification or deletion-dependent ion channel dysregulation.

Rare *PIEZO1-RSPO2* fusions are present in traditional serrated adenomas (TSA) (Hashimoto et al. 2019), although the functional consequence remains unknown. The fusion spans exon 1 of *PIEZO1* and exon 3 of *RSPO2*, comprising only 21 N-terminal amino acids of PIEZO1 while retaining critical functional domains of RSPO2. The relatively minor contribution of PIEZO1 amino acids suggest that *PIEZO1* promoter, rather than ion conductance, contributes to the fusion product to promote *RSPO2*-dependent WNT activation. Colon cancers harbor other recurrent fusions comprising *RSPO* genes (Seshagiri et al. 2012), highlighting a convergence on aberrant RSPO function in tumorigenesis. Many oncogenic fusions involving receptor tyrosine kinases display constitutive kinase activity due to truncation of regulatory domains. In contrast, there has been a paucity of fusion products comprising ion channel genes in cancer. One possible reason may be that ion channel function requires stereotyped membrane topology, and truncations abolishing gating or pore-forming domains may not be beneficial to the cancer cell.

2.3 Mutations

A subset of endocrine tumors (aldosterone-producing adenomas) harbor recurrent somatic and germline mutations in *KCNJ5*, an inwardly rectifying potassium channel, and *CACNA1D*, a voltage-gated calcium channel. *KCNJ5* mutations occur near the potassium selectivity filter and reduce potassium while increasing sodium conductance (Choi et al. 2011). *CACNA1D* mutations reside in the S6 pore-lining segment, which increases calcium influx (Scholl et al. 2013). Glioblastomas harbor sodium, calcium, and potassium channel mutations associated with poor prognosis (Joshi et al. 2011).

Metastatic urothelial carcinomas contain somatic missense mutations in ion channel genes including *CACNA1S*, *KCNK9*, and *SCN8A* (Sharma et al. 2019). As this data comes from the case report of a single patient, the prevalence and functional consequence of these ion channel mutations is unknown. There exist few reports of highly recurrent ion channel mutations spanning multiple cancer types, suggesting that ion channels may not simply be categorized as oncogenes or tumor suppressors. Rather, ion channel mutations in cancer are context and cell-type dependent.

2.4 Ion Channel Expression and Tumor Heterogeneity

To date, whether ion channel expression displays heterogeneity in cancer has been poorly explored. Expression heterogeneity can occur between patients (intertumoral), within different regions of the tumor (intratumoral), or change with disease progression (temporal). As an example of intertumoral heterogeneity, Group 4 medulloblastomas, which comprise one third of all medulloblastomas, display overexpression of potassium channel KCNA1. Immunohistochemical detection of KCNA1 is routinely used for molecular subgrouping (Northcott et al. 2011; Remke et al. 2011; Taylor et al. 2012), although KCNA1 function in medulloblastoma remains undetermined. Interestingly, integrated DNA methylation and gene expression analysis reveal enrichment of ion channel genes in a subset of SHH-activated medulloblastoma (Cavalli et al. 2017). In addition to being overexpressed in multiple tumor types, *TRPA1* is enriched in breast cancer with the exception of the normal-like subtype (Takahashi et al. 2018). Intratumor heterogeneity implies that distinct cells within the same tumor are dependent on different oncogenic events and that monotherapy will be insufficient. Likewise, temporal heterogeneity indicates that genetic alterations identified at diagnosis evolve as new mutations arise following treatment. Ion channel dysregulation in spatial or temporal tumor heterogeneity remains to be explored. We note that such studies are of great importance and will elucidate whether specific ion channels represent stable or dynamic therapeutic targets in cancer.

It is also important to note that genomic and proteomic analyses cannot entirely predict ionic events. Post-transcriptional and post-translational modifications regulate ion channel folding, stability, and trafficking to membrane domains. Various modes of stimulation govern the "on" or "off" state of ion channels. Furthermore, electrogenic proteins, including ion channels and transporters, work in concert to maintain ionic homeostasis and cellular resting membrane potential (Vmem), which can exert critical influence on cancer cell behavior (Yang and Brackenbury 2013; Payne et al. 2019). Therefore, it is crucial to consider how electrical signals propagate through tumors and holistically investigate ion channels at the genomic, proteomic, and physiological levels.

3 Ion Channel Functions in Tumorigenesis

3.1 Ion Channels in Tumor Initiation

A DNA transposon-based forward genetic screen implicates potassium channel *Kcnq1* and chloride channel *Cftr* loss-of-function in the initiation of murine gastrointestinal cancer (Starr et al. 2009). Intestinal-specific knockout of *Cftr* or global knockout of *Kcnq1* in the sensitized Apc^{min} mouse model of intestinal cancer increases tumor incidence (Than et al. 2014, 2016). These two channels are proposed to functionally interact in healthy intestinal epithelium, where Kcnq1-mediated basolateral potassium export provides the electrochemical drive for apical export of chloride by Cftr. Transcriptomic analysis of Kcnq1-or Cftr-deficient tumors displays enrichment for dysregulation of immune response and lipid metabolism. However, the mechanism by which loss of *Kcnq1* or *Cftr* initiates gastrointestinal cancer remains to be fully defined. Interestingly, *CFTR* mutations are causal for cystic fibrosis, and patients with cystic fibrosis have increased risk of digestive tract cancers (Neglia et al. 1995; Maisonneuve et al. 2013), providing relevance for the mouse data to human disease.

Somatic and germline gain-of-function mutations in calcium and potassium channel genes are suggested to initiate human endocrine tumors (Choi et al. 2011; Scholl et al. 2013). In approximately 40% of adrenal aldosterone-producing adenomas, recurrent mutations localize near the potassium selectivity filter of *KCNJ5*. Expression of mutant KCNJ5 in HEK293T and glomerulosa cells causes membrane depolarization attributable to reduced potassium selectivity and increased sodium conductance (Choi et al. 2011). Subsequent depolarization is proposed to activate voltage-gated calcium channels and increase intracellular calcium, thereby promoting aldosterone production and cell proliferation. A high-throughput screen in a HEK293 inducible expression system identified compounds which inhibit mutant KCNJ5. A series of clinically approved macrolides, bacteriostatic antibiotics with established safety profiles and oral bioavailability, suppresses mutant KCNJ5-dependent sodium conductance and normalizes aldosterone production in human adrenocortical cancer cells (Scholl et al. 2017). This study highlights the promise of drug repurposing to target ion channels in cancer.

Mutations in *CACNA1D*, which encodes a L-type voltage-gated calcium channel, occur in 11% of aldosterone-producing adenomas. All seven identified *CACNA1D* mutations affect conserved residues near the transmembrane S6 domain. Expression of mutant CACNA1D in HEK293 cells facilitates channel opening at less depolarized potentials. Through impaired inactivation, this leads to sustained channel activation and increased calcium influx. Thus, *CACNA1D* and *KCNJ5* mutations both lead to increased calcium influx to induce depolarization (Scholl et al. 2013). Given that *KCNJ5* and *CACNA1D* mutations are mutually exclusive, de novo germline mutations occur at the same locations as somatic mutations, and there is a lack of additional somatic mutations in these tumors, single ion channel mutations may be sufficient to initiate aldosterone-producing adenomas. Functional studies to

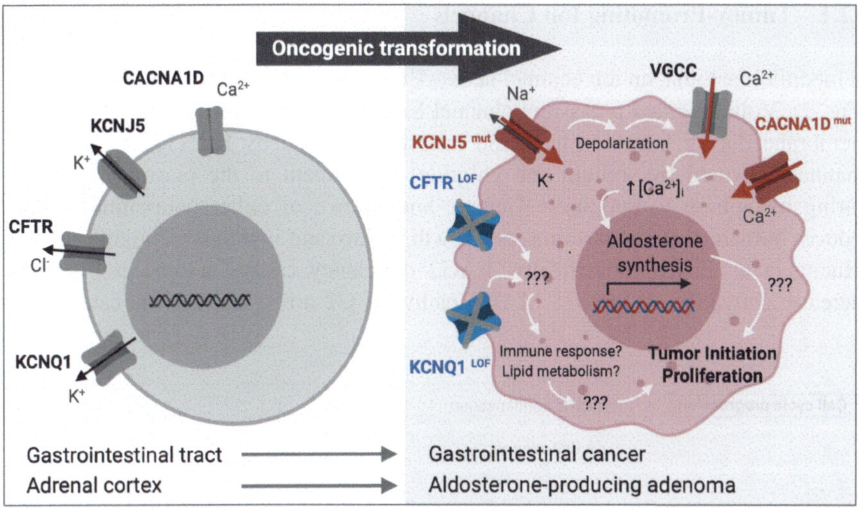

Fig. 1 Ion channels in tumor initiation. Dysregulated expression or function of several ion channels has been implicated in tumor initiation. Sequencing of human endocrine tumors identifies recurrent mutations in potassium channel *KCNJ5* and calcium channel *CACNA1D*. *KCNJ5* mutation promotes sodium over potassium conductance, depolarization, and activation of voltage-gated calcium channels. *CACNA1D* mutations impair channel inactivation and increase calcium influx. Mutations in these channels converge on increasing intracellular calcium and aldosterone production to promote cell proliferation. Through forward genetic screens, KCNQ1 and CFTR loss of function have been identified as tumor initiating events in mouse gastrointestinal cancers. Transcriptomic analysis of resulting tumors implicates perturbed immune response and lipid metabolism. The mechanism by which perturbed potassium and chloride homeostasis induce gastrointestinal cancers remains to be elucidated

induce these mutations in relevant cell types using genetically engineered animal models will fully define the role of these altered ion channels in tumor initiation. Overall, the small number of studies implicating ion channels in tumor initiation indicates that ion channels may be more often co-opted by cancer cells to regulate tumor progression and maintenance, following the oncogenic event that initiated the tumor (Fig. 1).

3.2 *Ion Channels in Tumor Progression*

As an evolving disease, cancer cells may no longer depend on initiating genes during tumor progression. The demonstration that EAG1 promotes tumor progression in subcutaneous xenograft models is among the first reports to implicate ion channels in tumor progression (Pardo et al. 1999).

3.2.1 Tumor-Promoting Ion Channels

In medulloblastoma, an ion channel network cooperates to regulate cell proliferation (Fig. 2). Voltage-gated potassium channel EAG2 (KCNH5) localizes to the plasma membrane during late G2 phase and mitosis, followed by KCNT2 (a potassium channel activated by sodium and chloride) enrichment to the plasma membrane during metaphase to telophase. Genetic knockdown of either potassium channel reduces human medulloblastoma cell growth in vitro and in mouse xenograft models (Huang et al. 2015). Specifically, EAG2 deficiency causes ectopic cell volume increase, activation of the p38 MAPK pathway, G2 arrest and mitotic catastrophe

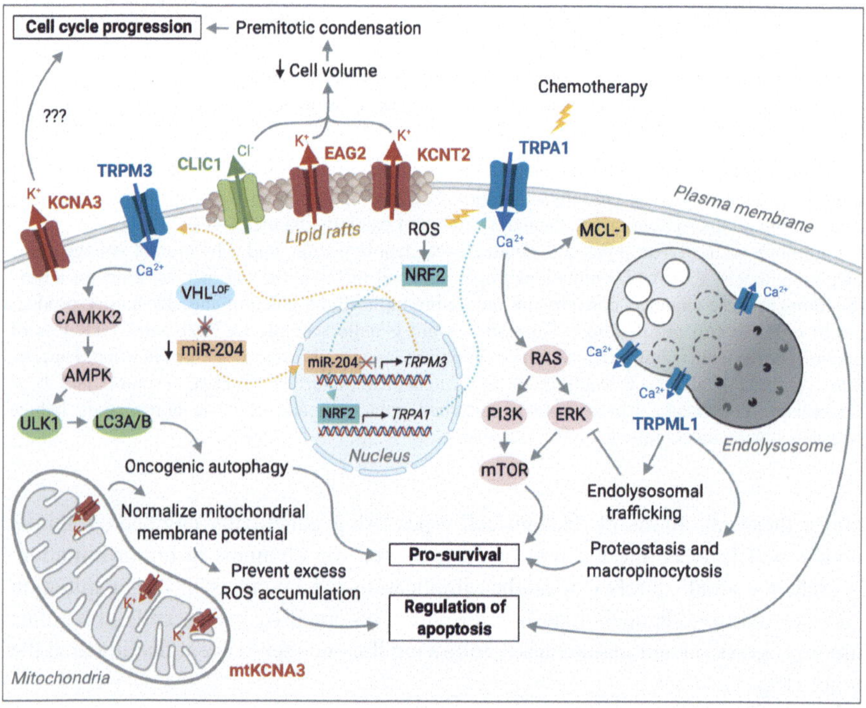

Fig. 2 Cell autonomous functions of ion channels in tumorigenesis. Ion channels in cancer display great diversity with regard to subcellular localization and intracellular signaling. Plasma membrane-localized ion channels CLIC1 and EAG2, both of which are present at lipid rafts, and KCNT2 function cooperatively. During late G2 and mitosis, chloride and potassium efflux reduces premitotic cell volume for cell cycle progression and tumor cell proliferation. Plasma membrane-localized TRP channels TRPM3 and TRPA1 engage autophagy and RAS/PI3K/mTOR signaling respectively to promote tumor cell survival. Potassium channel KCNA3 has localization-dependent functions in cancer. KCNA3 on the plasma membrane promotes cell cycle progression, and KCNA3 at the inner mitochondrial membrane regulates mitochondrial membrane potential and production of reactive oxidative species (ROS) to suppress apoptosis. Endolysosomal channel TRPML1 regulates proteostasis and macropinocytosis through ERK/mTOR signaling to promote tumor cell survival

(Huang et al. 2012). Screening of FDA-approved drugs identifies the antipsychotic thioridazine as an EAG2 channel inhibitor, which reduces tumor progression and prolongs survival in xenograft medulloblastoma mouse models. Furthermore, thioridazine displays efficacy in a human patient with relapsed metastatic medulloblastoma. Thioridazine treatment reduced tumor volume, providing the first proof-of-principle for using an ion channel blocker in treating brain tumor patients (Huang et al. 2015). Ultimately the patient did not tolerate prolonged treatment due to mood lability and depression, which is likely due to the inhibitory effect of thioridazine on dopaminergic and serotonergic receptors. These results demonstrate promise for clinical application of EAG2 inhibition in cancer and highlight the need to develop blockers with improved on-target specificity.

Chloride intracellular channel CLIC1 regulates cell volume homeostasis and cell cycle progression of rapidly dividing medulloblastoma cells (Fig. 2). CLIC1 colocalizes with EAG2 at lipid raft microdomains on the plasma membrane during mitosis. CLIC1 deficiency suppresses in vivo tumor growth in xenograft and genetic mouse models of medulloblastoma. EAG2, KCNT2, and CLIC1 mediate potassium and chloride efflux respectively, to synergistically regulate cell volume, premitotic cytoplasmic condensation, and cell proliferation. Loss of the orthologous clic and eag channels reduces brain tumor growth in *Drosophila* (Huang et al. 2015; Francisco et al. 2020). These results reveal an evolutionarily conserved role for CLIC1 and EAG2 in brain tumor growth and highlight functionally coupled ion channels as vulnerabilities in tumor progression.

Paralytic, which encodes the sole voltage-gated sodium channel in *Drosophila*, regulates the proliferative output of neuroblasts (*Drosophila* neural stem cells). Furthermore, loss of paralytic exerts tumor-suppressive effects in multiple *Drosophila* neuroblast-derived models of brain tumor (Piggott et al. 2019). Sodium channels with similar function in mammalian brain tumors remain unexplored.

Tissue stiffening frequently occurs during solid tumor progression. For example, glioma aggression and patient prognosis correlate with increasing tumor stiffness (Miroshnikova et al. 2016). Mechanosensitive cation channel PIEZO1 is overexpressed in human gliomas. PIEZO1 localizes at focal adhesions of glioblastoma stem cell processes, where its activation induces calcium influx and integrin-FAK signaling to promote extracellular matrix remodeling and tissue stiffening. The stiffer microenvironment elevates PIEZO1 expression to increase glioma cell proliferation. Therefore, PIEZO1 orchestrates a feedforward loop in which it promotes glioma stiffness and malignancy by sensing and responding to heightened tissue stiffness (Fig. 3). Targeting Piezo in *Drosophila* glioma models or PIEZO1 in glioblastoma xenograft mouse models suppresses tumor growth in vivo (Chen et al. 2018).

Potassium channel KCNA3 is expressed in several tumor types (Comes et al. 2013) and represents an actionable target to induce cancer cell death. In subcutaneous models of human lung adenocarcinoma and mouse melanoma, KCNA3 localizes to the plasma membrane and inner mitochondrial membrane, where it regulates cancer cell proliferation and apoptosis, respectively (Jang et al. 2011; Leanza et al. 2012) (Fig. 2). During lymphocyte apoptosis, apoptotic regulator BAX interacts

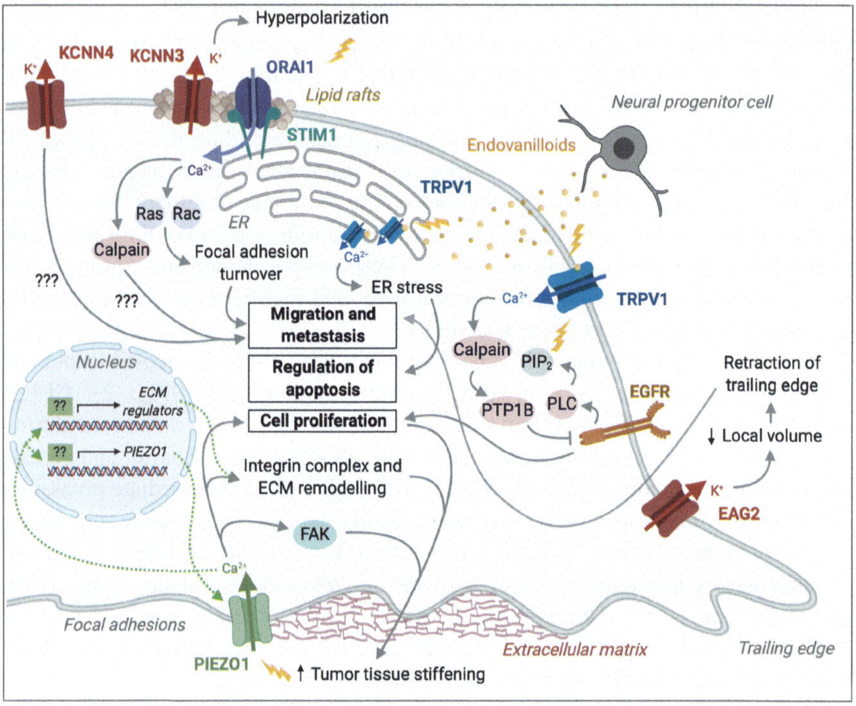

Fig. 3 Ion channel function in tumor cell invasiveness and non-autonomous interactions. Ion channels regulate cancer cell invasiveness and migration through distinct mechanisms. EAG2-mediated potassium efflux leads to local cell volume reduction and retraction of the trailing edge to facilitate cell migration. Potassium channel KCNN4 regulates cell invasiveness, while the mechanism has not been described. Several studies described store-operated calcium channel ORAI1 in cancer metastasis. ORAI1 activates in cooperation with co-localized KCNN3 and STIM1 at the endoplasmic reticulum (ER). ORAI1 activation and subsequent calcium influx enhances focal adhesion turnover. Human brain tumor cells express TRPV1, which can be activated by endovanilloids secreted by neural progenitor cells which home to the tumor. TRPV1 activation induces ER stress and tumor cell apoptosis. In gastric cancer models, TRPV1 activation inhibits EGFR-mediated cell proliferation. By localizing to glioma cell focal adhesions, PIEZO1 senses tumor stiffness to increases its own expression, cell proliferation, and integrin and extracellular matrix (ECM) remodeling to increase tumor stiffness and malignancy

with and inhibits mitochondrial KCNA3, causing mitochondrial membrane hyperpolarization, production of reactive oxidative species, and cytochrome C release (Szabo et al. 2008). Small molecule KCNA3 inhibitors, such as PAP-1, exhibit 23- to 125-fold selectivity over related Kv1 family members (Schmitz et al. 2005). Recently developed synthetic PAP-1 derivatives display increased specificity toward mitochondrial KCNA3 (Leanza et al. 2017). Administration of PAP-1 derivatives elicits tumor-specific apoptosis in leukemic, melanoma, and pancreatic cancer cells in vitro and in orthotopic xenograft models. Importantly, while KCNA3 is expressed in many organs, its inhibition does not seem to induce apoptosis in non-tumoral organs, including the brain, heart, liver, spleen, and kidney (Leanza et al. 2012, 2013, 2017). Interestingly, siRNA-mediated KCNA3 knockdown fails to induce cancer cell apoptosis (Leanza et al. 2012), highlighting a distinction between pharmacological versus genetic perturbation. In the latter case, other genes regulating ion homeostasis and mitochondrial membrane potential may compensate for KCNA3 deficiency. While the effects of KCNA3 inhibition on normal physiology besides inducing cell death require further examination, these findings highlight KCNA3 as a tumor-specific target.

As reactive oxidative species (ROS) readily accumulate in cancer cells, adaptation to oxidative stress is crucial for tumor progression. ROS signals through oxidant defense factor NRF2 to mediate an anti-oxidant defense program and induce expression of the redox-sensitive cation channel TRPA1. ROS elevation induces TRPA1-mediated calcium influx and upregulates RAS-ERK, PI3K/AKT, and mTOR-dependent pro-survival cues, as well as MCL-1-mediated anti-apoptotic signaling (Fig. 2). TRPA1 knockdown reduces tumor growth in patient-derived xenograft (PDX) breast cancer models. TRPA1 inhibition using AM-0902, an orally bioactive TRPA1 inhibitor, reduces PDX tumor growth in vivo, although the short plasma half-life of AM-0902 may pose challenges for bioavailability (Takahashi et al. 2018). As TRPA1 inhibitors are in clinical trials for pain and respiratory diseases, this study demonstrates the clinical prospect of TRPA1-based therapy in breast cancer.

Endolysosomal cation channel TRPML1 represents a critical dependency in human melanoma cells but not normal melanocytes. TRPML1 loss decreases melanoma cell growth in vitro and in xenograft models through elevating MAPK and mTORC1 activity, ultimately impairing macropinocytosis and inducing proteotoxic stress (Kasitinon et al. 2019). TRPML1 function in tumorigenesis may differ depending on cancer type. For example, TRPML1 inhibition has opposing effects on elevating and attenuating ERK signaling in melanoma cells (Kasitinon et al. 2019) and HRAS-driven human carcinoma cells (Jung et al. 2019), respectively. TRPML1 in HRAS-driven cancers mediates cholesterol distribution, HRAS nanoclustering, and ERK phosphorylation to maintain signaling and proliferation (Jung et al. 2019). These studies suggest a role for TRPML1 in endolysosomal fusion to create a permissive environment for oncogenic signaling. The exact mechanism by which TRPML1 activity in tumor cells regulates cholesterol transport and endolysosomes remains to be defined (Fig. 3). TRPML1 mutations underlie mucolipidosis-type IV, a neurodegenerative lysosomal storage disorder (Bargal et al. 2000; Sun et al. 2000). TRPML1 activation is neuroprotective (Tsunemi et al. 2019), and Trpml1 knockout female mice exhibit luteal cell degeneration, progesterone deficiency, and infertility (Wang et al. 2019). Thus, the consequences of TRPML1 loss in physiology should also be considered in developing TRPML1-based cancer therapies.

Inactivation of the tumor suppressor VHL is common in clear cell renal cell carcinoma (ccRCC) (The Cancer Genome Atlas Research Network 2013; Sato et al. 2013). In VHL-deficient ccRCCs, TRP channel TRPM3 regulates an oncogenic autophagy network. VHL regulates expression of tumor suppressive microRNA miR-204, derived from intron 6 of the *TRPM3* gene. miR-204 targets and inhibits TRPM3, as well as autophagy regulators LC3B and LC3B2. VHL inactivation and miR-204 downregulation elevate autophagy to sustain tumor growth (Mikhaylova et al. 2012). TRPM3 activation and calcium influx regulate autophagy through CAMKK2 and ULK1 signaling, which converge to positively modulate autophagic regulators LC3A/B (Fig. 2). Genetic knockdown or pharmacological inhibition of TRPM3 reduces tumor growth in orthotopic xenograft models (Hall et al. 2014). Mefenamic acid is an orally bioavailable, FDA-approved, non-steroidal anti-inflammatory drug (NSAID) that specifically inhibits TRPM3 over other TRP channels (Klose et al. 2011). Mefenamic acid reduces TRPM3 expression and autophagy in VHL-inactive human cancer cells in vitro and decreases tumor growth in subcutaneous xenograft models (Hall et al. 2014). These findings demonstrate the potential to target TRPM3 in cancers with VHL deficiency and increased TRPM3 expression. It remains to be determined whether anti-tumor effects of mefenamic acid, which also inhibits cyclooxygenase (COX) enzymes to suppress prostaglandin synthesis, may in part be attributed to a COX-dependent mechanism. Interestingly, upon VHL induction, miR-204 is co-expressed with two short but not full-length *TRPM3* transcripts (Mikhaylova et al. 2012). This may be due to a putative promoter upstream of the first exon of shorter transcripts. The contribution of short and full-length *TRPM3* transcripts to ccRCC tumor progression and its upstream regulatory machinery is intriguing. For example, do short *TRPM3* transcripts encode functional proteins with a role in tumorigenesis, or are they byproducts of miR-204 activation? Concomitant microRNA expression with shorter gene transcripts may be a general mechanism for intronic miRNAs from long genes. Many ion channels are encoded by long genes spanning multiple exons. The existence and function of microRNAs and alternative isoforms from ion channel genes, particularly in cancer, represent an interesting area for future study.

3.2.2 Tumor Suppressive Ion Channels

Neural progenitor cells possess tumor-suppressive function against high-grade astrocytomas (Stock et al. 2012). Neural progenitors display tropism for brain tumors and release endovanilloids, which act non-cell autonomously on TRPV1 channels expressed in human brain tumor cells to induce cell death via the endoplasmic reticulum stress pathway (Fig. 3). This raises the possibility that neural stem cell therapy or TRPV1 activation may be used to treat brain tumors. As proof of concept, the authors demonstrate that administering the blood-brain barrier permeable synthetic TRPV1 agonist arvanil prolongs survival of xenograft tumor-bearing mice. Interestingly, TRPV1 activation inhibits EGFR-mediated epithelial cell proliferation through calcium and PTP1B signaling, and loss of TRPV1 increases tumor formation in the Apc^{min} mouse model of intestinal tumorigenesis, suggesting that TRPV1 activation is also tumor suppressive in intestinal cancer (de Jong et al. 2014).

Given the evidence for ion channel involvement in both promoting and inhibiting tumor growth, most ion channels appear to act as neither oncogenes nor tumor suppressors. Rather, ion channel activity must remain within a proper window to facilitate tumorigenesis. Supra-physiological gain- or loss-of-function is likely detrimental to tumor growth.

3.3 Ion Channels in Metastasis

Tumor metastasis comprises a sequence of events that include dissemination of tumor cells from the primary site, survival in the circulatory system, and invasion and colonization of distant locations for neoplastic growth. Most reported functions of ion channels in tumor metastasis are pertinent to cell migration.

In gliomas, genetic knockdown or pharmacological inhibition of calcium-activated potassium channel KCNN4 reduces tumor cell migration in vitro. KCNN4 knockdown in glioma xenografts reduces invasive growth in vivo (Turner et al. 2014). Therefore, KCNN4 may contribute to the diffuse nature of malignant gliomas (Fig. 3).

Expression of potassium channel EAG2 is upregulated in a subset of metastatic medulloblastoma compared to matched primary tumors. EAG2 localizes to the trailing edge of medulloblastoma cells to promote local potassium efflux and rear cell retraction, which are essential for medulloblastoma cell motility (Fig. 3). Targeting EAG2 reduces medulloblastoma metastasis in mouse xenograft models and decreases metastatic burden in a patient with metastatic medulloblastoma (Huang et al. 2015).

STIM and Orai proteins mediate store-operated calcium entry, a calcium influx mechanism in non-excitable cells. Second messenger signaling induces rapid and transient release of calcium from the endoplasmic reticulum (ER). Decreased ER calcium concentration is sensed by STIM proteins, which cluster near the plasma membrane to complex with the pore-forming Orai for calcium entry. In human breast cancer cells, calcium influx through Orai1 and STIM1 promotes cell migration in vitro through small GTPase-mediated focal adhesion turnover. Store-operated calcium influx may increase FAK tyrosine kinase activity and calcium-dependent protease calpain to regulate focal adhesion dynamics (Fig. 3). Administration of SKF96365, an inhibitor of store-operated calcium entry, reduces breast cancer metastasis from mammary gland to lung in xenograft mouse models (Yang et al. 2009). As SKF96365 also blocks TRPC and low-voltage-activated T-type calcium channels (Singh et al. 2010), additional studies are needed to determine whether the phenotypes can be solely attributed to Orai1. Since genetic or pharmacological manipulations of Orai1/STIM1 also perturbed focal adhesions in mouse embryonic fibroblasts (Yang et al. 2009), future investigation is necessary to determine the full spectrum of phenotypes from pharmacologically targeting this mode of calcium entry.

Orai colocalizes and functionally associates with calcium-activated potassium channel KCNN3 in cancer. In an orthotopic breast cancer xenograft model, KCNN3 knockdown reduces bone metastases without affecting primary tumor growth or lung metastases. KCNN3-Orai1 complexes localize to lipid rafts to regulate calcium entry, calpain activation, and cell migration (Fig. 3). Disruption of KCNN3-Orai1 lipid localization using alkyl-lipid KCNN3 blocker Ohmline impairs calcium influx, migration, and bone metastases (Chantome et al. 2013). Ohmline does not cause systemic toxicity in vivo (Girault et al. 2011). These results suggest that ion channel function depends on precise subcellular localization, which can be leveraged for cancer treatment.

As tumor metastasis involves tumor cell dissemination, intravasation, survival through systemic circulation, extravasation, colonization at a distant site, and eventual metastatic growth, future studies are required to specify which exact steps depend on ion channel function. In addition, it is important to determine whether a particular ion channel is uniquely required for metastasis. If targeting an ion channel reduces the growth of primary tumor and metastasis, it becomes difficult to ascertain whether impaired metastasis is due to a reduction in primary tumor burden or a defect in the metastatic cascade.

3.4 Ion Channels in Therapeutic Resistance and Tumor Recurrence

As described above, TRPA1 overexpression promotes tumor cell survival in environments of high oxidative stress. The anti-cancer agent carboplatin, which generates reactive chemicals to cause cell death, induces TRPA1-dependent increases in intracellular calcium concentration and oscillatory calcium responses. In carboplatin-resistant breast cancer cells, TRPA1 inhibition restores carboplatin sensitivity in vitro and in vivo, highlighting the therapeutic potential of TRPA1-based combination therapy (Takahashi et al. 2018). The contribution of ion channels toward tumor recurrence and resistance has not been sufficiently studied in vivo. Some malignancies, such as glioblastoma, recur in nearly all patients despite first-line treatment (Nabors et al. 2017). Future study to target ion channels in both primary and recurrent tumors will provide the foundation to develop new therapies to improve patient outcome.

3.5 Therapeutic Potential of Targeting Ion Channels

As small molecules and FDA-approved drugs have been used to target ion channels in cancer, we note that stringent criteria must be upheld in defining on-target versus off-target effects. Pharmacodynamic and pharmacokinetic data, such as the maximum tolerated dose, minimum effective concentration, bioavailability, half-life, and systemic toxicity, must be evaluated. Finally, the IC_{50}/EC_{50} concentrations at which a molecule acts on ion channels should be consistent with the achievable therapeutic concentrations in vivo.

4 Emergent Areas of Ion Channels in Cancer and Outlook

4.1 Ion Channels in Governing Local Ion Milieu

Most studies in the field focus on how ion channels regulate cancer biology through cell autonomous mechanisms. Intratumoral heterogeneity and cell non-autonomous mechanisms are crucial in fueling malignant growth. For example, reciprocal signaling between stem-like and differentiated tumor cells through growth factor secretion creates a tumor ecosystem to drive glioma progression (Wang et al. 2018). Metabolic adaptation of tumor cells, which reside in either hypoxic or oxygenated niches, induces distinct gene expression programs (Allen et al. 2016; Jin et al. 2017). In this regard, one can imagine that tumor heterogeneity in the local ionic milieu and changes in extracellular ion dynamics may exert cell non-autonomous effects (Fig. 4). Indeed, in mouse and human melanomas, dying tumor cells release intracellular potassium into the extracellular environment. Elevated extracellular

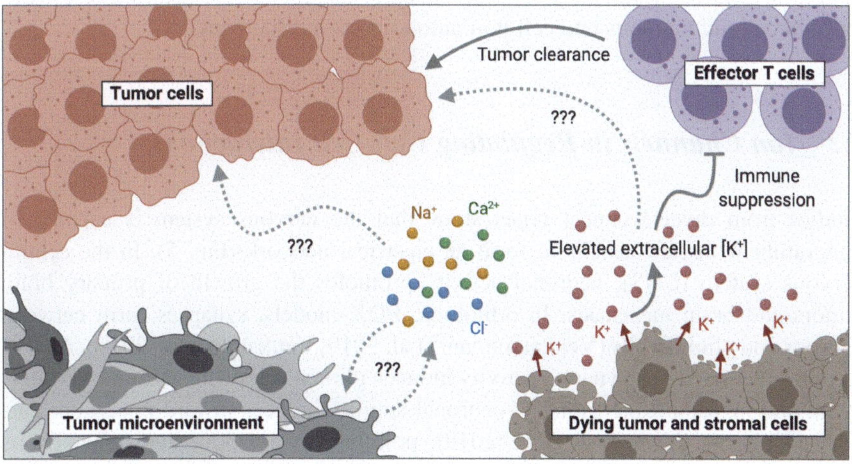

Fig. 4 Ion channels govern local ion milieu and tumor-immune cell interaction. The contribution of ion channels and the extracellular ionic environment to cell non-autonomous interactions in cancer is largely under-explored. In melanoma, tumor cell death releases potassium ions into the extracellular space. Elevation of extracellular potassium suppresses anti-tumor functions of resident T cells through inhibition of Akt/mTOR signaling, dysregulated metabolism, and altered histone acetylation. Whether perturbing the homeostasis of other ion classes has effects in the tumor microenvironment remains to be determined. *Dotted arrows denote hypothetical functions*

potassium suppresses the anti-tumor function of T cells through inhibiting Akt/mTOR and PP2A signaling (Eil et al. 2016). In addition, elevated extracellular potassium limits nutrient uptake and reduces the level of metabolic intermediate acetyl-CoA. Decreased acetyl-CoA suppresses histone acetylation of T-cell effector and exhaustion loci to attenuate their anti-tumor activity. These altered T cells display increased stemness and improved persistence in vivo (Vodnala et al. 2019). These data suggest that manipulation of extracellular potassium levels is a strategy to modulate T-cell-based immunotherapy. In addition to melanoma, it would be of interest to determine the effect of extracellular potassium on infiltrated T cells in other types of tumors and normal organs.

Altered extracellular ion concentrations may shift the membrane potential of cancer cells, activate voltage-gated ion channels, and induce bioelectrical signaling (Tuszynski et al. 2017; Payne et al. 2019). Early studies in chick spinal cord neurons implicate membrane potential in cell cycle progression (Cone and Cone 1976). A correlation exists between membrane potential and differentiation status, where more proliferative cells possess depolarized membrane potentials, while terminally differentiated non-dividing cells reside at more hyperpolarized potentials (Cone 1971). In the developing mouse brain, the membrane potential of neural progenitor cells becomes progressively hyperpolarized during corticogenesis. Inducing hyperpolarization in vivo shifts the neurogenic output and transcriptome of progenitor cells toward a later developmental stage (Vitali et al. 2018). Thus, tumor-specific variation in extracellular ionic balance may explain why certain cancers are more malignant than others (Fig. 4). Future study of the functional impact of local ion milieu in cancer may uncover new ion channel targets to modulate tumor growth through cell-autonomous and cell non-autonomous mechanisms.

4.2 Ion Channels in Regulating Cell-Cell Interactions

Studies from diverse cancer types show that the nervous system is capable of integrating with cancer cells to form an electrical network (Fig. 5). In the central nervous system (CNS), neuronal activity promotes the growth of primary brain tumors and brain metastasis. In orthotopic PDX models, synapses form between neurons and glioma cells (Venkataramani et al. 2019; Venkatesh et al. 2019). AMPA receptor-mediated postsynaptic activity induces glioma cell depolarization, proliferation, and invasion. Interestingly, neuronal activity induces inward potassium currents in glioma (Venkatesh et al. 2019), pointing toward potassium channels as mediators of this process.

In mouse PDX intracardiac injection models of breast-to-brain metastasis, neoplastic cells form functional neuronal synapses, which promote metastatic colonization in an NMDA signaling-dependent manner. In human brain metastases, components of NMDA receptor signaling are elevated when compared to paired primary breast tumors (Zeng et al. 2019). Outside of brain malignancies, innervation of peripheral nerves promotes tumorigenesis in PDX and transgenic mouse models

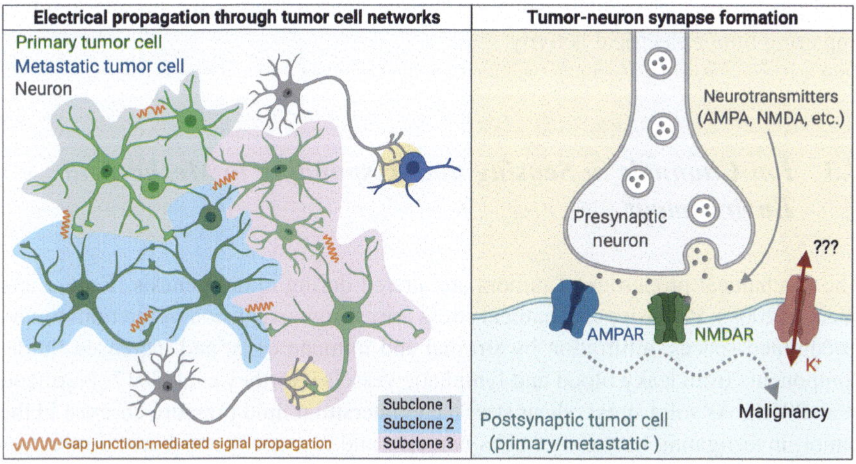

Fig. 5 Ion channels mediate tumor-neuron synapses and propagate electrical signaling. In a network of interconnected tumor cells, electrical activity can occur through gap junction or ion channel-mediated signaling. Propagation of electrical signals mediates interactions between different subclones and buffers tumor cells against therapeutic stress. Surrounding neurons are capable of forming synapses with primary tumor or metastatic tumor cells (*left*). Tumor-neuron synaptic transmission activates neurotransmitter signaling and induces potassium currents and depolarization. Synaptic activity increases tumor proliferation and metastasis (*right*). *Dotted arrows denote hypothetical functions*

of gastric, prostate, and pancreatic cancer in a paracrine manner (Magnon et al. 2013; Hayakawa et al. 2017; Renz et al. 2018). It would be interesting to determine whether such interactions depend on ion channel-mediated signal propagation to proliferate and metastasize.

Bioelectric signaling in cancer is not a single-cell phenomenon. Altered electrical activity in cancer can propagate through an interconnected network. Gap junctions are intercellular channels formed by connexin proteins. Gap junctions between adjacent cells enable rapid cellular communication through passage of ions, second messengers, and other small molecules. PDX models demonstrate that glioma cells form networks through gap junction coupling (Osswald et al. 2015; Venkataramani et al. 2019; Venkatesh et al. 2019). As gap junctions possess distinct ion permeability, differential gap junction expression determines differences in membrane potential and cancer stem cell self-renewal (Hitomi et al. 2015). Such connectivity can buffer interconnected glioma cells against therapeutic stress (Osswald et al. 2015). Furthermore, rapid propagation of electrical state may enable cooperativity among different tumor subclones. We refer readers to excellent reviews on gap junctions in cancer (Aasen et al. 2016; Sinyuk et al. 2018) and the nascent field of cancer neuroscience (Monje et al. 2020).

In summary, ion channel-mediated electrical activities of tumor-tumor interactions and tumor-neuronal synapses should be considered in order to gain a comprehensive view of the dynamic tumor microenvironment. Pharmacologically manipulating ion channel dependencies may impede malignant growth through impairing tumor electrical activity.

4.3 Ion Channels in Sensing and Responding to Mechanical Environment

The mechanical properties of tumors are altered during tumorigenesis (Kumar and Weaver 2009). Pervasive mechanical stress arises from expansion of the tumor mass in confined spaces, infiltration by stromal and immune cells, and increased fluidic components from leaky blood and lymphatic vessels (Northey et al. 2017; Northcott et al. 2018). As solid stress, shear stress, and interstitial fluid pressure co-exist in the tumor, investigating how tumor cells perceive and respond to mechanical microenvironment through mechanosensitive ion channels will uncover new therapeutic opportunities (Fig. 6).

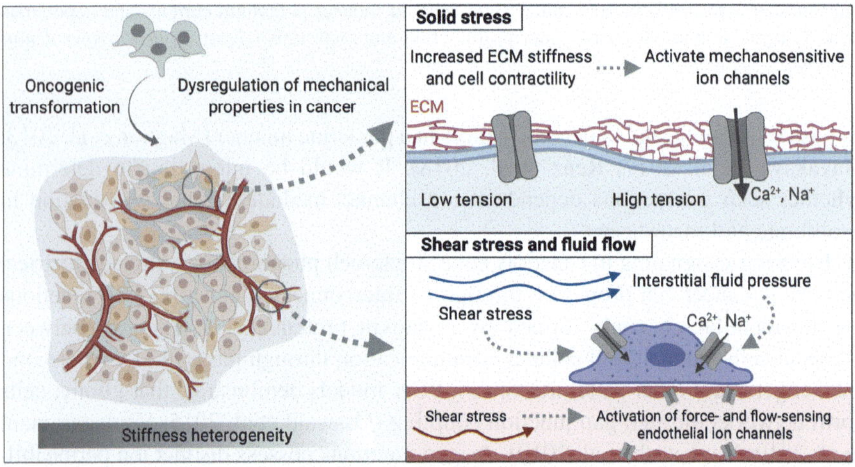

Fig. 6 Mechanosensitive ion channels sense tissue mechanics and regulate tumor malignancy. Aberrant tissue mechanics is a common feature of solid tumors. Tumor stiffness can be sensed by mechanosensitive ion channels. Distinct tumor regions are mechanically heterogeneous (*left*). In a region of high mechanical stress, mechanosensitive ion channels activate intracellular signaling to drive cell proliferation (*top right*). In regions with elevated interstitial fluid pressure or shear stress, flow-sensitive ion channels expressed in tumor cells or tumor vasculature may be activated to promote tumor growth and microenvironmental remodeling (*bottom right*). *Dotted arrows denote hypothetical functions*

4.3.1 Solid Stress

Solid stress arises from expansion of the tumor mass and non-fluid tumor components which compress and distend cells and tissues (Jain et al. 2014). Through mechanoreciprocity, solid stress may propagate to the surrounding stromal tissue to increase extracellular matrix (ECM) tension and induce changes to ECM material properties to drive tumor growth (Butcher et al. 2009). In glioma and pancreatic cancer, increased ECM stiffening and elevated epithelial tension are associated with increased malignancy and shorter patient survival. In both types of tumors, elevated tenascin C and STAT3-mediated mechanosignaling form positive feedback loops by further increasing ECM stiffness and tumor aggression (Laklai et al. 2016; Miroshnikova et al. 2016). Recurrent tumors display further stiffening compared to the primary tumors (Miroshnikova et al. 2016). Mechanical heterogeneity in distinct tumor regions may contribute to the growth of heterogeneous tumor cell populations. Mechanosensitive ion channels Piezo and PIEZO1 are evolutionarily conserved regulators of tumor growth in *Drosophila* and xenograft mouse models of glioma, respectively. Glioma cells perceive microenvironmental stiffness through PIEZO1. PIEZO1 signaling elevates integrin and FAK signaling to increase ECM production, tumor proliferation, and its own expression, forming a feedforward circuit to promote tumor malignancy (Chen et al. 2018).

4.3.2 Shear Stress and Fluid Flow

Dysregulated tumor mechanics impair blood vessel integrity and reduce lymphatic drainage, which lead to increased shear stress and perturbed fluid flow. Decreased circulation can hinder nutrient distribution, drug delivery, and therapeutic response (Tong et al. 2004; Winkler et al. 2004). Fluid mechanics may activate flow-sensitive ion channels, although this notion is yet to be fully elucidated in cancer. In adult mouse neural stem cells of the subependymal zone, flow-sensitive sodium channel ENaC senses ventricular fluid flow to promote proliferation and neurogenic output through Na^+ and Ca^{2+} signaling, activation of calcium-release activated channels, and ERK signaling (Petrik et al. 2018). ENaC knockdown reduces glioma cell migration in vitro (Kapoor et al. 2009), while its in vivo function in cancer remains to be explored. Tumor angiogenesis, a hallmark of cancer, is the ability for cancers to generate or recruit vasculature to provide oxygen and nutrients to the tumor (Hanahan and Weinberg 2000, 2011). Ion channels have been described in endothelial cells, including the shear stress-activated ENaC (Wang et al. 2009; Guo et al. 2016), and mechanosensitive ion channels such as TRPV4 and PIEZOs (Gerhold and Schwartz 2016). Whether ion channel function in tumor vasculature can be therapeutically exploited remains to be investigated.

In summary, ion channels are expressed in a tissue- or cancer-specific manner that can create actionable therapeutic windows (Table 1). The challenges of cancer plasticity and heterogeneity must be considered when investigating context-dependent ion channel functions. It is important to investigate ion channel function using in vivo model systems that recapitulate physiological features of cancer. Defining the particular stage of tumorigenesis that an ion channel regulates and elucidating its mechanism of action should lead to improved design of targeted therapies. Future studies of non-cell autonomous ion channel functions will uncover how electrical signals propagate through a network of cancer and stromal cells and how ion channels perceive mechanical cues to control tumor malignancy. As this field makes strides forward, we look forward to unlocking the potential to target ion channels in cancer.

Acknowledgment We thank Ali Momin, Hyun-Kee Min, Michelle Francisco, Weifan Dong, and Xin Chen for constructive suggestions on the manuscript. This work is supported by SickKids Foundation, Arthur and Sonia Labatt Brain Tumour Research Centre, Garron Family Cancer Centre, b.r.a.i.n.child, Meagan's Walk, Natural Sciences and Engineering Research Council (NSERC) Discovery Grant, Ontario Institute for Cancer Research (OICR) Translational Research Initiative, Canadian Institute of Health Research (CIHR) Project Grants, Concern Foundation Conquer Cancer Now Award, and Sontag Foundation Distinguished Scientist Award to Xi Huang. Jerry J. Fan is supported by SickKids Restracomp Scholarship. Xi Huang thanks the love and support from Lucas Huang and Liam Huang. Xi Huang is a Catalyst Scholar at The Hospital for Sick Children and Canada Research Chair in Cancer Biophysics.

Declaration of Interest The authors declare no competing financial interests.

References

Aasen T, Mesnil M, Naus CC et al (2016) Gap junctions and cancer: communicating for 50 years. Nat Rev Cancer 16:775–788. https://doi.org/10.1038/nrc.2016.105

Abdul M, Hoosein N (2006) Reduced Kv1.3 potassium channel expression in human prostate cancer. J Membr Biol 214:99–102. https://doi.org/10.1007/s00232-006-0065-7

Agarwal JR, Griesinger F, Stuhmer W, Pardo LA (2010) The potassium channel ether a go-go is a novel prognostic factor with functional relevance in acute myeloid leukemia. Mol Cancer 9:18. https://doi.org/10.1186/1476-4598-9-18

Alexander SPH, Kelly E, Mathie A et al (2019) The concise guide to pharmacology 2019/20: introduction and other protein targets. Br J Pharmacol 176:S1–S20. https://doi.org/10.1111/bph.14747

Allen E, Miéville P, Warren CM et al (2016) Metabolic symbiosis enables adaptive resistance to anti-angiogenic therapy that is dependent on mTOR signaling. Cell Rep 15:1144–1160. https://doi.org/10.1016/j.celrep.2016.04.029

Bargal R, Avidan N, Ben-Asher E et al (2000) Identification of the gene causing mucolipidosis type IV. Nat Genet 26:118–122. https://doi.org/10.1038/79095

Bates E (2015) Ion channels in development and cancer. Annu Rev Cell Dev Biol 31:231–247. https://doi.org/10.1146/annurev-cellbio-100814-125338

Bonnet S, Archer SL, Allalunis-Turner J et al (2007) A mitochondria-K+ channel Axis is suppressed in cancer and its normalization promotes apoptosis and inhibits cancer growth. Cancer Cell 11:37–51. https://doi.org/10.1016/j.ccr.2006.10.020

Brevet M, Haren N, Sevestre H et al (2009) DNA methylation of $K_v1.3$ potassium channel gene promoter is associated with poorly differentiated breast adenocarcinoma. Cell Physiol Biochem 24:25–32. https://doi.org/10.1159/000227810

Bulk E, Ay A-S, Hammadi M et al (2015) Epigenetic dysregulation of K $_{Ca}$ 3.1 channels induces poor prognosis in lung cancer: K $_{Ca}$ 3.1 channel and poor lung cancer prognosis. Int J Cancer 137:1306–1317. https://doi.org/10.1002/ijc.29490

Butcher DT, Alliston T, Weaver VM (2009) A tense situation: forcing tumour progression. Nat Rev Cancer 9:108–122. https://doi.org/10.1038/nrc2544

Cavalli FMG, Remke M, Rampasek L et al (2017) Intertumoral heterogeneity within medulloblastoma subgroups. Cancer Cell 31:737–754.e6. https://doi.org/10.1016/j.ccell.2017.05.005

Chantome A, Potier-Cartereau M, Clarysse L et al (2013) Pivotal role of the lipid raft SK3-Orai1 complex in human cancer cell migration and bone metastases. Cancer Res 73:4852–4861. https://doi.org/10.1158/0008-5472.CAN-12-4572

Chen C-D, Wang C-S, Huang Y-H et al (2007) Overexpression of CLIC1 in human gastric carcinoma and its clinicopathological significance. Proteomics 7:155–167. https://doi.org/10.1002/pmic.200600663

Chen X, Wanggou S, Bodalia A et al (2018) A feedforward mechanism mediated by mechanosensitive ion channel PIEZO1 and tissue mechanics promotes Glioma aggression. Neuron 100:799–815.e7. https://doi.org/10.1016/j.neuron.2018.09.046

Choi M, Scholl UI, Yue P et al (2011) K+ channel mutations in adrenal aldosterone-producing adenomas and hereditary hypertension. Science 331:768–772. https://doi.org/10.1126/science.1198785

Comes N, Bielanska J, Vallejo-Gracia A et al (2013) The voltage-dependent K+ channels Kv1.3 and Kv1.5 in human cancer. Front Physiol 4. https://doi.org/10.3389/fphys.2013.00283

Cone CD (1971) Unified theory on the basic mechanism of normal mitotic control and oncogenesis. J Theor Biol 30:151–181. https://doi.org/10.1016/0022-5193(71)90042-7

Cone C, Cone C (1976) Induction of mitosis in mature neurons in central nervous system by sustained depolarization. Science 192:155–158. https://doi.org/10.1126/science.56781

Crottès D, Lin Y-HT, Peters CJ et al (2019) TMEM16A controls EGF-induced calcium signaling implicated in pancreatic cancer prognosis. Proc Natl Acad Sci U S A 116:13026–13035. https://doi.org/10.1073/pnas.1900703116

Cuddapah VA, Sontheimer H (2011) Ion channels and transporters in cancer. 2. ion channels and the control of cancer cell migration. Am J Phys Cell Phys 301:C541–C549. https://doi.org/10.1152/ajpcell.00102.2011

de Jong PR, Takahashi N, Harris AR et al (2014) Ion channel TRPV1-dependent activation of PTP1B suppresses EGFR-associated intestinal tumorigenesis. J Clin Invest 124:3793–3806. https://doi.org/10.1172/JCI72340

den Uil SH, Coupé VMH, Linnekamp JF et al (2016) Loss of KCNQ1 expression in stage II and stage III colon cancer is a strong prognostic factor for disease recurrence. Br J Cancer 115:1565–1574. https://doi.org/10.1038/bjc.2016.376

Ding Q, Li M, Wu X et al (2015) CLIC1 overexpression is associated with poor prognosis in gallbladder cancer. Tumor Biol 36:193–198. https://doi.org/10.1007/s13277-014-2606-5

Djamgoz MBA, Onkal R (2012) Persistent current blockers of voltage-gated sodium channels: a clinical opportunity for controlling metastatic disease. PRA 8:66–84. https://doi.org/10.2174/157489281130801006

Eil R, Vodnala SK, Clever D et al (2016) Ionic immune suppression within the tumour microenvironment limits T cell effector function. Nature 537:539–543. https://doi.org/10.1038/nature19364

Francisco MA, Wanggou S, Fan JJ et al (2020) Chloride intracellular channel 1 cooperates with potassium channel EAG2 to promote medulloblastoma growth. J Exp Med 217:e20190971. https://doi.org/10.1084/jem.20190971

Fraser SP, Ozerlat-Gunduz I, Brackenbury WJ et al (2014) Regulation of voltage-gated sodium channel expression in cancer: hormones, growth factors and auto-regulation. Philos Trans R Soc B 369:20130105. https://doi.org/10.1098/rstb.2013.0105

Gerhold KA, Schwartz MA (2016) Ion channels in endothelial responses to fluid shear stress. Physiology 31:359–369. https://doi.org/10.1152/physiol.00007.2016

Girault A, Haelters J-P, Potier-Cartereau M et al (2011) New alkyl-lipid blockers of SK3 channels reduce Cancer cell migration and occurrence of metastasis. CCDT 11:1111–1125. https://doi.org/10.2174/156800911798073069

Guo D, Liang S, Wang S et al (2016) Role of epithelial Na^+ channels in endothelial function. J Cell Sci 129:290–297. https://doi.org/10.1242/jcs.168831

Hall DP, Cost NG, Hegde S et al (2014) TRPM3 and miR-204 establish a regulatory circuit that controls oncogenic autophagy in clear cell renal cell carcinoma. Cancer Cell 26:738–753. https://doi.org/10.1016/j.ccell.2014.09.015

Hanahan D, Weinberg RA (2000) The hallmarks of cancer. Cell 100:57–70. https://doi.org/10.1016/S0092-8674(00)81683-9

Hanahan D, Weinberg RA (2011) Hallmarks of cancer: the next generation. Cell 144:646–674. https://doi.org/10.1016/j.cell.2011.02.013

Hartung F, Stühmer W, Pardo LA (2011) Tumor cell-selective apoptosis induction through targeting of KV10.1 via bifunctional TRAIL antibody. Mol Cancer 10:109. https://doi.org/10.1186/1476-4598-10-109

Hashimoto T, Ogawa R, Yoshida H et al (2019) EIF3E-RSPO2 and PIEZO1-RSPO2 fusions in colorectal traditional serrated adenoma. Histopathol His:13867. https://doi.org/10.1111/his.13867

Hayakawa Y, Sakitani K, Konishi M et al (2017) Nerve growth factor promotes gastric tumorigenesis through aberrant cholinergic signaling. Cancer Cell 31:21–34. https://doi.org/10.1016/j.ccell.2016.11.005

Hemmerlein B, Weseloh RM, Mello de Queiroz F et al (2006) Overexpression of Eag1 potassium channels in clinical tumours. Mol Cancer 5:41. https://doi.org/10.1186/1476-4598-5-41

Hille B (2001) Ion channels of excitable membranes, 3rd edn. Sinauer Associates is an imprint of Oxford University Press, Sunderland

Hitomi M, Deleyrolle LP, Mulkearns-Hubert EE et al (2015) Differential connexin function enhances self-renewal in glioblastoma. Cell Rep 11:1031–1042. https://doi.org/10.1016/j.celrep.2015.04.021

Huang X, Jan LY (2014) Targeting potassium channels in cancer. J Cell Biol 206:151–162. https://doi.org/10.1083/jcb.201404136

Huang X, Dubuc AM, Hashizume R et al (2012) Voltage-gated potassium channel EAG2 controls mitotic entry and tumor growth in medulloblastoma via regulating cell volume dynamics. Genes Dev 26:1780–1796. https://doi.org/10.1101/gad.193789.112

Huang X, He Y, Dubuc AM et al (2015) EAG2 potassium channel with evolutionarily conserved function as a brain tumor target. Nat Neurosci 18:1236–1246. https://doi.org/10.1038/nn.4088

Innamaa A, Jackson L, Asher V et al (2013) Expression and prognostic significance of the oncogenic K2P potassium channel KCNK9 (TASK-3) in ovarian carcinoma. Anticancer Res 8

Jain RK, Martin JD, Stylianopoulos T (2014) The role of mechanical forces in tumor growth and therapy. Annu Rev Biomed Eng 16:321–346. https://doi.org/10.1146/annurev-bioeng-071813-105259

Jang SH, Choi SY, Ryu PD, Lee SY (2011) Anti-proliferative effect of Kv1.3 blockers in A549 human lung adenocarcinoma in vitro and in vivo. Eur J Pharmacol 651:26–32. https://doi.org/10.1016/j.ejphar.2010.10.066

Jia N, Dong S, Zhao G et al (2016) CLIC1 overexpression is associated with poor prognosis in pancreatic ductal adenocarcinomas. J Cancer Res Ther 12:892–896. https://doi.org/10.4103/0973-1482.154057

Jin X, Kim LJY, Wu Q et al (2017) Targeting glioma stem cells through combined BMI1 and EZH2 inhibition. Nat Med 23:1352–1361. https://doi.org/10.1038/nm.4415

Joshi AD, Parsons D, Velculescu VE, Riggins GJ (2011) Sodium ion channel mutations in glioblastoma patients correlate with shorter survival. Mol Cancer 10:17. https://doi.org/10.1186/1476-4598-10-17

Jung J, Cho K, Naji AK et al (2019) HRAS-driven cancer cells are vulnerable to TRPML1 inhibition. EMBO Rep 20. https://doi.org/10.15252/embr.201846685

Kapoor N, Bartoszewski R, Qadri YJ et al (2009) Knockdown of ASIC1 and epithelial sodium channel subunits inhibits glioblastoma whole cell current and cell migration. J Biol Chem 284:24526–24541. https://doi.org/10.1074/jbc.M109.037390

Kasitinon SY, Eskiocak U, Martin M et al (2019) TRPML1 promotes protein homeostasis in melanoma cells by negatively regulating MAPK and mTORC1 signaling. Cell Rep 28:2293–2305.e9. https://doi.org/10.1016/j.celrep.2019.07.086

Kim CJ, Cho YG, Jeong SW et al (2004) Altered expression of KCNK9 in colorectal cancers. APMIS 112:588–594. https://doi.org/10.1111/j.1600-0463.2004.apm1120905.x

Klose C, Straub I, Riehle M et al (2011) Fenamates as TRP channel blockers: mefenamic acid selectively blocks TRPM3: mefenamic acid selectively blocks TRPM3. Br J Pharmacol 162:1757–1769. https://doi.org/10.1111/j.1476-5381.2010.01186.x

Kumar S, Weaver VM (2009) Mechanics, malignancy, and metastasis: the force journey of a tumor cell. Cancer Metastasis Rev 28:113–127. https://doi.org/10.1007/s10555-008-9173-4

Laklai H, Miroshnikova YA, Pickup MW et al (2016) Genotype tunes pancreatic ductal adenocarcinoma tissue tension to induce matricellular fibrosis and tumor progression. Nat Med 22:497–505. https://doi.org/10.1038/nm.4082

Lang F, Föller M, Lang KS et al (2005) Ion channels in cell proliferation and apoptotic cell death. J Membr Biol 205:147–157. https://doi.org/10.1007/s00232-005-0780-5

Lastraioli E, Guasti L, Crociani O et al (2004) herg1 gene and HERG1 protein are overexpressed in colorectal cancers and regulate cell invasion of tumor cells. Cancer Res 64:606–611. https://doi.org/10.1158/0008-5472.CAN-03-2360

Leanza L, Henry B, Sassi N et al (2012) Inhibitors of mitochondrial Kv1.3 channels induce Bax/Bak-independent death of cancer cells. EMBO Mol Med 4:577–593. https://doi.org/10.1002/emmm.201200235

Leanza L, Trentin L, Becker KA et al (2013) Clofazimine, Psora-4 and PAP-1, inhibitors of the potassium channel Kv1.3, as a new and selective therapeutic strategy in chronic lymphocytic leukemia. Leukemia 27:1782–1785. https://doi.org/10.1038/leu.2013.56

Leanza L, Romio M, Becker KA et al (2017) Direct pharmacological targeting of a mitochondrial ion channel selectively kills tumor cells in vivo. Cancer Cell 31:516–531.e10. https://doi.org/10.1016/j.ccell.2017.03.003

Li C, Rezania S, Kammerer S et al (2015) Piezo1 forms mechanosensitive ion channels in the human MCF-7 breast cancer cell line. Sci Rep 5:8364. https://doi.org/10.1038/srep08364

Li B, Mao Y, Wang Z et al (2018) CLIC1 promotes the progression of gastric cancer by regulating the MAPK/AKT pathways. Cell Physiol Biochem 46:907–924. https://doi.org/10.1159/000488822

Lu J, Dong Q, Zhang B et al (2015) Chloride intracellular channel 1 (CLIC1) is activated and functions as an oncogene in pancreatic cancer. Med Oncol 32:171. https://doi.org/10.1007/s12032-015-0616-9

Magnon C, Hall SJ, Lin J et al (2013) Autonomic nerve development contributes to prostate cancer progression. Science 341:1236361–1236361. https://doi.org/10.1126/science.1236361

Maisonneuve P, Marshall BC, Knapp EA, Lowenfels AB (2013) Cancer risk in cystic fibrosis: a 20 year nationwide study from the United States. JNCI J Natl Cancer Inst 105:122–129. https://doi.org/10.1093/jnci/djs481

Mello de Queiroz F, Suarez-Kurtz G, Stühmer W, Pardo LA (2006) Ether à go-go potassium channel expression in soft tissue sarcoma patients. Mol Cancer 5:42. https://doi.org/10.1186/1476-4598-5-42

Mikhaylova O, Stratton Y, Hall D et al (2012) VHL-regulated MiR-204 suppresses tumor growth through inhibition of LC3B-mediated autophagy in renal clear cell carcinoma. Cancer Cell 21:532–546. https://doi.org/10.1016/j.ccr.2012.02.019

Miroshnikova YA, Mouw JK, Barnes JM et al (2016) Tissue mechanics promote IDH1-dependent HIF1α–tenascin C feedback to regulate glioblastoma aggression. Nat Cell Biol 18:1336–1345. https://doi.org/10.1038/ncb3429

Monje M, Borniger JC, D'Silva NJ et al (2020) Roadmap for the emerging field of cancer neuroscience. Cell 181:219–222. https://doi.org/10.1016/j.cell.2020.03.034

Monteith GR, Prevarskaya N, Roberts-Thomson SJ (2017) The calcium–cancer signalling nexus. Nat Rev Cancer 17:373–380. https://doi.org/10.1038/nrc.2017.18

Morelli MB, Nabissi M, Amantini C et al (2016) Overexpression of transient receptor potential mucolipin-2 ion channels in gliomas: role in tumor growth and progression. Oncotarget 7. https://doi.org/10.18632/oncotarget.9661

Mu D, Chen L, Zhang X et al (2003) Genomic amplification and oncogenic properties of the KCNK9 potassium channel gene. Cancer Cell 3:297–302. https://doi.org/10.1016/S1535-6108(03)00054-0

Nabors LB, Portnow J, Ammirati M et al (2017) NCCN guidelines insights: central nervous system cancers, version 1.2017. J Natl Compr Cancer Netw 15:1331–1345. https://doi.org/10.6004/jnccn.2017.0166

Napp J, Pardo LA, Hartung F et al (2016) In vivo imaging of tumour xenografts with an antibody targeting the potassium channel Kv10.1. Eur Biophys J 45:721–733. https://doi.org/10.1007/s00249-016-1152-z

Neglia JP, FitzSimmons SC, Maisonneuve P et al (1995) The risk of cancer among patients with cystic fibrosis. N Engl J Med 332:494–499. https://doi.org/10.1056/NEJM199502233320803

Northcott PA, Korshunov A, Witt H et al (2011) Medulloblastoma comprises four distinct molecular variants. JCO 29:1408–1414. https://doi.org/10.1200/JCO.2009.27.4324

Northcott JM, Dean IS, Mouw JK, Weaver VM (2018) Feeling stress: the mechanics of cancer progression and aggression. Front Cell Dev Biol 6:17. https://doi.org/10.3389/fcell.2018.00017

Northey JJ, Przybyla L, Weaver VM (2017) Tissue force programs cell fate and tumor aggression. Cancer Discov 7:1224–1237. https://doi.org/10.1158/2159-8290.CD-16-0733

Osswald M, Jung E, Sahm F et al (2015) Brain tumour cells interconnect to a functional and resistant network. Nature 528:93–98. https://doi.org/10.1038/nature16071

Ouadid-Ahidouch H, Dhennin-Duthille I, Gautier M et al (2013) TRP channels: diagnostic markers and therapeutic targets for breast cancer? Trends Mol Med 19:117–124. https://doi.org/10.1016/j.molmed.2012.11.004

Overington JP, Al-Lazikani B, Hopkins AL (2006) How many drug targets are there? Nat Rev Drug Discov 5:993–996. https://doi.org/10.1038/nrd2199

Pardo LA, Stühmer W (2014) The roles of K+ channels in cancer. Nat Rev Cancer 14:39–48. https://doi.org/10.1038/nrc3635

Pardo LA, del Camino D, Sánchez A et al (1999) Oncogenic potential of EAG K(+) channels. EMBO J 18:5540–5547. https://doi.org/10.1093/emboj/18.20.5540

Payne SL, Levin M, Oudin MJ (2019) Bioelectric control of metastasis in solid tumors. Bioelectricity 1:114–130. https://doi.org/10.1089/bioe.2019.0013

Peretti M, Angelini M, Savalli N et al (2015) Chloride channels in cancer: focus on chloride intracellular channel 1 and 4 (CLIC1 AND CLIC4) proteins in tumor development and as novel therapeutic targets. Biochim Biophys Acta Biomembr 1848:2523–2531. https://doi.org/10.1016/j.bbamem.2014.12.012

Petrik D, Myoga MH, Grade S et al (2018) Epithelial sodium channel regulates adult neural stem cell proliferation in a flow-dependent manner. Cell Stem Cell 22:865–878.e8. https://doi.org/10.1016/j.stem.2018.04.016

Piggott BJ, Peters CJ, He Y et al (2019) Paralytic, the *Drosophila* voltage-gated sodium channel, regulates proliferation of neural progenitors. Genes Dev 33:1739–1750. https://doi.org/10.1101/gad.330597.119

Pollak J, Rai KG, Funk CC et al (2017) Ion channel expression patterns in glioblastoma stem cells with functional and therapeutic implications for malignancy. PLoS One 12:e0172884. https://doi.org/10.1371/journal.pone.0172884

Preußat K, Beetz C, Schrey M et al (2003) Expression of voltage-gated potassium channels Kv1.3 and Kv1.5 in human gliomas. Neurosci Lett 346:33–36. https://doi.org/10.1016/S0304-3940(03)00562-7

Prevarskaya N, Skryma R, Shuba Y (2011) Calcium in tumour metastasis: new roles for known actors. Nat Rev Cancer 11:609–618. https://doi.org/10.1038/nrc3105

Prevarskaya N, Skryma R, Shuba Y (2018) Ion channels in cancer: are cancer hallmarks oncochannelopathies? Physiol Rev 98:559–621. https://doi.org/10.1152/physrev.00044.2016

Rabjerg M, Oliván-Viguera A, Hansen LK et al (2015) High expression of KCa3.1 in patients with clear cell renal carcinoma predicts high metastatic risk and poor survival. PLoS One 10: e0122992. https://doi.org/10.1371/journal.pone.0122992

Remke M, Hielscher T, Northcott PA et al (2011) Adult medulloblastoma comprises three major molecular variants. JCO 29:2717–2723. https://doi.org/10.1200/JCO.2011.34.9373

Renz BW, Tanaka T, Sunagawa M et al (2018) Cholinergic signaling via muscarinic receptors directly and indirectly suppresses pancreatic tumorigenesis and cancer stemness. Cancer Discov 8:1458–1473. https://doi.org/10.1158/2159-8290.CD-18-0046

Roger S, Gillet L, Le Guennec J-Y, Besson P (2015) Voltage-gated sodium channels and cancer: is excitability their primary role? Front Pharmacol 6. https://doi.org/10.3389/fphar.2015.00152

Santoni G, Farfariello V (2011) TRP channels and cancer: new targets for diagnosis and chemotherapy. EMIDDT 11:54–67. https://doi.org/10.2174/187153011794982068

Sato Y, Yoshizato T, Shiraishi Y et al (2013) Integrated molecular analysis of clear-cell renal cell carcinoma. Nat Genet 45:860–867. https://doi.org/10.1038/ng.2699

Schmitz A, Sankaranarayanan A, Azam P et al (2005) Design of PAP-1, a selective small molecule Kv1.3 blocker, for the suppression of effector memory T cells in autoimmune diseases. Mol Pharmacol 68:1254–1270. https://doi.org/10.1124/mol.105.015669

Schneider SW, Pagel P, Rotsch C et al (2000) Volume dynamics in migrating epithelial cells measured with atomic force microscopy. Pflugers Arch - Eur J Physiol 439:297–303. https://doi.org/10.1007/s004249900176

Scholl UI, Goh G, Stölting G et al (2013) Somatic and germline CACNA1D calcium channel mutations in aldosterone-producing adenomas and primary aldosteronism. Nat Genet 45:1050–1054. https://doi.org/10.1038/ng.2695

Scholl UI, Abriola L, Zhang C et al (2017) Macrolides selectively inhibit mutant KCNJ5 potassium channels that cause aldosterone-producing adenoma. J Clin Investig 127:2739–2750. https://doi.org/10.1172/JCI91733

Schwab A, Gabriel K, Finsterwalder F et al (1995) Polarized ion transport during migration of transformed Madin-Darby canine kidney cells. Pflugers Arch 430:802–807. https://doi.org/10.1007/bf00386179

Schwab A, Fabian A, Hanley PJ, Stock C (2012) Role of ion channels and transporters in cell migration. Physiol Rev 92:1865–1913. https://doi.org/10.1152/physrev.00018.2011

Seshagiri S, Stawiski EW, Durinck S et al (2012) Recurrent R-spondin fusions in colon cancer. Nature 488:660–664. https://doi.org/10.1038/nature11282

Setti M, Savalli N, Osti D et al (2013) Functional role of CLIC1 Ion channel in glioblastoma-derived stem/progenitor cells. JNCI J Natl Cancer Inst 105:1644–1655. https://doi.org/10.1093/jnci/djt278

Sharma J, Deb B, George IA et al (2019) Somatic mutations profile of a young patient with metastatic Urothelial carcinoma reveals mutations in genes involved in ion channels. Front Oncol 9:435. https://doi.org/10.3389/fonc.2019.00435

Singh A, Hildebrand M, Garcia E, Snutch T (2010) The transient receptor potential channel antagonist SKF96365 is a potent blocker of low-voltage-activated T-type calcium channels: SKF96365 block of T-type calcium channels. Br J Pharmacol 160:1464–1475. https://doi.org/10.1111/j.1476-5381.2010.00786.x

Sinyuk M, Mulkearns-Hubert EE, Reizes O, Lathia J (2018) Cancer connectors: connexins, gap junctions, and communication. Front Oncol 8. https://doi.org/10.3389/fonc.2018.00646

Starr TK, Allaei R, Silverstein KAT et al (2009) A transposon-based genetic screen in mice identifies genes altered in colorectal cancer. Science 323:1747–1750. https://doi.org/10.1126/science.1163040

Stock K, Kumar J, Synowitz M et al (2012) Neural precursor cells induce cell death of high-grade astrocytomas through stimulation of TRPV1. Nat Med 18:1232–1238. https://doi.org/10.1038/nm.2827
Sun M, Goldin E, Stahl S et al (2000) Mucolipidosis type IV is caused by mutations in a gene encoding a novel transient receptor potential channel. Hum Mol Genet 9:2471–2478. https://doi.org/10.1093/hmg/9.17.2471
Szabo I, Bock J, Grassme H et al (2008) Mitochondrial potassium channel Kv1.3 mediates Bax-induced apoptosis in lymphocytes. Proc Natl Acad Sci 105:14861–14866. https://doi.org/10.1073/pnas.0804236105
Takahashi N, Chen H-Y, Harris IS et al (2018) Cancer cells co-opt the neuronal redox-Sensing Channel TRPA1 to promote oxidative-stress tolerance. Cancer Cell 33:985–1003.e7. https://doi.org/10.1016/j.ccell.2018.05.001
Taylor MD, Northcott PA, Korshunov A et al (2012) Molecular subgroups of medulloblastoma: the current consensus. Acta Neuropathol 123:465–472. https://doi.org/10.1007/s00401-011-0922-z
Than BLN, Goos JACM, Sarver AL et al (2014) The role of KCNQ1 in mouse and human gastrointestinal cancers. Oncogene 33:3861–3868. https://doi.org/10.1038/onc.2013.350
Than BLN, Linnekamp JF, Starr TK et al (2016) CFTR is a tumor suppressor gene in murine and human intestinal cancer. Oncogene 35:4191–4199. https://doi.org/10.1038/onc.2015.483
The Cancer Genome Atlas Network (2012a) Comprehensive molecular characterization of human colon and rectal cancer. Nature 487:330–337. https://doi.org/10.1038/nature11252
The Cancer Genome Atlas Network (2012b) Comprehensive molecular portraits of human breast tumours. Nature 490:61–70. https://doi.org/10.1038/nature11412
The Cancer Genome Atlas Research Network (2008) Comprehensive genomic characterization defines human glioblastoma genes and core pathways. Nature 455:1061–1068. https://doi.org/10.1038/nature07385
The Cancer Genome Atlas Research Network (2011) Integrated genomic analyses of ovarian carcinoma. Nature 474:609–615. https://doi.org/10.1038/nature10166
The Cancer Genome Atlas Research Network (2012) Comprehensive genomic characterization of squamous cell lung cancers. Nature 489:519–525. https://doi.org/10.1038/nature11404
The Cancer Genome Atlas Research Network (2013) Comprehensive molecular characterization of clear cell renal cell carcinoma. Nature 499:43–49. https://doi.org/10.1038/nature12222
Tong RT, Boucher Y, Kozin SV et al (2004) Vascular normalization by vascular endothelial growth factor receptor 2 blockade induces a pressure gradient across the vasculature and improves drug penetration in tumors. Cancer Res 64:3731–3736. https://doi.org/10.1158/0008-5472.CAN-04-0074
Tsunemi T, Perez-Rosello T, Ishiguro Y et al (2019) Increased Lysosomal exocytosis induced by Lysosomal Ca $^{2+}$ channel agonists protects human dopaminergic neurons from α-synuclein toxicity. J Neurosci 39:5760–5772. https://doi.org/10.1523/JNEUROSCI.3085-18.2019
Turner KL, Honasoge A, Robert SM et al (2014) A proinvasive role for the Ca $^{2+}$-activated K $^{+}$ channel KCa3.1 in malignant glioma. Glia 62:971–981. https://doi.org/10.1002/glia.22655
Tuszynski J, Tilli TM, Levin M (2017) Ion channel and neurotransmitter modulators as electroceutical approaches to the control of cancer. Curr Pharm Des 23:4827–4841. https://doi.org/10.2174/1381612823666170530105837
Venkataramani V, Tanev DI, Strahle C et al (2019) Glutamatergic synaptic input to glioma cells drives brain tumour progression. Nature 573:532–538. https://doi.org/10.1038/s41586-019-1564-x
Venkatesh HS, Morishita W, Geraghty AC et al (2019) Electrical and synaptic integration of glioma into neural circuits. Nature 573:539–545. https://doi.org/10.1038/s41586-019-1563-y
Vitali I, Fièvre S, Telley L et al (2018) Progenitor hyperpolarization regulates the sequential generation of neuronal subtypes in the developing Neocortex. Cell 174:1264–1276.e15. https://doi.org/10.1016/j.cell.2018.06.036
Vodnala SK, Eil R, Kishton RJ et al (2019) T cell stemness and dysfunction in tumors are triggered by a common mechanism. Science 363:eaau0135. https://doi.org/10.1126/science.aau0135

Wang S, Meng F, Mohan S et al (2009) Functional ENaC channels expressed in endothelial cells: a new candidate for mediating shear force. Microcirculation 16:276–287. https://doi.org/10.1080/10739680802653150
Wang W, Xu X, Wang W et al (2011) The expression and clinical significance of CLIC1 and HSP27 in lung adenocarcinoma. Tumor Biol 32:1199–1208. https://doi.org/10.1007/s13277-011-0223-0
Wang R, Gurguis CI, Gu W et al (2015) Ion channel gene expression predicts survival in glioma patients. Sci Rep 5:11593. https://doi.org/10.1038/srep11593
Wang H-Y, Wang W, Liu Y-W et al (2017a) Role of KCNB1 in the prognosis of gliomas and autophagy modulation. Sci Rep 7:14. https://doi.org/10.1038/s41598-017-00045-7
Wang L, Hao J, Zhang Y et al (2017b) Orai1 mediates tumor-promoting store-operated Ca2+ entry in human gastrointestinal stromal tumors via c-KIT and the extracellular signal–regulated kinase pathway. Tumour Biol 39:1010428317691426. https://doi.org/10.1177/1010428317691426
Wang X, Prager BC, Wu Q et al (2018) Reciprocal signaling between Glioblastoma stem cells and differentiated tumor cells promotes malignant progression. Cell Stem Cell 22:514–528.e5. https://doi.org/10.1016/j.stem.2018.03.011
Wang Z, El Zowalaty AE, Li Y et al (2019) Association of luteal cell degeneration and progesterone deficiency with lysosomal storage disorder mucolipidosis type IV in Mcoln1−/− mouse model†. Biol Reprod 101:782–790. https://doi.org/10.1093/biolre/ioz126
Winkler F, Kozin SV, Tong RT et al (2004) Kinetics of vascular normalization by VEGFR2 blockade governs brain tumor response to radiation: role of oxygenation, angiopoietin-1, and matrix metalloproteinases. Cancer Cell 6:553–563. https://doi.org/10.1016/j.ccr.2004.10.011
Yang M, Brackenbury WJ (2013) Membrane potential and cancer progression. Front Physiol 4. https://doi.org/10.3389/fphys.2013.00185
Yang S, Zhang JJ, Huang X-Y (2009) Orai1 and STIM1 are critical for breast tumor cell migration and metastasis. Cancer Cell 15:124–134. https://doi.org/10.1016/j.ccr.2008.12.019
Yang H, Zhang Q, He J, Lu W (2010) Regulation of calcium signaling in lung cancer. J Thorac Dis 2:52–56
Zeng Q, Michael IP, Zhang P et al (2019) Synaptic proximity enables NMDAR signalling to promote brain metastasis. Nature 573:526–531. https://doi.org/10.1038/s41586-019-1576-6
Zhang J, Zhou Y, Huang T et al (2018) PIEZO1 functions as a potential oncogene by promoting cell proliferation and migration in gastric carcinogenesis. Mol Carcinog 57:1144–1155. https://doi.org/10.1002/mc.22831

Rev Physiol Biochem Pharmacol (2022) 183: 135–156
https://doi.org/10.1007/112_2021_62

Published online: 23 July 2021

Potassium and Chloride Ion Channels in Cancer: A Novel Paradigm for Cancer Therapeutics

Umberto Banderali, Luigi Leanza, Najmeh Eskandari, and Saverio Gentile

Contents

Abstract Cancer is a collection of diseases caused by specific changes at the genomic level that support cell proliferation indefinitely. Traditionally, ion channels are known to control a variety of cellular processes including electrical signal generation and transmission, secretion, and contraction by controlling ionic gradients. However, recent studies had brought to light important facts on ion channels in cancer biology.

In this review we discuss the mechanism linking potassium or chloride ion channel activity to biochemical pathways controlling proliferation in cancer cells and the potential advantages of targeting ion channels as an anticancer therapeutic option.

U. Banderali
Human Health Therapeutics Research Centre, National Research Council of Canada, Ottawa, ON, Canada

L. Leanza
Department of Biology, University of Padova, Padova, Italy

N. Eskandari and S. Gentile (✉)
Department of Hematology and Oncology, University of Illinois Chicago, Chicago, IL, USA
e-mail: sgentile@uic.edu

Keywords Cancer biology · Chloride channel · Potassium channel · Proliferative signaling · Therapeutic approach

Abbreviations

AVD	Apoptotic volume decrease
ER/PR	Estrogen/progesterone receptors
GPER	G protein estrogen receptor
HER2	Human epidermal growth factor receptor 2
K2P	Two-pore-domain potassium channel
Kir	Inward-rectifier potassium channel
Kv	Voltage-gated potassium channel
LRRC	Leucine-rich repeat-containing
Vm	Voltage membrane
VRAC	Volume regulated anion channels

1 Introduction

All living cells under resting conditions present an intracellular negative electrical potential called the transmembrane potential or resting membrane potential (Vm). Vm arises from the concerted activities of a variety of ion channels and transporters (Wright 2004) that maintain unequal distribution of ions (Na^+, K^+, Ca^{2+}, Cl^-) across the membranes of all cells. Changes of ionic gradients as a function of time are determined by the specific assortment of ion channels that are expressed in a specific cell. In turn, fluctuations of the Vm can activate/inactivate a variety of ion channels. The nominal cell Vm value is a fundamental factor for the appropriate life activities and the ionic pumps and channels are designed to work around this potential (Levin 2012, 2014). Cells that are highly differentiated and traditionally known as "excitable cells" such as neurons, cardiac myocytes, or pancreatic *beta*-cells are designed to operate around very negative Vm (−70 mV). In these cells a series of highly selective ion channels generates action potentials by controlling rapid changes of ionic fluxes that cross the surface membrane according to their electrochemical gradients. The shape and frequency of action potentials can finely regulate electrical, chemical, and/or mechanical signal transmission in a time scale ranging from a fraction of milliseconds to several seconds and underlie brain function, heart contraction, and/or secretion. Nevertheless, a dividing cell is designed to function throughout the division process around more depolarized Vm and ion channels are critical factors in controlling important pathways underlying the proliferation processes (Ng et al. 2008). These events can develop over a more extended time scale when compared with those in excitable tissues and include the intricate and well-synchronized process of cell proliferation (O'Grady and Lee 2005; Pardo 2004;

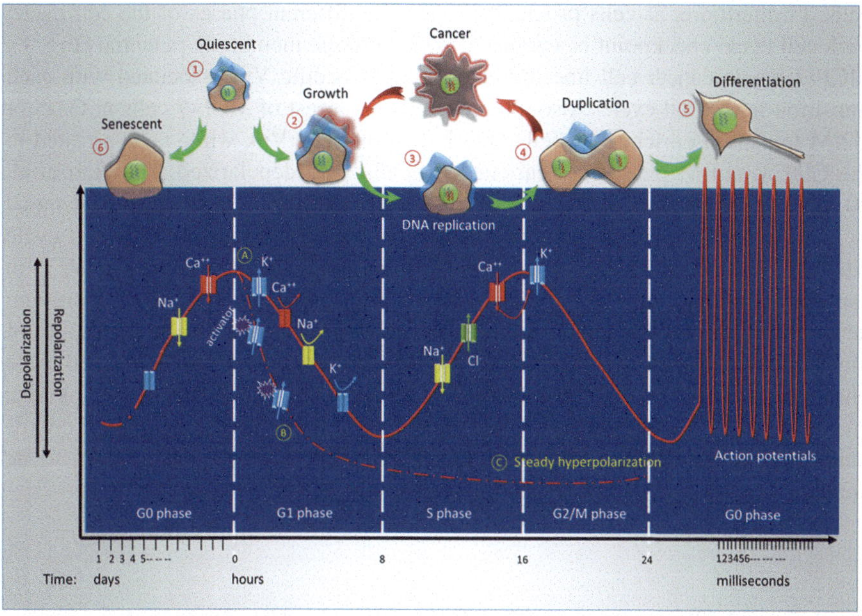

Fig. 1 Schematic representation of the relationship between changes in membrane potential and cell cycle. After exposure to specific conditions including hormones or growth factors, quiescent cells (1) progress through the different phases of the cell cycle to grow (2; G1-phase), duplicate the DNA (3; S-phase), divide (4; G2-Phase) and eventually differentiate into specific cell type (5; G0/G1-phase) or becoming senescent (6; G0-phase). Concerted activity of ion channels determines changes variations of membrane potential (a) which is necessary to stimulate the passage through the different phases of the cell cycle. Eventually in cells that stop duplicating and differentiate, selected groups of ion channels will underlie the specialized function (i.e., generation of action potentials). Transient alteration of channel density/activity in a specific cell cycle phase can produce a faster depolarization/repolarization (example in (b)) that would promote proliferative cells to pass from that cell cycle phase to the next. This can result in an increased proliferation rate and tumor growth. In contrast, chronic alteration of a channel activity by using activator molecules produces a steady hyperpolarization which inhibits the passage through cell cycle phases

Wang et al. 2000; Gray et al. 2004; Munaron et al. 2004; Kapur et al. 2007; DeCoursey et al. 1984) in both physiological and pathological status such as cancer.

Cancer develops when the process of cell replacement becomes dysregulated, damaged cells escape cell death and proliferate indefinitely. Therefore, cancer is traditionally considered a disease of proliferation.

The first evidence that ion channels play a fundamental role in cell proliferation came from the analysis of basic electrogenic differences between differentiated and highly proliferative cells like cancer cells. Terminally differentiated cells such as neurons and myocytes present a highly hyperpolarized resting membrane potential. In contrast, cells showing high mitotic activity such as those undergoing malignant transformation present a more depolarized membrane potential when compared to normal cells from the same tissue (Cone 1971). These observations suggest that proliferative cells tend to regulate ionic gradients differently from differentiated cells. Furthermore, as cells progress through the different phases of the cell cycle, each cell cycle checkpoint is associated with a unique membrane potential (Fig. 1). MCF-7 breast cancer cell line presents a phase-specific Vm associated with each transition in the cell cycle. Also, pharmacological arrest of MCF-7 cells in G1/S or G2/M transition enriches for cells with hyperpolarized Vm, while cells arrested in the G0/G1 and M phases are enriched for cells with depolarized Vm (Rao et al. 2015). Furthermore, several studies have shown that depolarization is a potent signal to initiate DNA synthesis. This event can cause ectopic reentry into the cell cycle which is pivotal for the malignant proliferation of undifferentiated cells (such as in cancer) as well as for differentiated cells such as neurons undergoing neurodegeneration (Lobikin et al. 2012; Cone 1970). This suggests that changes of ion channel activities can determine the passage of the cell from a specific cell cycle phase to another (Cone 1969, 1970, 1971; Binggeli and Weinstein 1986; Tokuoka and Morioka 1957; Johnstone 1959; Marino et al. 1994). Nevertheless, the contribution of a specific ion channel to the process of carcinogenesis appears to be contextual to a tissue or a specific cancer type.

2 K^+ and Cl^- Ions as a Key Factor in the Bioelectrical Signaling in Cancer Cell Proliferation

Generally, it is clear that the coordinated activity of ion channels (Wright 2004; Rao et al. 2015; Accardi 2015) produces a spatial and temporal variation of intracellular signaling that can contribute to cell proliferation (Ozkucur et al. 2011; Wei et al. 2008; Schwab et al. 2006; Goswami and Hucho 2007; Djamgoz et al. 2001; Morokuma et al. 2008).

K^+ ion is the major intracellular cation reaching a concentration up to 150 mM versus the 4 mM in the extracellular space. Loss of K^+ ion through opening of K^+ channels is the major contributor for the falling of membrane potential to negative values; a process known as hyperpolarization. To control this process, at least 80 genes encode for different basic transmembrane alpha subunits of K^+ channels which forms the pore that transports K^+ ions making K^+ channel the largest class of ion channels. Based on their structure and/or function, K^+ channels can be grouped into three major classes: the voltage-gated (Kv), the inwardly-rectifying (Kir), and the two-pore domain (K2P) channels which can assemble as dimers or tetramers. These proteins can respond to a variety of stimuli including changes in Vm, second messengers (e.g., ATP; Ca^{2+}), post-translational modification and by associating with additional soluble or membrane-bound cofactors to determine the fine control of K^+ ion fluxes across surface cell membranes as well as on the membranes of intracellular organelles.

Several studies have underlined the concept that high K^+ channels activity is characterized as a potent limiting factor for cancer cell proliferation. For example, ectopic expression of a constitutively open Kir channel in metastatic cancer can produce hyperpolarization which has been associated with inhibition of oncogene-induced carcinogenesis (Lobikin et al. 2012; Chernet and Levin 2013). Also, upregulation of $K_{Ca}1.1$ expression is associated with the arrest of breast cancer cells proliferation (Lallet-Daher et al. 2013). In contrast, low expression of the Kv7.1 channel has been associated with carcinogenesis of colorectal cancer (Than et al. 2014) and expression of Kv1.5 inversely correlates with tumor aggressiveness in lymphoma (Vallejo-Gracia et al. 2013).

However, it also has been found that increased expression and activity of K^+ channels can promote cancer. For example, exogenous expression of each of Kv1.3, Kv10.1, Kv11.1, or $K_{2p}9.1$ channels produces carcinogenicity in healthy cells (Mu et al. 2003; Pardo et al. 1999; Serrano-Novillo et al. 2019; Bianchi et al. 1998; Fukushiro-Lopes et al. 2018).

Also, expression levels of several K^+ channels such as Kv1.3, Kv10.1, Kv11.1, $K_{2p}9.1$, Kv1.4 were found to be augmented in several cancers, including breast, colon, and glioma, when compared with normal tissues from which the cancer has been generated (Pardo et al. 1999). In myeloid leukemia, expression of Kv10.1 correlates with higher relapse rates and a significantly shorter overall survival (Agarwal et al. 2010).

Chloride is the most abundant anion in the intra- and extra-cellular fluids and the mechanisms of chloride transport across the membrane participate in many important cellular functions such as: fine-tuning of the resting potential, pH homeostasis, control of cellular volume, and modulation of signaling pathways. In the setting of cancer, these functions can be in some way altered: highly proliferating cells have depolarized membrane potential, intra- vs. extra-cellular pH gradients are modified in migrating cells and apoptotic cells reduce their volume. There are different families of chloride selective channels and members of these families have been reported to play important roles in cancer cells.

Anoctamins (ANOs, also referred to as TMEM16s) is a family of calcium activated chloride selective channels comprising 10 members: ANO-1 to ANO-10. This family of channels plays important functions such as transepithelial ion transport, smooth muscle contraction, olfaction, phototransduction, nociception, and control of neuronal excitability (Pedemonte and Galietta 2014). Some members of this family have also been reported to play important roles in cancer development. In particular, ANO-1 has been reported to be highly expressed in numerous different types of cancer, in which it plays active roles in malignancy. ANO-1 is overexpressed, for instance, in colorectal cancer, prostate cancer, and lung cancer (Guan et al. 2016; Sui et al. 2014) and its genetic inhibition strongly reduces proliferation, metastasis, and invasiveness both in in vitro and in vivo models (Jia et al. 2015). A therapeutic approach in cancer could consist in reducing cancer cells proliferation, and this can be achieved in two different ways: (1) slowing down the proliferation rate or (2) increasing cell death. Interestingly, in an extensive study on colon and prostate cell lines (SW480, HCT116, HT-29, and PC-3), Guan et al. (Guan et al. 2016) showed that silencing the ANO-1 gene results both in arresting the cell cycle to the G1 phase, thus retarding cell cycle progression, and in inducing apoptosis, indicating an important role played by ANO-1 in both these processes. In the intestinal cell line CaCo-2, it was also shown that ANO-1 plays an important role in the balance between cell differentiation and proliferation: knocking down ANO-1 significantly reduces differentiation in favor of proliferation (Yang et al. 2013).

Changes in volume are a very common feature in cell life and ion channels and transporters are essential elements in the control of cellular volume. Volume or shape changes are also critical processes in cancer cells. For instance, the formation of lamellipodia at the leading edge of a moving cell and the retraction of the cell body at the trailing edge are triggered by ion exchanges across channels and transporters which in turn promote osmotically driven transmembrane water fluxes (Schwab et al. 2012). Invading cells dramatically change their shape and volume while crawling through the extracellular matrix or crossing blood vessel walls. Cells shrink during apoptosis undergoing the apoptotic volume decrease (AVD).

After the discovery of anion channels capable of mediating the fluxes of organic and inorganic anions during cellular volume changes (Banderali and Roy 1992) (volume regulated anion channels VRAC), the search for their molecular identity has been subject to many controversies. Difficulties in identifying these channels may have arisen in part from the heterogeneity of functional phenotypes they show (Banderali and Ehrenfeld 1996; Shennan 2008). In 2014, VRAC channels were discovered to be the result of the heteromeric assembly (likely a hexamer) of proteins belonging to the family LRRC8 (leucine-rich repeat–containing 8). This family counts five different members: LRRC8-A, B, C, D, and E (Voss et al. 2014; Jentsch et al. 2016; Planells-Cases et al. 2015). Due to the variety of possible combinations of the five LCCR8 subunits to form a functional hexamer, it is possible to envisage that different assemblies of LRRC8 isoforms may give rise to the heterogeneity of biophysical behaviors of VRAC channels observed in nature (Jentsch et al. 2016).

The importance of VRAC has been documented in various cancers. In endometrial cancer, VRAC was shown to participate in cell migration and invasion via its function as volume regulator and mediating intracellular calcium increase (Li et al. 2013). In hepatocellular carcinoma, progressive inhibition of VRAC with increasing doses of tamoxifen closely correlates with decreased migration of estrogen receptor negative tumor cells (Mao et al. 2013). In atypical teratoid/rhabdoid (AT/RT) brain tumor cells, the degree of inhibition of VRAC by different chloride channel blockers positively correlates with a decrease in cancer cell survival (Banderali et al. 2012). Interestingly, tumor cell disruption after swelling induced by hypotonic shock is used in hepatocellular carcinoma (HCC) therapy however, ways to improve efficacy of this kind of treatment are still needed (Kudou et al. 2016). In this context, an in vitro study demonstrated that application of a hypotonic shock combined with the inhibition of VRAC, which is responsible for volume control in HCC cells, results to be a more effective approach than the hypotonic shock alone (Banderali et al. 2012; Planells-Cases et al. 2015). Similarly, an important risk factor in gastric and pancreatic cancer resection is represented by viable and potentially tumorigenic cancer cells exfoliated into the peritoneal cavity. Exposure to distilled water via lavage of the peritoneal cavity is used to disrupt the exfoliated cancer cells during surgery of gastric cancers. In this case too, in vitro studies showed that block of VRAC enhances the cytocidal effect of the hypotonic shock induced by distilled water alone, and this approach could be successfully applied in surgeries of gastric and pancreatic resection (Iitaka et al. 2012; Nako et al. 2012).

3 Mechanisms Controlling Expression of Ion Channels in Cancer Cells

Carcinogenesis is a biological process characterized by alterations at the cellular, genetic, and epigenetic levels that produce abnormal cell proliferation. Therefore, atypical expression of ion channels can result as a consequence of mechanisms that become abnormal in cancer cells.

Gene amplification is a critical mechanism in carcinogenesis and it is often associated with overexpression of genes that are considered tumor markers (Albertson 2006). Remarkably, the KCNK9 gene which encodes for the $K_{2p}9.1$ channel has been found to be overexpressed in a novel amplicon of the chromosomal region 8q24.3 in breast cancer (Mu et al. 2003). Interestingly, multiple cancers of different histogenesis present high incidence of copy number gains that affects the 8q chromosome (Struski et al. 2002). This suggests that the increased level of the $K_{2p}9.1$ channel in some cancer cells can be explained by the mechanism of amplification of the KCNK9 gene as a part of the oncogenic signature. Similarly, genomic amplification of the KCNH1 gene, encoding for the Kv10.1 channel, has been observed in head and neck carcinomas and in colorectal adenocarcinoma (Menendez et al. 2012a, b). These events suggest that the specific K^+ channels could be considered as "driver" genes to promote carcinogenesis.

It has been also shown that the abundance of several K^+ channels such as Kv10.1, Kv11.1, Kv1.3, or Kv1.5 in cancer cells oscillates during the cell cycle. Interestingly, some channels reach the highest concentration in the G1 phase (Crociani et al. 2003; Villalonga et al. 2008). For example, in neuroblastoma cells, the transition from the G1 to S phase and the transition from the G2 to M phase are characterized by an increased outflow of K^+ and hyperpolarization (Boonstra et al. 1981).

Interestingly, hypermethylation of the promoter regions of genes encoding for the Kv1.3 or Kv1.5 channels downregulates the transcription of these channels, and this has been linked to progression of breast cancer (Brevet et al. 2009a, b). Also, promoter hypermethylation of the KCNH2 gene encoding for the Kv11.1 channel has been associated with downregulation of the channel in ovarian cancer (Cicek et al. 2013). Inversely, loss of DNA methylation of the KCNN4 promoter has been associated with high expression of the encoded $K_{Ca}3.1$ channel and poor survival in non-small cell lung cancer patients (Bulk et al. 2015). Therefore, disruption of epigenetic processes can lead to altered K^+ channels gene function and contribute to carcinogenesis in different ways.

The effect of ion channels on proliferation-related signaling pathways can be finely modulated by a variety of mechanisms including transcriptional, translational, post-translational, and epigenetic events. The activity of several ion channels is regulated by mitogen-activated biochemical signaling. Sex hormones that are mitogenic in hormone-dependent cancers (e.g., breast and prostate) can regulate ion channel activity through a variety of mechanisms that increase channel synthesis or activate membrane signaling pathways in which ion channels are effectors.

The sex hormone estrogen can induce transcription of potassium, calcium, and sodium channels via the novel membrane-bound G protein estrogen receptor (GPER) (Sun et al. 2010). In breast cancer, ANO-1 interacts with important receptors that regulate cell proliferation, including the estrogen (ER) and progesterone receptors (PR) and human epidermal growth factor receptor 2 (HER2) in a subtype-dependent manner (Loibl et al. 2021). ANO-1 activity promotes proliferation in ER/PR-positive cells but inhibits it in ER/PR-negative cells. In addition, ANO-1 overexpression in PR-positive/HER2-negative patients is a favorable prognostic factor following tamoxifen treatment Finally, ANO-1 acts as a HER2 transcriptional factor in YMB-1 breast cancer cells in which siRNA-mediated inhibition of ANO-1 significantly impairs HER2 transcription (Fujimoto et al. 2017). Other work demonstrated that exposure of ERα-positive breast cancer cells to 17β-estradiol (17β-E2) elicits a significant transmembrane current mediated by the chloride channel 3 (CLC-3) (Yang et al. 2018). The interplay between ERα and CLC-3 in breast cancer cells seems to be essential to elicit estrogen-induced tumor growth in these cells.

Activity of K^+ channels is also regulated by hormone receptors. Hormone binding to G protein-coupled receptors (GPCR) releases the active βγ subunit of the heteromeric GTPase complex, which directly binds to and activates K^+ channels (e.g., GIRK) (Plummer et al. 2005). Mass spectrometry investigations revealed that the Kv11.1 potassium ion channel is among the 10 most phosphorylated proteins expressed in mammary tumors arising an MMTV-PyMT transgenic mouse model (Ali et al. 2014). Although the specific effect of this post-translational modification has not been characterized yet, others have shown that reversible phosphorylation of Kv11.1 induced by hormone-activated biochemical cascades dramatically changes channel activity. These events appear to be related to a non-genomic mechanism linking the effect of steroid hormone receptor activation to the regulation of surface membrane ion channel activities (Gentile 2012; Gentile et al. 2006). The findings led investigators to propose that phosphorylated Kv11.1 is part of a putative oncogenic signature (Ali et al. 2014). Finally, the presence of signaling modules such as the oxygen sensor Per-Arnt-Sim (PAS) or the cyclic nucleotide binding site on intracellular domains of EAG family members suggests that ion channels can be activated by specific metabolic states in a cancer cell. For example, it has been shown that the PAS domain significantly regulates the biophysical properties of Kv11.1 (Morais Cabral et al. 1998). Although the stimulus detected by the domain has not been determined, it is possible to speculate that the Kv11.1 channel could play an essential role during hypoxia, which is a critical factor in tumorigenesis.

4 Ion Channel Activity Controls Proliferative Signaling in Cancer

Throughout the cell cycle, cells undergo checkpoints in which a specific group of proteins controls the health status of the cell (e.g., DNA damage). If the cells do not present the essential conditions, the cell cycle is arrested and cells eventually die. This process becomes dysfunctional in cancer cells. Consequently, these cells are not capable to differentiate or die and undergo a short circuit for which the proliferative process virtually never ends (Fig. 1).

Constitutive blockage or opening of the ion channel by pharmacological approach would impair the normal cell cycle progression by affecting depolarization or repolarization (Fig. 1).

Recent work is beginning to elucidate a variety of ways that ion channels regulate key growth and survival pathways in cancer. Inhibition or removal of the Kv11.1 protein in cancer cells activates apoptosis suggesting a direct link between alteration of K^+ gradients and a key process involved in cell death (Bianchi et al. 1998; Wang et al. 2002; Lansu and Gentile 2013; Zheng et al. 2009). In contrast, hyperactivity of Kv11.1 arrests breast tumor growth by activating a "cellular senescent program" (Bentzen et al. 2011) which is defined as a permanent arrest of the cell cycle induced by a progressive increase in cellular stress (Munoz-Espin and Serrano 2014). This data raises the possibility that induction of senescence may be a potent therapeutic strategy in cancer. Additionally, Kv11.1 activity triggers a series of regulatory proteins that control progression of the cell cycle and cell metabolism. For example, cyclin E2 and p21waf are important cell cycle regulators. Cyclin E2 promotes progression from the G1 to S phase of the cell cycle and is overexpressed in several cancers. In contrast, the tumor suppressor protein p21waf arrests the cell cycle in G0/G1. Expression of p21waf is often lost in cancers of different histogenesis. Interestingly, pharmacological activation of Kv11.1 produced a rapid proteasome-mediated degradation of cyclin E2 (Perez-Neut et al. 2015a). Also, prolonged stimulation of Kv11.1 activity produced de-*novo* synthesis of p21waf (Perez-Neut et al. 2016b). These events contributed significantly to the arrest of the cell cycle in the G0/G1 phase. Notably, this occurred in cells that were resistant to anticancer therapy.

Interestingly, both degradation of cyclin E2 and p21waf synthesis were driven by a Kv11.1-dependent increase in intracellular Ca^{2+}. This observation can be explained by the fact that variations of K^+ fluxes (which is the most abundant intracellular cation) significantly contribute to changes of Vm in all cells. The opening of K^+ channels allows K^+ to leave the cell, resulting in cell hyperpolarization. In contrast to "excitable cells," where hyperpolarization inhibits the rapid opening of specific Ca^{2+} channels (E.g., L-type), the increasing intracellular negative potential in non-excitable cells provides a driving force for extracellular Ca^{2+} to cross the membrane and enter into the cytoplasm (Lepple-Wienhues et al. 1996; Wang 2004) through Ca^{2+} channels. Consequently, changes of intracellular Ca^{2+}

concentration activate several biochemical cascades that control a range of events during different phases of proliferation.

Loss of K^+ ion can activate intricate signaling independently of Ca^{2+}. Autophagy is a homeostatic mechanism for cellular quality control and a critical cell stress response mechanism. The role of autophagy in cancer is still debated as it can play a dual role of enhancing survival and activating apoptosis. Stimulation of the Kv11.3 (hERG3) in melanoma cells produced a rapid activation of AMP-Activated Kinase (AMPK) which initiated autophagy in a Ca^{2+}-independent manner (Perez-Neut et al. 2015b). Ultimately, Kv11.3-dependent autophagy protected cancer cells from death and contributed to the senescent phenotype, indicating that K^+ channel activity alone can control a variety of biochemical cascades regulating cell survival.

The CLC-3 chloride channel protein can be expressed both in the plasma membrane and intracellularly, and both situations occur in cancer. CLC-3 function in malignancy is dependent on protein localization (Zhang et al. 2014). Control of cellular volume appears to be one of the most critical functions of CLC-3, which is fundamental to maintaining cell survival. It was reported that membrane-bound CLC-3 can form a complex with aquaporins (AQP) and the concerted action of these two proteins facilitates the exchange of anionic osmolytes and water across the plasma membrane in response to osmotic challenges. When located intracellularly, CLC-3 influences the Akt signaling pathway. CLC-3 knockdown in osteosarcoma cells reduces cell proliferation by suppressing phosphorylation of the Akt/GSK3β pathway, leading to downregulation of cyclin D1 and cyclin E, thus arresting the cell cycle in G0/G1 phase (Du and Yang 2015).

5 Ion Channels as Protein Partners in Cancer Biology

The anti-proliferative effects of ion channel activity can in part be ascribed to their function in ion transport. However, direct interaction with or regulation of other proteins and/or signaling pathways also occurs. The epidermal growth factor receptor (EGFR) is an essential factor in the progression of head and neck squamous cell carcinoma (HNSCC). Bill et al. (2015) reported that EGFR and ANO-1 form a functional complex in HNSCC controlling cell proliferation and that co-inhibition of the two components has additive inhibitory effects on proliferation. In a recent work, Godse et al. (2017) showed that ANO-1 expression has a strong negative correlation with the pro-apoptotic protein Bim in HNSCC. This correlation was found both in vitro and in vivo, and it was shown to account for both the low rate of apoptosis in untreated tumor cells and for the suppression of apoptosis activation in response to cisplatin. A positive correlation between ANO-1 and EGFR has also been reported in breast cancer (Wu et al. 2017). Knockdown of ANO-1 decreases EGFR expression and attenuates the activation of AKT, Src, and ERK via the calmodulin-dependent protein kinase II (CaMKII) signaling pathway.

Potassium ion channels also appear to directly interact with proteins involved in cellular processes that are key to the malignant phenotype. Conformational changes of the voltage-gated domain in the potassium channels Kv10.1 and Kv11.1 can directly interact with β1-integrin and FAK in focal adhesions or lipid rafts and contribute to cell migration, which is a critical factor for cancer metastasis (Fang et al. 2007). Mutations that eliminate the ion flux of the Kv10.1 channel (a member of the superfamily of ether-a-go-go channels) did not alter its ability to activate p38/MAPK (another autophagy regulator) (Hegle et al. 2006). In contrast, mutations that forced the channel in an open state inhibited p38/MAPK activity suggesting that Kv channels can function as voltage sensors to indirectly control the activity of key players of autophagy as well as other biochemical pathways.

6 Organellar Ion Channels in Cancer

The surface area of the plasma membrane is estimated to be only 2–5% of the total cell membrane area and it is established that intracellular membranes contain a variety of ion channels (Xu et al. 2015). Lysosomes are membrane-enclosed intracellular vesicles that contain hydrolases for the degradation and recycling of macromolecules to preserve cellular homeostasis (Xu and Ren 2015). Ion channels play fundamental roles in controlling the activity of the lysosomes ranging from detection of nutrient status to the generation of an acidic environment to maintain the activity of digestive enzymes. Cancer cells can adapt to scarce nutrient condition by increasing lysosomal function which appears at least in part to be due to dysregulation of their ion channels. For example, cancer cells ongoing nutrient starvation and/or inhibition of the cell growth regulator mammalian target of rapamycin complex 1 (mTORC1) can upregulate the lysosomal cation channel ML1 (Mucilipin-1). Consequently, this event resulted in an increased lysosomal proteolytic activity (Wang et al. 2015). Interestingly, lysosomal two-pore channel (TPC) dysfunction was also found to be involved in cancer cell migration and invasion. The pharmacological inhibition of TPC1 or TPC2 produced an accumulation of the adhesive protein b1-integrin in endocytic vesicles which resulted in an impaired formation of lamellipodia (Grimm et al. 2018).

Mitochondria control ATP production, and they are involved in several pathological processes including cancer. Overexpression of mitochondrial uncoupling protein 2 (UCP2) has been documented in numerous tumor types. Mitochondrial uncoupling regulates mitochondrial proton motive force, preventing hyperpolarization and thermogenesis (Baffy 2010; Baffy et al. 2011). UCP2 has been shown to protect cells from oxidative stress (Arsenijevic et al. 2000) and to block the apoptosis-inducing effects of chemotherapeutic drugs (Gentile 2012). Defective mitochondrial calcium signaling is considered a hallmark of cancer (Delierneux et al. 2020; Grasso et al. 2020). Mitochondrial permeability to calcium is mediated by the permeability transition pore (mPTP) whose molecular identity is debated, but the most recent model suggests that is formed by dimers of F0F1 ATP synthase (Giorgio et al. 2013). mPTP activation, mainly mediated by mitochondrial calcium overload and production of reactive oxygen species (ROS), has been linked to cancer (Biasutto et al. 2016).

Voltage-gated potassium channels are the most abundant ion channels in mitochondria (Choe 2002). Mitochondrial Kv1.3 is located on the inner mitochondrial membrane and is the most studied member of this family. One of its functions seems to be regulation of organelle-mediated apoptosis (Leanza et al. 2015; Szabo et al. 2005, 2008, 2011). This channel is overexpressed in several tumor types. In vitro, ex vivo, and in vivo experiments reveal that pharmacological inhibition of the channel by membrane permeant inhibitors (Psora-4, PAP-1, and clofazimine) can selectively kill cancer cells and spare healthy cells in several tumors, such as melanoma, chronic lymphocytic leukemia, glioma, and pancreatic ductal adenocarcinoma (Leanza et al. 2015; Szabo et al. 2005, 2008, 2011). After the potassium flux to the matrix is blocked, mitochondrial membrane hyperpolarization occurs, which increases production of ROS. ROS in turn can activate mPTP, causing mitochondrial membrane depolarization and cytochrome c mobilization from the cristae and its release from the inter-membrane space to the cytosol, where it contributes to activating the apoptosome (Leanza et al. 2012, 2013, 2017; Szabo et al. 2015).

Another mitochondrial voltage-gated potassium channel, mitoKv1.5, has been implicated in cancer. Interestingly, this channel is one of the two potassium channels that are under-expressed rather than overexpressed in tumor cells. The functional significance and consequences of this finding are not yet determined, but this finding provides further evidence that mitochondrial-associated ion channels are dysregulated and are likely biologically important aberrations in neoplasia (Ryland et al. 2016).

7 Targeting Ion Channels as Therapeutic Approach Against Cancer

Although cancer treatments have dramatically improved in the past decades, cancer is still a leading cause of death worldwide. Several factors contribute to limit the effectiveness of treatment for patients with clinically challenging cancers including the difficulty in identifying proteins governing crucial biochemical signaling, the modest efficacy and significant toxicity of available chemotherapeutic agents, and the lack of approved targeted therapies. These therapeutic limitations can be fulfilled by pharmacological targeting ion channels for at least three reasons: (1) ion channels control critical biochemical pathways that underline all hallmarks of cancer. (2) Drug discovery focused on ion channels has generated an abundance of medicines targeting potassium channels that are of critical importance in the pharmacopeia for treating human diseases (Kaczorowski et al. 2008). (3) The unique localization of ion channels on the cell membrane provides a therapeutic advantage as it makes this protein an ideal target for safe pharmacologic manipulation and repurposing drugs.

Importantly, several drugs have been discovered or developed in order to target ion channels in cancer cells. Several of these compounds have demonstrated an efficacy in in vivo pre-clinical models (Leanza et al. 2016) .Clinical data demonstrate that expression levels of specific ion channels can be associated with improved overall survival and, therefore, it can inform whether to use channel activators or blockers. An example of the beneficial use of specific ion channels activators is provided by the following study: In ovarian cancer patients who underwent tumor resection, high expression levels of the Kir6.2 channel associate with prolonged overall survival (Innamaa et al. 2013), suggesting that high activity of these channels is a positive prognostic factor and does not favor cancer recurrence. Works in our lab demonstrated that pharmacological stimulation of Kir6.2 inhibited ovarian tumor growth by arresting the cell cycle in a specific cell cycle phase (Fukushiro-Lopes et al. 2018, 2020; Perez-Neut et al. 2015a, 2016a, b; Lansu and Gentile 2013) and by causing cell death selectively in cancer cells. Interestingly, the Kir6.2 channel activator used in this study is the small molecule minoxidil. Minoxidil was initially approved by the FDA as antihypertensive drug (Loniten®) and later approved for its properties to arrest hair loss (Rogaine®). Therefore, these studies indicate the benefit of repurposing drugs that target ion channels as an effective strategy against cancer. Similar studies (Fukushiro-Lopes et al. 2018, 2020; Perez-Neut et al. 2015a, 2016a, b; Lansu and Gentile 2013) in which activator molecules of specific potassium channels such as Kv11.1 were used in vitro, in vivo, and ex vivo testing also demonstrated that this pharmacological strategy appears to be promising in treating other clinically challenging cancer such as triple negative breast cancers. Remarkably, chronic use of a Kv11.1 activator such as NS1643 arrested tumor growth by producing a senescent-like phenotype. While activating the senescence program, cells that were exposed to NS1643 underwent autophagy. The combination of NS1643 with the FDA autophagy blocker, hydroxychloroquine (Plaquenil®) produced cancer cell death rather than senescence. Notably, no significant side effects were associated with the use of these K^+ channels activators. In contrast, pharmacological blockades of the Kv11.1 channels for therapeutic strategy are generally discouraged due to the increased risk of severe side effects including cardiac ventricular fibrillation. However, it is interesting to note that several anticancer drugs currently in use present a Kv11.1 blocking activity suggesting that their anticancer effect might be due, at least in part, to their action on ion channels (Gentile 2016).

Targeting chloride channels in cancer therapy may prove to be a difficult task. In fact, despite some important advances in the development of inhibitors specific to a particular channel (Koster et al. 2018), pharmacology of these proteins remains still very challenging. Lack of specificity, instability, or unpredictable behaviors make chloride channels difficult therapeutic targets at this point. Nonetheless, due to the continuously growing evidence of their physiological importance in cancer as well as in many other pathologies, the need to develop efficacious and selective modulators is primordial.

An interesting strategy for targeting ion channels in cancer is the use of antibodies. The anticancer antibody-based therapy proposed by Pardo exploits the unique property of some cancers to express the Kv10.1 channel which in a healthy body is found only in the central nervous system. An initial approach included the use of functional antibodies targeting the Kv10.1 channel with antagonistic activity. This method has been successfully used to arrest a human-derived breast cancer (Gomez-Varela et al. 2007) in in vivo testing. More recently, antibodies with dual specificity (Antibody-drug-conjugates, ADCs) were tested. When cancer cells were exposed to a single-domain antibody (nanobody) with high affinity against the Kv10.1 channel and fused to a cytokine (ligand of a tumor necrosis factor-related apoptosis inducing ligand or TRAIL) (Hartung et al. 2020), apoptosis is rapidly activated. Although these studies are still at the early stage and important in vivo experiments, pharmacokinetic and pharmacodynamic studies are still pending, the use of ADC nanobodies targeting ion channels and surface membrane tumor markers offers an impressive range of opportunities ranging from diagnostic to personalized medicine.

Another Kv channel Kv1.3 can be druggable in cancer treatment. Apart from membrane permeant inhibitors Psora-4, PAP-1, and clofazimine, new mitochondria targeted (PAPTP and PCARBTP) or more soluble (PAP-1-MHEG) PAP-1 derivatives have been recently developed (Leanza et al. 2017; Peruzzo et al. 2020) These compounds demonstrated their efficacy in vitro against different cancer cell lines as well as in vivo in pre-clinical models leading to 90% or 60% reduction of tumor volume in melanoma or pancreatic ductal adenocarcinoma orthotopic mouse models respectively, without inducing any side effect (Leanza et al. 2017). Unfortunately, these new drugs were not able to cross the blood brain barrier to reach brain tumors in vivo; nevertheless, they showed a great ability to eliminate glioblastoma cells in in vitro experiments (Venturini et al. 2017). Importantly, it has been recently shown that conjugating PAPTP to small peptides could be a new possible strategy to reach brain tumors (Parrasia et al. 2021). Importantly, altered expression of ion channels has been associated with resistance to therapeutic compounds used for cancer treatments (a topic recently well discussed in other reviews (Hoffmann and Lambert 2014; Kischel et al. 2019)). Anticancer drug resistance is determined by multifactorial events including reduction of drug cellular import, alteration of gene expression which can be driven by the therapeutic agent (i.e., tamoxifen), variation of cancer cell metabolism and more (Housman et al. 2014; Mansoori et al. 2017). In addition, in view of the opportunistic nature of cancer cells, each of these events can occur alone in different types of cancer cells or in combination to confer resistance to one anticancer agent rather than another. Therefore, although drug resistance in cancer cells has been associated with expression of specific ion channels, still little is known about the role of these channels in each event regulating drug resistance. However, due to the complexity of this phenomenon, it is possible to speculate that these cells change the ion channel assets rather than relying on a specific member of this class of proteins.

8 Conclusion

Research into the role of ion channels in the pathophysiology of human disease has focused extensively on "excitable" tissues such as neurons and myocytes as well as on epithelial transport in the past. However, a rapidly growing body of literature indicates that ion channels can play important roles in cancer as well. At this time the role of specific channels appears to be cancer context dependent; however, the notion that that abnormal expression of ion channels in cancer cells may be an adaptive tool by which malignant cells acquire and/or maintain accelerated proliferation appears to be convincing. It is also clear that the activity of ion channels in cancer is finely controlled by a variety of temporally activated mechanisms. Chronic stimulation or inhibition of ion channels activity (e.g., Kv11.1) produces arrest of the cell cycle which can occur in the same cell cycle phase independently from the use of an activator or inhibitor. This suggests that in order to contribute to tumorigenesis, activity of ion channels is transient in nature. We can speculate that the ability of cancer cells to transiently amplify a specific ion channel function in a specific cell cycle phase (for example, by altering protein expression level or current activity) can result in changes in membrane potentials which in turn accelerate the transitioning of cells from that cycle phase to the next. Consequently, the duplication time of cancer cells becomes shorter when compared to normal cells resulting in tumor growth.

A better understanding of the molecular mechanisms involved should lead to development of more effective channel activators and inhibitors that will enable clinical investigation of a novel paradigm for cancer therapeutics.

References

Accardi A (2015) Cell signaling. Lipids link ion channels and cancer. Science 349(6250):789–790

Agarwal JR, Griesinger F, Stuhmer W, Pardo LA (2010) The potassium channel ether a go-go is a novel prognostic factor with functional relevance in acute myeloid leukemia. Mol Cancer 9:18

Albertson DG (2006) Gene amplification in cancer. Trends Genet 22(8):447–455

Ali NA, Wu J, Hochgrafe F, Chan H, Nair R, Ye S et al (2014) Profiling the tyrosine phosphoproteome of different mouse mammary tumour models reveals distinct, model-specific signalling networks and conserved oncogenic pathways. Breast Cancer Res 16(5):437

Arsenijevic D, Onuma H, Pecqueur C, Raimbault S, Manning BS, Miroux B et al (2000) Disruption of the uncoupling protein-2 gene in mice reveals a role in immunity and reactive oxygen species production. Nat Genet 26(4):435–439

Baffy G (2010) Uncoupling protein-2 and cancer. Mitochondrion 10(3):243–252

Baffy G, Derdak Z, Robson SC (2011) Mitochondrial recoupling: a novel therapeutic strategy for cancer? Br J Cancer 105(4):469–474

Banderali U, Ehrenfeld J (1996) Heterogeneity of volume-sensitive chloride channels in basolateral membranes of A6 epithelial cells in culture. J Membr Biol 154(1):23–33

Banderali U, Roy G (1992) Anion channels for amino-acids in Mdck cells. Am J Physiol 263(6): C1200–C1207

Banderali U, Jayanthan A, Hoeksema KA, Narendran A, Giles WR (2012) Ion channels in pediatric CNS atypical teratoid/rhabdoid tumor (AT/RT) cells: potential targets for novel therapeutic agents. J Neurooncol 107(1):111–119
Bentzen BH, Bahrke S, Wu K, Larsen AP, Odening KE, Franke G et al (2011) Pharmacological activation of Kv11.1 in transgenic long QT-1 rabbits. J Cardiovasc Pharmacol 57(2):223–230
Bianchi L, Wible B, Arcangeli A, Taglialatela M, Morra F, Castaldo P et al (1998) Herg encodes a K+ current highly conserved in tumors of different histogenesis: a selective advantage for cancer cells? Cancer Res 58(4):815–822
Biasutto L, Azzolini M, Szabo I, Zoratti M (2016) The mitochondrial permeability transition pore in AD 2016: an update. Biochim Biophys Acta 1863(10):2515–2530
Bill A, Gutierrez A, Kulkarni S, Kemp C, Bonenfant D, Voshol H et al (2015) ANO1/TMEM16A interacts with EGFR and correlates with sensitivity to EGFR-targeting therapy in head and neck cancer. Oncotarget 6(11):9173–9188
Binggeli R, Weinstein RC (1986) Membrane potentials and sodium channels: hypotheses for growth regulation and cancer formation based on changes in sodium channels and gap junctions. J Theor Biol 123(4):377–401
Boonstra J, Mummery CL, Tertoolen LG, Van Der Saag PT, De Laat SW (1981) Cation transport and growth regulation in neuroblastoma cells. Modulations of K+ transport and electrical membrane properties during the cell cycle. J Cell Physiol 107(1):75–83
Brevet M, Haren N, Sevestre H, Merviel P, Ouadid-Ahidouch H (2009a) DNA methylation of K(v) 1.3 potassium channel gene promoter is associated with poorly differentiated breast adenocarcinoma. Cell Physiol Biochem 24(1–2):25–32
Brevet M, Fucks D, Chatelain D, Regimbeau JM, Delcenserie R, Sevestre H et al (2009b) Deregulation of 2 potassium channels in pancreas adenocarcinomas: implication of KV1.3 gene promoter methylation. Pancreas 38(6):649–654
Bulk E, Ay AS, Hammadi M, Ouadid-Ahidouch H, Schelhaas S, Hascher A et al (2015) Epigenetic dysregulation of KCa 3.1 channels induces poor prognosis in lung cancer. Int J Cancer 137 (6):1306–1317
Chernet BT, Levin M (2013) Transmembrane voltage potential is an essential cellular parameter for the detection and control of tumor development in a Xenopus model. Dis Model Mech 6 (3):595–607
Choe S (2002) Potassium channel structures. Nat Rev Neurosci 3(2):115–121
Cicek MS, Koestler DC, Fridley BL, Kalli KR, Armasu SM, Larson MC et al (2013) Epigenome-wide ovarian cancer analysis identifies a methylation profile differentiating clear-cell histology with epigenetic silencing of the HERG K+ channel. Hum Mol Genet 22(15):3038–3047
Cone CD Jr (1969) Electroosmotic interactions accompanying mitosis initation in sarcoma cells in vitro. Trans N Y Acad Sci 31(4):404–427
Cone CD Jr (1970) Variation of the transmembrane potential level as a basic mechanism of mitosis control. Oncology 24(6):438–470
Cone CD Jr (1971) Unified theory on the basic mechanism of normal mitotic control and oncogenesis. J Theor Biol 30(1):151–181
Crociani O, Guasti L, Balzi M, Becchetti A, Wanke E, Olivotto M et al (2003) Cell cycle-dependent expression of HERG1 and HERG1B isoforms in tumor cells. J Biol Chem 278(5):2947–2955
DeCoursey TE, Chandy KG, Gupta S, Cahalan MD (1984) Voltage-gated K+ channels in human T lymphocytes: a role in mitogenesis? Nature 307(5950):465–468
Delierneux C, Kouba S, Shanmughapriya S, Potier-Cartereau M, Trebak M, Hempel N (2020) Mitochondrial calcium regulation of redox signaling in cancer. Cells (2):9
Djamgoz MBA, Mycielska M, Madeja Z, Fraser SP, Korohoda W (2001) Directional movement of rat prostate cancer cells in direct-current electric field: involvement of voltagegated Na+ channel activity. J Cell Sci 114(Pt 14):2697–2705
Du S, Yang L (2015) ClC-3 chloride channel modulates the proliferation and migration of osteosarcoma cells via AKT/GSK3beta signaling pathway. Int J Clin Exp Pathol 8 (2):1622–1630

Fang D, Hawke D, Zheng Y, Xia Y, Meisenhelder J, Nika H et al (2007) Phosphorylation of beta-catenin by AKT promotes beta-catenin transcriptional activity. J Biol Chem 282 (15):11221–11229
Fujimoto M, Inoue T, Kito H, Niwa S, Suzuki T, Muraki K et al (2017) Transcriptional repression of HER2 by ANO1 Cl(−) channel inhibition in human breast cancer cells with resistance to trastuzumab. Biochem Biophys Res Commun 482(1):188–194
Fukushiro-Lopes DF, Hegel AD, Rao V, Wyatt D, Baker A, Breuer EK et al (2018) Preclinical study of a Kv11.1 potassium channel activator as antineoplastic approach for breast cancer. Oncotarget 9(3):3321–3337
Fukushiro-Lopes D, Hegel AD, Russo A, Senyuk V, Liotta M, Beeson GC et al (2020) Repurposing Kir6/SUR2 channel activator minoxidil to arrests growth of gynecologic cancers. Front Pharmacol 11:577
Gentile S (2012) Ion channel phosphorylopathy: a link between genomic variation and human disease. ChemMedChem 7(10):1757–1761
Gentile S (2016) hERG1 potassium channel in cancer cells: a tool to reprogram immortality. Eur Biophys J 45(7):649–655
Gentile S, Darden T, Erxleben C, Romeo C, Russo A, Martin N et al (2006) Rac GTPase signaling through the PP5 protein phosphatase. Proc Natl Acad Sci U S A 103(13):5202–5206
Giorgio V, von Stockum S, Antoniel M, Fabbro A, Fogolari F, Forte M et al (2013) Dimers of mitochondrial ATP synthase form the permeability transition pore. Proc Natl Acad Sci U S A 110(15):5887–5892
Godse NR, Khan N, Yochum ZA, Gomez-Casal R, Kemp C, Shiwarski DJ et al (2017) TMEM16A/ANO1 inhibits apoptosis via downregulation of Bim expression. Clin Cancer Res 23 (23):7324–7332
Gomez-Varela D, Zwick-Wallasch E, Knotgen H, Sanchez A, Hettmann T, Ossipov D et al (2007) Monoclonal antibody blockade of the human Eag1 potassium channel function exerts antitumor activity. Cancer Res 67(15):7343–7349
Goswami C, Hucho T (2007) TRPV1 expression-dependent initiation and regulation of filopodia. J Neurochem 103(4):1319–1333
Grasso D, Zampieri LX, Capeloa T, Van de Velde JA, Sonveaux P (2020) Mitochondria in cancer. Cell Stress 4(6):114–146
Gray LS, Perez-Reyes E, Gomora JC, Haverstick DM, Shattock M, McLatchie L et al (2004) The role of voltage gated T-type Ca2+ channel isoforms in mediating "capacitative" Ca2+ entry in cancer cells. Cell Calcium 36(6):489–497
Grimm C, Bartel K, Vollmar AM, Biel M (2018) Endolysosomal cation channels and cancer – a link with great potential. Pharmaceuticals (Basel) 11(1)
Guan LZ, Song Y, Gao J, Gao JJ, Wang KW (2016) Inhibition of calcium-activated chloride channel ANO1 suppresses proliferation and induces apoptosis of epithelium originated cancer cells. Oncotarget 7(48):78619–78630
Hartung F, Kruwel T, Shi X, Pfizenmaier K, Kontermann R, Chames P et al (2020) A novel anti-Kv10.1 nanobody fused to single-chain TRAIL enhances apoptosis induction in cancer cells. Front Pharmacol 11:686
Hegle AP, Marble DD, Wilson GF (2006) A voltage-driven switch for ion-independent signaling by ether-a-go-go K+ channels. Proc Natl Acad Sci U S A 103(8):2886–2891
Hoffmann EK, Lambert IH (2014) Ion channels and transporters in the development of drug resistance in cancer cells. Philos Trans R Soc Lond B Biol Sci 369(1638):20130109
Housman G, Byler S, Heerboth S, Lapinska K, Longacre M, Snyder N et al (2014) Drug resistance in cancer: an overview. Cancers (Basel) 6(3):1769–1792
Iitaka D, Shiozaki A, Ichikawa D, Kosuga T, Komatsu S, Okamoto K et al (2012) Blockade of chloride ion transport enhances the cytocidal effect of hypotonic solution in gastric cancer cells. J Surg Res 176(2):524–534

Innamaa A, Jackson L, Asher V, Van Shalkwyk G, Warren A, Hay D et al (2013) Expression and prognostic significance of the oncogenic K2P potassium channel KCNK9 (TASK-3) in ovarian carcinoma. Anticancer Res 33(4):1401–1408

Jentsch TJ, Lutter D, Planells-Cases R, Ullrich F, Voss FK (2016) VRAC: molecular identification as LRRC8 heteromers with differential functions. Pflugers Archiv Eur J Physiol 468 (3):385–393

Jia LH, Liu W, Guan LZ, Lu M, Wang KW (2015) Inhibition of calcium-activated chloride channel ANO1/TMEM16A suppresses tumor growth and invasion in human lung cancer. Plos One 10 (8)

Johnstone BM (1959) Micro-electrode penetration of ascites tumour cells. Nature 183(4658):411

Kaczorowski GJ, McManus OB, Priest BT, Garcia ML (2008) Ion channels as drug targets: the next GPCRs. J Gen Physiol 131(5):399–405

Kapur N, Mignery GA, Banach K (2007) Cell cycle-dependent calcium oscillations in mouse embryonic stem cells. Am J Physiol Cell Physiol 292(4):C1510–C1518

Kischel P, Girault A, Rodat-Despoix L, Chamlali M, Radoslavova S, Abou Daya H et al (2019) Ion channels: new actors playing in chemotherapeutic resistance. Cancers (Basel) 11(3)

Koster AK, Wood CAP, Thomas-Tran R, Chavan TS, Almqvist J, Choi KH et al (2018) A selective class of inhibitors for the CLC-Ka chloride ion channel. Proc Natl Acad Sci U S A 115(21): E4900–E49E9

Kudou M, Shiozaki A, Kosuga T, Ichikawa D, Konishi H, Morimura R et al (2016) Inhibition of regulatory volume decrease enhances the cytocidal effect of hypotonic shock in hepatocellular carcinoma. J Cancer 7(11):1524–1533

Lallet-Daher H, Wiel C, Gitenay D, Navaratnam N, Augert A, Le Calve B et al (2013) Potassium channel KCNA1 modulates oncogene-induced senescence and transformation. Cancer Res 73 (16):5253–5265

Lansu K, Gentile S (2013) Potassium channel activation inhibits proliferation of breast cancer cells by activating a senescence program. Cell Death Dis 4:e652

Leanza L, Henry B, Sassi N, Zoratti M, Chandy KG, Gulbins E et al (2012) Inhibitors of mitochondrial Kv1.3 channels induce Bax/Bak-independent death of cancer cells. EMBO Mol Med 4(7):577–593

Leanza L, Trentin L, Becker KA, Frezzato F, Zoratti M, Semenzato G et al (2013) Clofazimine, Psora-4 and PAP-1, inhibitors of the potassium channel Kv1.3, as a new and selective therapeutic strategy in chronic lymphocytic leukemia. Leukemia 27(8):1782–1785

Leanza L, Venturini E, Kadow S, Carpinteiro A, Gulbins E, Becker KA (2015) Targeting a mitochondrial potassium channel to fight cancer. Cell Calcium 58(1):131–138

Leanza L, Manago A, Zoratti M, Gulbins E, Szabo I (2016) Pharmacological targeting of ion channels for cancer therapy: In vivo evidences. Biochim Biophys Acta 1863(6 Pt B):1385–1397

Leanza L, Romio M, Becker KA, Azzolini M, Trentin L, Manago A et al (2017) Direct pharmacological targeting of a mitochondrial ion channel selectively kills tumor cells in vivo. Cancer Cell 31(4):516–531.e10

Lepple-Wienhues A, Berweck S, Bohmig M, Leo CP, Meyling B, Garbe C et al (1996) K+ channels and the intracellular calcium signal in human melanoma cell proliferation. J Membr Biol 151 (2):149–157

Levin M (2012) Molecular bioelectricity in developmental biology: new tools and recent discoveries: control of cell behavior and pattern formation by transmembrane potential gradients. Bioessays 34(3):205–217

Levin M (2014) Molecular bioelectricity: how endogenous voltage potentials control cell behavior and instruct pattern regulation in vivo. Mol Biol Cell 25(24):3835–3850

Li M, Wu DB, Wang J (2013) Effects of volume-activated chloride channels on the invasion and migration of human endometrial cancer cells. Eur J Gynaecol Oncol 34(1):60–64

Lobikin M, Chernet B, Lobo D, Levin M (2012) Resting potential, oncogene-induced tumorigenesis, and metastasis: the bioelectric basis of cancer in vivo. Phys Biol 9(6):065002

Loibl S, Marme F, Martin M, Untch M, Bonnefoi H, Kim SB et al (2021) Palbociclib for residual high-risk invasive HR-positive and HER2-negative early breast cancer-the penelope-B trial. J Clin Oncol:JCO2003639

Mansoori B, Mohammadi A, Davudian S, Shirjang S, Baradaran B (2017) The different mechanisms of cancer drug resistance: a brief review. Adv Pharm Bull 7(3):339–348

Mao JW, Yuan J, Wang LW, Zhang HF, Jin XB, Zhu JY et al (2013) Tamoxifen inhibits migration of estrogen receptor-negative hepatocellular carcinoma cells by blocking the swelling-activated chloride current. J Cell Physiol 228(5):991–1001

Marino AA, Morris DM, Schwalke MA, Iliev IG, Rogers S (1994) Electrical potential measurements in human breast cancer and benign lesions. Tumour Biol 15(3):147–152

Menendez ST, Villaronga MA, Rodrigo JP, Alvarez-Teijeiro S, Garcia-Carracedo D, Urdinguio RG et al (2012a) Frequent aberrant expression of the human ether a go-go (hEAG1) potassium channel in head and neck cancer: pathobiological mechanisms and clinical implications. J Mol Med (Berl) 90(10):1173–1184

Menendez ST, Rodrigo JP, Alvarez-Teijeiro S, Villaronga MA, Allonca E, Vallina A et al (2012b) Role of HERG1 potassium channel in both malignant transformation and disease progression in head and neck carcinomas. Mod Pathol 25(8):1069–1078

Morais Cabral JH, Lee A, Cohen SL, Chait BT, Li M, Mackinnon R (1998) Crystal structure and functional analysis of the HERG potassium channel N terminus: a eukaryotic PAS domain. Cell 95(5):649–655

Morokuma J, Blackiston D, Adams DS, Seebohm G, Trimmer B, Levin M (2008) Modulation of potassium channel function confers a hyperproliferative invasive phenotype on embryonic stem cells. Proc Natl Acad Sci U S A 105(43):16608–16613

Mu D, Chen L, Zhang X, See LH, Koch CM, Yen C et al (2003) Genomic amplification and oncogenic properties of the KCNK9 potassium channel gene. Cancer Cell 3(3):297–302

Munaron L, Antoniotti S, Lovisolo D (2004) Intracellular calcium signals and control of cell proliferation: how many mechanisms? J Cell Mol Med 8(2):161–168

Munoz-Espin D, Serrano M (2014) Cellular senescence: from physiology to pathology. Nat Rev Mol Cell Biol 15(7):482–496

Nako Y, Shiozaki A, Ichikawa D, Komatsu S, Konishi H, Iitaka D et al (2012) Enhancement of the cytocidal effects of hypotonic solution using a chloride channel blocker in pancreatic cancer cells. Pancreatology 12(5):440–448

Ng EY, Sree SV, Ng KH, Kaw G (2008) The use of tissue electrical characteristics for breast cancer detection: a perspective review. Technol Cancer Res Treat 7(4):295–308

O'Grady SM, Lee SY (2005) Molecular diversity and function of voltage-gated (Kv) potassium channels in epithelial cells. Int J Biochem Cell Biol 37(8):1578–1594

Ozkucur N, Perike S, Sharma P, Funk RH (2011) Persistent directional cell migration requires ion transport proteins as direction sensors and membrane potential differences in order to maintain directedness. BMC Cell Biol 12:4

Pardo LA (2004) Voltage-gated potassium channels in cell proliferation. Physiology (Bethesda) 19:285–292

Pardo LA, del Camino D, Sanchez A, Alves F, Bruggemann A, Beckh S et al (1999) Oncogenic potential of EAG K(+) channels. EMBO J 18(20):5540–5547

Parrasia S, Rossa A, Varanita T, Checchetto V, De Lorenzi R, Zoratti M et al (2021) An Angiopep2-PAPTP construct overcomes the blood-brain barrier. New perspectives against brain tumors. Pharmaceuticals (Basel) 14(2)

Pedemonte N, Galietta LJ (2014) Structure and function of TMEM16 proteins (anoctamins). Physiol Rev 94(2):419–459

Perez-Neut M, Shum A, Cuevas BD, Miller R, Gentile S (2015a) Stimulation of hERG1 channel activity promotes a calcium-dependent degradation of cyclin E2, but not cyclin E1, in breast cancer cells. Oncotarget 6(3):1631–1639

Perez-Neut M, Rao VR, Haar L, Jones KW, Gentile S (2015b) Current and potential antiarrhythmic drugs targeting voltage-gated cardiac ion channels. Cardiol Pharmacol 4(139)

Perez-Neut M, Haar L, Rao V, Santha S, Lansu K, Rana B et al (2016a) Activation of hERG3 channel stimulates autophagy and promotes cellular senescence in melanoma. Oncotarget 7 (16):21991–22004

Perez-Neut M, Rao VR, Gentile S (2016b) hERG1/Kv11.1 activation stimulates transcription of p21waf/cip in breast cancer cells via a calcineurin-dependent mechanism. Oncotarget 7 (37):58893–58902

Peruzzo R, Mattarei A, Azzolini M, Becker-Flegler KA, Romio M, Rigoni G et al (2020) Insight into the mechanism of cytotoxicity of membrane-permeant psoralenic Kv1.3 channel inhibitors by chemical dissection of a novel member of the family. Redox Biol 37:101705

Planells-Cases R, Lutter D, Guyader C, Gerhards NM, Ullrich F, Elger DA et al (2015) Subunit composition of VRAC channels determines substrate specificity and cellular resistance to Pt-based anti-cancer drugs. EMBO J 34(24):2993–3008

Plummer HK 3rd, Dhar MS, Cekanova M, Schuller HM (2005) Expression of G-protein inwardly rectifying potassium channels (GIRKs) in lung cancer cell lines. BMC Cancer 5:104

Rao VR, Perez-Neut M, Kaja S, Gentile S (2015) Voltage-gated ion channels in cancer cell proliferation. Cancers (Basel) 7(2):849–875

Ryland KE, Hawkins AG, Weisenberger DJ, Punj V, Borinstein SC, Laird PW et al (2016) Promoter methylation analysis reveals that KCNA5 ion channel silencing supports Ewing sarcoma cell proliferation. Mol Cancer Res 14(1):26–34

Schwab A, Wulf A, Schulz C, Kessler W, Nechyporuk-Zloy V, Romer M et al (2006) Subcellular distribution of calcium-sensitive potassium channels (IK1) in migrating cells. J Cell Physiol 206 (1):86–94

Schwab A, Fabian A, Hanley PJ, Stock C (2012) Role of ion channels and transporters in cell migration. Physiol Rev 92(4):1865–1913

Serrano-Novillo C, Capera J, Colomer-Molera M, Condom E, Ferreres JC, Felipe A (2019) Implication of voltage-gated potassium channels in neoplastic cell proliferation. Cancers (Basel) 11(3)

Shennan DB (2008) Swelling-induced taurine transport: relationship with chloride channels, anion-exchangers and other swelling-activated transport pathways. Cell Physiol Biochem 21 (1–3):15–28

Struski S, Doco-Fenzy M, Cornillet-Lefebvre P (2002) Compilation of published comparative genomic hybridization studies. Cancer Genet Cytogenet 135(1):63–90

Sui YJ, Sun MY, Wu F, Yang LF, Di WH, Zhang GZ et al (2014) Inhibition of TMEM16A expression suppresses growth and invasion in human colorectal cancer cells. Plos One 9(12)

Sun J, Chu Z, Moenter SM (2010) Diurnal in vivo and rapid in vitro effects of estradiol on voltage-gated calcium channels in gonadotropin-releasing hormone neurons. J Neurosci 30 (11):3912–3923

Szabo I, Bock J, Jekle A, Soddemann M, Adams C, Lang F et al (2005) A novel potassium channel in lymphocyte mitochondria. J Biol Chem 280(13):12790–12798

Szabo I, Bock J, Grassme H, Soddemann M, Wilker B, Lang F et al (2008) Mitochondrial potassium channel Kv1.3 mediates Bax-induced apoptosis in lymphocytes. Proc Natl Acad Sci U S A 105(39):14861–14866

Szabo I, Soddemann M, Leanza L, Zoratti M, Gulbins E (2011) Single-point mutations of a lysine residue change function of Bax and Bcl-xL expressed in Bax- and Bak-less mouse embryonic fibroblasts: novel insights into the molecular mechanisms of Bax-induced apoptosis. Cell Death Differ 18(3):427–438

Szabo I, Trentin L, Trimarco V, Semenzato G, Leanza L (2015) Biophysical characterization and expression analysis of Kv1.3 potassium channel in primary human leukemic B cells. Cell Physiol Biochem 37(3):965–978

Than BL, Goos JA, Sarver AL, O'Sullivan MG, Rod A, Starr TK et al (2014) The role of KCNQ1 in mouse and human gastrointestinal cancers. Oncogene 33(29):3861–3868

Tokuoka S, Morioka H (1957) The membrane potential of the human cancer and related cells. I Gan 48(4):353–354

Vallejo-Gracia A, Bielanska J, Hernandez-Losa J, Castellvi J, Ruiz-Marcellan MC, Ramon y Cajal S et al (2013) Emerging role for the voltage-dependent K+ channel Kv1.5 in B-lymphocyte physiology: expression associated with human lymphoma malignancy. J Leukoc Biol 94 (4):779–789

Venturini E, Leanza L, Azzolini M, Kadow S, Mattarei A, Weller M et al (2017) Targeting the potassium channel Kv1.3 kills glioblastoma cells. Neurosignals 25(1):26–38

Villalonga N, Martinez-Marmol R, Roura-Ferrer M, David M, Valenzuela C, Soler C et al (2008) Cell cycle-dependent expression of Kv1.5 is involved in myoblast proliferation. Biochim Biophys Acta 1783(5):728–736

Voss FK, Ullrich F, Munch J, Lazarow K, Lutter D, Mah N et al (2014) Identification of LRRC8 heteromers as an essential component of the volume-regulated anion channel VRAC. Science 344(6184):634–638

Wang Z (2004) Roles of K+ channels in regulating tumour cell proliferation and apoptosis. Pflugers Arch 448(3):274–286

Wang XT, Nagaba Y, Cross HS, Wrba F, Zhang L, Guggino SE (2000) The mRNA of L-type calcium channel elevated in colon cancer: protein distribution in normal and cancerous colon. Am J Pathol 157(5):1549–1562

Wang H, Zhang Y, Cao L, Han H, Wang J, Yang B et al (2002) HERG K+ channel, a regulator of tumor cell apoptosis and proliferation. Cancer Res 62(17):4843–4848

Wang W, Gao Q, Yang M, Zhang X, Yu L, Lawas M et al (2015) Up-regulation of lysosomal TRPML1 channels is essential for lysosomal adaptation to nutrient starvation. Proc Natl Acad Sci U S A 112(11):E1373–E1381

Wei JF, Wei L, Zhou X, Lu ZY, Francis K, Hu XY et al (2008) Formation of Kv2.1-FAK complex as a mechanism of FAK activation, cell polarization and enhanced motility. J Cell Physiol 217 (2):544–557

Wright SH (2004) Generation of resting membrane potential. Adv Physiol Educ 28(1–4):139–142

Wu H, Wang H, Guan S, Zhang J, Chen Q, Wang X et al (2017) Cell-specific regulation of proliferation by Ano1/TMEM16A in breast cancer with different ER, PR, and HER2 status. Oncotarget 8(49):84996–85013

Xu H, Ren D (2015) Lysosomal physiology. Annu Rev Physiol 77:57–80

Xu HX, Martinoia E, Szabo I (2015) Organellar channels and transporters. Cell Calcium 58(1):1–10

Yang B, Cao L, Liu B, McCaig CD, Pu J (2013) The transition from proliferation to differentiation in colorectal cancer is regulated by the calcium activated chloride channel A1. PLos One 8(4)

Yang H, Ma L, Wang Y, Zuo W, Li B, Yang Y et al (2018) Activation of ClC-3 chloride channel by 17beta-estradiol relies on the estrogen receptor alpha expression in breast cancer. J Cell Physiol 233(2):1071–1081

Zhang H, Li H, Liu E, Guang Y, Yang L, Mao J et al (2014) The AQP-3 water channel and the ClC-3 chloride channel coordinate the hypotonicity-induced swelling volume in nasopharyngeal carcinoma cells. Int J Biochem Cell Biol 57:96–107

Zheng F, Li H, Du W, Huang S (2009) Role of hERG1 K(+) channels in leukemia cells as a positive regulator in SDF-1a-induced proliferation. Hematology 16(3):177–184

Rev Physiol Biochem Pharmacol (2022) 183: 157–176
https://doi.org/10.1007/112_2020_24

Published online: 8 August 2020

Potassium and Calcium Channel Complexes as Novel Targets for Cancer Research

Marie Potier-Cartereau, William Raoul, Gunther Weber, Karine Mahéo, Raphael Rapetti-Mauss, Maxime Gueguinou, Paul Buscaglia, Caroline Goupille, Nelig Le Goux, Souleymane Abdoul-Azize, Thierry Lecomte, Gaëlle Fromont, Aurélie Chantome, Olivier Mignen, Olivier Soriani, and Christophe Vandier

Contents

M. Potier-Cartereau, W. Raoul, G. Weber, K. Mahéo, A. Chantome, and C. Vandier (✉)
N2C UMR 1069, University of Tours, INSERM, Tours, France
e-mail: christophe.vandier@univ-tours.fr

R. Rapetti-Mauss and O. Soriani
iBV, INSERM, CNRS, University of the Côte d'Azur, Nice, France

M. Gueguinou
EA 7501 GICC, University of Tours, Tours, France

P. Buscaglia, N. Le Goux, S. Abdoul-Azize, and O. Mignen
LBAI UMR 1227, University of Brest, INSERM, Brest, France

C. Goupille
N2C UMR 1069, University of Tours, INSERM, CHRU de Tours, Tours, France

T. Lecomte
EA 7501 GICC, University of Tours, CHRU de Tours, Department of Hepato-Gastroenterology and Digestive Oncology, Tours, France

G. Fromont
N2C UMR 1069, University of Tours, INSERM, CHRU de Tours, Department of Pathology, Tours, France

Abstract The intracellular Ca^{2+} concentration is mainly controlled by Ca^{2+} channels. These channels form complexes with K^+ channels, which function to amplify Ca^{2+} flux. In cancer cells, voltage-gated/voltage-dependent Ca^{2+} channels and non-voltage-gated/voltage-independent Ca^{2+} channels have been reported to interact with K^+ channels such as Ca^{2+}-activated K^+ channels and voltage-gated K^+ channels. These channels are activated by an increase in cytosolic Ca^{2+} concentration or by membrane depolarization, which induces membrane hyperpolarization, increasing the driving force for Ca^{2+} flux. These complexes, composed of K^+ and Ca^{2+} channels, are regulated by several molecules including lipids (ether lipids and cholesterol), proteins (e.g. STIM), receptors (e.g. S1R/SIGMAR1), and peptides (e.g. LL-37) and can be targeted by monoclonal antibodies, making them novel targets for cancer research.

Keywords Ca^{2+} channels · Cancer · K^+ channels · Lipids · LL-37 · SIGMAR1 · STIM

Abbreviations

AMP	Antimicrobial peptide
AQP5	Aquaporin 5
ARC	Arachidonic acid-regulated Ca^{2+} channels
BCR	B cell receptor
BiP	Binding immunoglobulin protein
BKCa	Big conductance calcium-activated potassium channel
CaM	Calmodulin
CaV	Voltage-gated/voltage-dependent Ca^{2+} channel
CLL	Chronic lymphocytic leukemia
CRC	Colorectal cancer
DHA	Docosahexaenoic acid
DRM	Detergent-resistant membrane
EAG1	Ether-à-go-go K^+ channel 1
ECM	Extracellular matrix
EGF	Epidermal growth factor
EGFR	Epidermal growth factor receptor
EMT	Epithelial-to-mesenchymal transition
ENaC	Epithelial Na channel
EPA	Eicosapentaenoic acid
ER	Endoplasmic reticulum
FAK	Focal adhesion kinase
FRET	Fluorescence resonance energy transfer
HCC	Hepatocellular carcinoma
hERG	Human ether-à-go-go-related gene
IKCa	Intermediate conductance calcium-activated potassium channel
KCa	Calcium-activated potassium channel

KCNE2	Potassium voltage-gated channel subfamily E member 2
KCNE3	Potassium voltage-gated channel subfamily E member 3
KCNH1	Potassium voltage-gated channel subfamily H member 1
KCNH2	Potassium voltage-gated channel subfamily H member 2
KCNN4	Potassium calcium-activated channel subfamily N member 4
KCNQ1	Potassium voltage-gated channel subfamily Q member 1
Kv	Voltage-gated potassium channel
LB	Lymphocyte B
LL-37	Cathelicidin antimicrobial peptides
mAb56	Monoclonal antibody 56
mAb62	Monoclonal antibody 62
mSTIM1	Membrane stromal interaction molecule 1
Ohmline	1-O-hexadecyl-2-O-methyl-sn-glycero-3-lactose
PI3K	Phosphoinositide 3-kinase
PIP2	Phosphatidylinositol bisphosphate
PRL-3	Phosphatase of regenerating liver-3
PUFA	Polyunsaturated fatty acid
S1R	Sigma-1 receptor
SICE	Store-independent calcium entry
SIGMAR1	Sigma-1 receptor
SK3	Small conductance calcium-activated potassium channel type 3
SKCa	Small conductance calcium-activated potassium channel
SOCE	Store-operated calcium entry
SPCA2	Secretory pathway Ca^{2+} ATPase 2
STIM1	Stromal interacting molecule 1
$STIM1_{PM}$	Stromal interacting molecule 1 plasma membrane
TGFβ	Transforming growth factor β
TNM	Tumor/node/metastasis
TRAAK	TWIK-related arachidonic acid-stimulated K+ channel
TRAIL	Tumor necrosis factor-related apoptosis-inducing ligand
TREK-1	TWIK1-related K^+ channel
TRP	Transient receptor potential
TRPC1	Transient receptor potential canonical 1
TRPV2	Transient receptor potential cation channel subfamily V member 2
TWIK	Two pore domain weak inward rectifying K^+

1 Introduction

The intracellular Ca^{2+} concentration (cytosolic and within intracellular organelles) is mainly controlled by ion channels and transporters. Ca^{2+} channels are protein molecules that span the cell membrane, allowing the passage of Ca^{2+} from one side of the membrane to the other. Other ion channels like K^+ channels act in cooperation to amplify this Ca^{2+} flux. The formation of such ion channel complexes in cancer cells represents the gain of a new biological function that did not exist for the individual channel. This suggests that, in order to evolve, the cancer cell could take advantage of the association between K^+ and Ca^{2+} channels as complexes. Voltage-gated/voltage-dependent Ca^{2+} channels (CaV) and non-voltage-gated/voltage-independent Ca^{2+} channels (Transient Receptor Potential (TRP) Orai and their partners, including Stromal Interacting Molecule [STIM]) have been reported to control various biological functions of tumor cells, including proliferation and migration/invasion, and have been proposed as drug targets to inhibit cancer progression (Bong and Monteith 2018; Buchanan and McCloskey 2016; Déliot and Constantin 2015; Mignen et al. 2017). In addition, various K^+ channels, including voltage-gated K^+ channels (Kv) and Ca^{2+}-activated K^+ channels (KCa), have been reported to act as Ca^{2+} channel partners, acting as amplifiers of Ca^{2+} entry. Among these K^+ channels, Kv10.1 (also known as ether-à-go-go K^+ channel 1 [EAG1] and member 1 of the K^+ voltage-gated channel subfamily H (EAG-related) [KCNH1]), Kv11.1 (also known as hERG or KCNH2), the big conductance BKCa (also known as KCa1.1), the intermediate conductance IKCa (also known as KCa3.1 or SK4) and the small conductance SK3 (also known as KCa2.3) channels were found to form functional complexes with Ca^{2+} channels. Indeed, following activation by an increase in cytosolic Ca^{2+} concentration for KCa channels or by membrane depolarization for Kv channels, these K^+ channels induce membrane hyperpolarization, increasing the driving force for Ca^{2+} entry. In addition, these channel complexes are regulated by lipids, receptors, and peptides and can be targeted by monoclonal antibodies, making them novel targets for cancer research (Fig. 1).

2 Calcium and Potassium Channel Complexes

A few years ago, we reviewed the role of Ca^{2+} and K^+ complexes in cancer, as well as their role in controlling constitutive Ca^{2+} entry (Gueguinou et al. 2014; Mignen et al. 2017). Since these reviews were published, it was revealed that these complexes are not limited to two ion channels, and multicomplexes of ion channels can involve more than two channels. Indeed, in colon cancer cells, Gueguinou et al. (2016) detailed the role of SK3 and Ca^{2+} channels in colon cancer cell migration. In these complexes, Orai1 and TRPC1 were found to be associated with SK3, which localized in nanodomains only after phosphorylation of reticular STIM1. These complexes not only control constitutive Ca^{2+} entry but also store-operated Ca^{2+} entry (SOCE), which is activated by the depletion of endoplasmic reticulum (ER) Ca^{2+} stores and store-independent Ca^{2+} entry (SICE). Indeed, ion channels form a complex triggered by STIM1 and regulating a singular mode for SK3 in regulating Orai1/TRPC1-dependent SOCE. Interestingly, we found that epidermal growth factor (EGF) activated these complexes, SOCE, and cell migration and that anti-epidermal growth factor receptor (EGFR) monoclonal antibodies act on EGFR

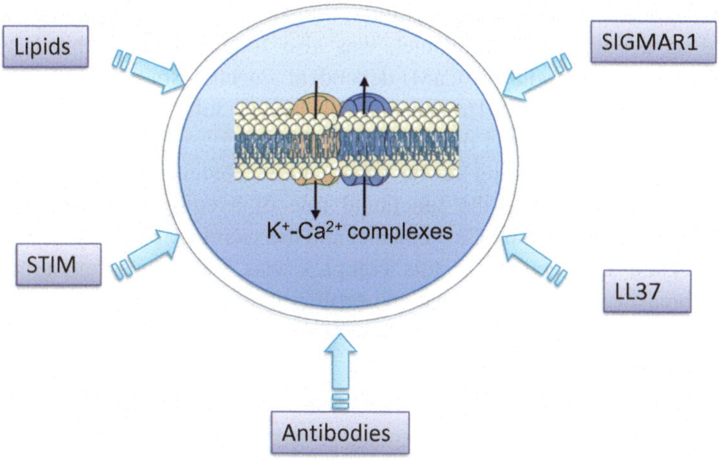

Fig. 1 The intracellular Ca^{2+} concentration is primarily controlled by Ca^{2+} channels that form complexes with K^+ channels, which act as amplifiers of Ca^{2+} flux through Ca^{2+} channels. In cancer cells, voltage-gated/voltage-dependent Ca^{2+} channels (CaV) and non-voltage-gated/voltage-independent Ca^{2+} channels have been reported to interact with Ca^{2+}-activated K^+ channels (KCa). Following activation by an increase in cytosolic Ca^{2+} concentration, these channels induce membrane hyperpolarization, increasing the driving force for Ca^{2+} flux. In addition, the entry of Ca^{2+} is regulated by various molecules including lipids (e.g. ether lipids), proteins (e.g. STIM), receptors with SIGMAR1 and peptides (e.g. LL-37), and are the target of monoclonal antibodies, making these channels promising novel targets for cancer research

to modulate SOCE activated by these complexes, leading to induction or a reduction in cancer cell migration (depending on the antibody tested). More recently, the role of Orai1 and TRPC1 channels in the constitutive Ca^{2+} entry into B lymphocytes was demonstrated (Debant et al. 2019; Garaud et al. 2018). This Ca^{2+} influx is regulated by the pool of STIM1 located at the plasma membrane (mSTIM1, also named $STIM1_{PM}$), which is increased in B cells of patients with chronic lymphocytic leukemia (CLL) with a pejorative clinical score and high lymphoproliferation rate. This constitutive Ca^{2+} entry represents an innovative therapeutic target for cancer, which is completely independent from SOCE entry supported by Orai1 channels and activated by B cell receptor (BCR) engagement in B cells.

$K_V10.1$ is a voltage-gated K^+ channel belonging to the superfamily of KCNH channels. These channels have well-known roles in cardiac physiology, cell proliferation, and neuronal excitability. The mRNA expression of this channel is mostly restricted to the brain, but expression has also been found in the testis and adrenal gland. Under physiological conditions, protein expression is strictly detected in the brain, as specified in the human atlas project (Uhlen et al. 2015). However, $K_V10.1$ has been found to be expressed in gastric and colorectal cancers and in esophageal squamous cell carcinomas (Ding et al. 2007a, b, 2008). In these cancers, $K_V10.1$ mRNA and protein were detected in over 70% of tumor samples and in adenomas, but not in adjacent matched tissues, suggesting a link between $K_V10.1$ expression and cancer onset or development. Interestingly, $K_V10.1$, which was initially described as a voltage-gated channel, was also found to be inhibited by Ca^{2+} signaling via a calmodulin (CaM)-dependent mechanism (Schonherr 2000). Marques-Carvalho et al. (2016) showed that the C lobe of CaM binds to the cytoplasmic BDC2 fragments of $K_V10.1$. In this complex, the channel adopts an unusual conformation leading to its deactivation. Recent research by Ouadid-Ahidouch's team described the functional role of a complex composed of the Kv10.1 K^+ channel, the Orai1 Ca^{2+} channel and the secretory pathway Ca^{2+} ATPase (SPCA2). Their cooperation in this complex was found to promote collagen I-induced breast cancer cell survival and proliferation (Badaoui et al. 2018; Peretti et al. 2019), as SPCA2 enhances membrane expression of both Kv10.1 and Orai1, leading to SICE.

3 Proteins Associated with Potassium and Calcium Channels

Numerous dysregulated signaling pathways are involved in cancer progression, and those involving plasma membrane proteins are good drug candidates in terms of their accessibility. Among the plasma membrane proteins that interact with Ca^{2+} channels and are deregulated in cancer, STIM1 appears to be a promising target. In fact, in the large body of available literature, STIM1 is known to be primarily located in the ER membrane and acts as a Ca^{2+} sensor, linking store depletion to store-operated Ca^{2+} channels. Many recent reviews have reported the mechanistic processes involved in STIM1 activation and its role in SOCE, as well as its deregulation and involvement in cancer (for review, Nelson and Roe 2018; Qiu and Lewis 2019). However, STIM1 was initially identified as a protein located at the plasma membrane that is involved in rhabdoid tumor growth suppression and myoblastic cell division regulation (Manji et al. 2000; Williams et al. 2002). In fact, $STIM1_{PM}$ was first thought to be a tumor suppressor protein. If the constitutive presence of STIM1 at the plasma membrane is no longer demonstrated, the role of $STIM1_{PM}$ is far from being elucidated. The role of $STIM1_{PM}$ in the activation of store-independent arachidonic acid-regulated Ca^{2+} (ARC) channels, supported by Orai3, has been known for more than 10 years (Shuttleworth et al. 2007; Thompson et al. 2013). This store-independent Ca^{2+} entry was later described as Ca^{2+} influx favoring prostate cancer cell proliferation (Dubois et al. 2014). As previously mentioned, recent work highlighted the role of $STIM1_{PM}$ in the regulation of constitutive Ca^{2+} entry of lymphocyte B (LB). An increase in $STIM1_{PM}$ together with enhanced constitutive Ca^{2+} entry is observed in patients with a poor prognosis and high LB doubling time (Debant et al. 2019). Further studies should explore the possible enhanced expression of proteins that interact with K^+ or Ca^{2+} channels in cancer cells to uncover new potential therapeutic targets.

Dysregulated signaling pathways, originating from mutation or from the interaction of cancer cells with their microenvironment, represent a driving force for cancer progression. Cell adhesion to the extracellular matrix (ECM) is primarily triggered by the binding of ECM components to integrins. These transmembrane proteins form clusters at focal adhesion sites, which promote substrate cell anchoring through the binding of mediator proteins to actin filaments. This process further activates signaling pathways to promote growth, survival or invasion initiated by focal adhesion kinase and Src protein activation (Cooper and Giancotti 2019). A growing number of studies demonstrate the involvement of K^+ channels in integrin signaling macrocomplexes in various cancers. In particular, hERG (also known as KCNH2 or Kv11.1) plays a key role in cell adhesion to the ECM; integrin activates hERG current and increases the density of hERG channels at the plasma membrane (Fiore et al. 2013; Pillozzi et al. 2007). Activation of these channels stimulates focal adhesion *kinase* (FAK)-dependent growth, angiogenesis, and survival in leukemia and colorectal cancer (CRC) via the PI3K/AKT pathway. The subsequent formation of the hERG/integrin complex stimulates cytoskeleton reorganization and cell migration (Becchetti et al. 2019). Interestingly, the association between hERG and β1-integrin occurs in cancer cells but not in the heart. Triggering this specific interaction may therefore represent an interesting strategy to unlock hERG signaling complexes in cancer cells.

Another example of the tight regulation of signaling pathways by ion channels is illustrated by the interplay between KCNQ1 and the Wnt/β-catenin pathway. The function of this channel in epithelial physiology has been largely described (Jespersen et al. 2005). In association with the β subunit of the KCNE family, especially KCNE3 and KCNE2, KCNQ1 regulates the transepithelial transport of electrolytes, solutes, and water (Heitzmann and Warth 2008). Surprisingly, KCNQ1 was recently identified as a tumor suppressor gene in mouse and human CRC (Than et al. 2014). In this study, the authors found that KCNQ1 knockout mice exhibit enhanced intestinal tumor multiplicity (number of tumors) and progression. Also, the loss of KCNQ1 expression in human CRC liver metastases has been observed to be associated with poor prognosis. In line with these results, another study demonstrated that the loss of KCNQ1 protein expression is a strong prognostic factor for an increased likelihood of recurrence and reduced survival in patients with stages II and III colon cancer (den Uil et al. 2016). Both of these reports confirmed the function of KCNQ1 as a tumor suppressor in CRC. However, the molecular mechanism underlying this role of KCNQ1 remained unknown until the identification of KCNQ1 as a component of the Wnt pathway (Rapetti-Mauss et al. 2017). In fact, KCNQ1 physically associates with β-catenin and E-cadherin at the plasma membrane to stabilize the adherens junctions (AJ) complex and control β-catenin localization. This association promotes epithelial integrity by preventing the epithelial-to-mesenchymal transition (EMT) and repressing Wnt signaling activity. Moreover, the same study showed that KCNQ1 expression itself is repressed by Wnt/β-catenin pathway activation through a direct interaction between the TCF-4/β-catenin transcription complex and the promoter region of KCNQ1. This bidirectional interaction between KCNQ1 and β-catenin highlights the function of this channel as a fine regulator of the Wnt signaling pathway. Recently, the physical interaction between KCNQ1 and β-catenin was observed in hepatocellular carcinoma (HCC) (Fan et al. 2018). The authors found that KCNQ1 expression is downregulated in HCC, and in line with the observations in CRC, patients with reduced KCNQ1 expression have lower overall survival. Furthermore, in HCC, the expression of KCNQ1 suppresses Wnt/β-catenin signaling pathway activity by interacting with β-catenin at the plasma membrane. These data suggest that KCNQ1, by sequestering β-catenin at the AJ, restricts the activation of the Wnt signaling pathway and acts as a tumor suppressor in numerous epithelial cancers. The mechanism seems to be conserved, suggesting a key role of KCNQ1 in epithelial homeostasis.

4 Role of Calcium and Potassium Channel Complexes in Epithelial-to-Mesenchymal Transition

Changes in plasma membrane ion channel expression have been reported during the EMT process, which converts epithelial cells to a mesenchymal-like phenotype and increases cancer cell invasion and migration (Azimi and Monteith 2016). Several studies have reported a critical role of Ca^{2+} as a key signaling transduction pathway regulating the induction of EMT. Major inducers of EMT, such as TGF-β (Cheng et al. 2016; Schaar et al. 2016), hypoxia, and EGF (Davis et al. 2014), lead to a transient increase in cytosolic Ca^{2+} concentration. In breast cancer, some Ca^{2+} channels have been identified to be involved in EMT, such as TRPC1/STIM1 (Schaar et al. 2016), TRPM7 (Davis et al. 2014) and Orai1/STIM1 (Hu et al. 2011). In colon cancer, the KCNN4 channel, which was found to be induced in tumor tissues compared to normal tissues (Ibrahim et al. 2019), participates in EMT induced by phosphatase of regenerating liver-3 (PRL-3). Moreover, *KCNN4* expression is positively correlated with the tumor/node/metastasis (TNM) stage of colorectal cancer (Lai et al. 2013). Recently, we identified a new signaling pathway involving a positive feedback loop between the EMT transcription factor Zeb1 and the SK3 channel, which leads to the amplification of Ca^{2+} entry and cellular migration (Figiel et al. 2019). Cytosolic Ca^{2+} is known to be involved in the expression of several EMT-associated genes. Indeed, intracellular Ca^{2+} chelation or blocking Ca^{2+} influx has been reported to reduce the expression of vimentin, Twist, Snail, and N-cadherin (Davis et al. 2014; Lai et al. 2013; Schaar et al. 2016).

5 Regulation of Potassium and Calcium Complexes

5.1 Regulation by Lipids

To be fully activated, such complexes were shown to be integrated into cholesterol-enriched nanodomains, also known as lipid rafts. This was demonstrated for Orai1-SK3 complexes in breast cancer cells (Chantome et al. 2013; Gueguinou et al. 2017) and for Orai1-TRPC1-SK3 complexes in colon cancer cells (Gueguinou et al. 2016). This suggests that the formation of ion channel complexes in cancer cells represents the gain of a new biological function, which only occurs when the complex is integrated into nanodomains. These channels may interact physically, as observed between SK3 and Orai1 channels (Chantome et al. 2013; Gueguinou et al. 2017), or may colocalize without physical interaction. These interactions between channels should be favored by their localization in caveolae and probably also by the presence of specific lipids like cholesterol in nanodomains. Several mechanisms have been proposed to explain the regulation of ion channels and interactions between channels by lipids (Fig. 2). One possible mechanism could involve a change in the biophysical properties of the membrane, as exemplified by the effect of cholesterol on membrane fluidity and bilayer thickness (Lundbaek et al. 1996; Schagina et al. 1989, 1992). More recently, the Piezo1 mechanosensitive cation channel was found to be regulated by fatty acids following a change in the membrane bending stiffness (Romero et al. 2019). In this study, margaric acid, a saturated fatty acid, was found to inhibit Piezo1 currents by increasing membrane bending stiffness, whereas polyunsaturated fatty acids (PUFA; arachidonic acid, eicosapentaenoic acid [EPA], and docosahexaenoic acid [DHA]) were found to decrease it (Romero et al. 2019). Other mechanosensitive channels such as TREK-1/TRAAK channels have been reported to be activated by inverted-conical-shaped lipids, and these lipids were found to experimentally induce convex deformation of the plasma membrane (Maingret et al. 2000). We found that the SK3 channel is inhibited by the synthetic ether-lipid 1-O-hexadecyl-2-O-methyl-sn-glycero-3-lactose (Ohmline), and we hypothesize that Ohmline interacts with cholesterol in nanodomains by removing the cholesterol OH moieties away from their main binding sites. This interaction would force new rearrangements with other lipid groups, leading to reorganization of the lipid phases and consequently SK3 channel inhibition (Herrera et al. 2017). In addition, the SK3 channel was found within cholesterol-enriched nanodomains, and MβCD was sufficient to abrogate the SK3 current, suggesting that the SK3 channel is activated by cholesterol (Gueguinou et al. 2017; Herrera et al. 2017). Another mechanism that could explain the regulation of ion channels by lipids is the involvement of specific lipid-channel interactions, as exemplified by the interaction between cholesterol and the KirBac1.1 channel (Singh et al. 2009) or fatty acids with the IKCa channel (Hamilton et al. 2003; Kacik et al. 2014). Indeed, arachidonic acid was found to inhibit IKCa through a direct interaction with the pore-lining amino acids Thr(250) and Val(275) in IKCa (Hamilton et al. 2003). In addition, the inhibition of IKCa by 14,15-epoxyeicotrienoic acids, 20-hydroxyeicosatetraeonic

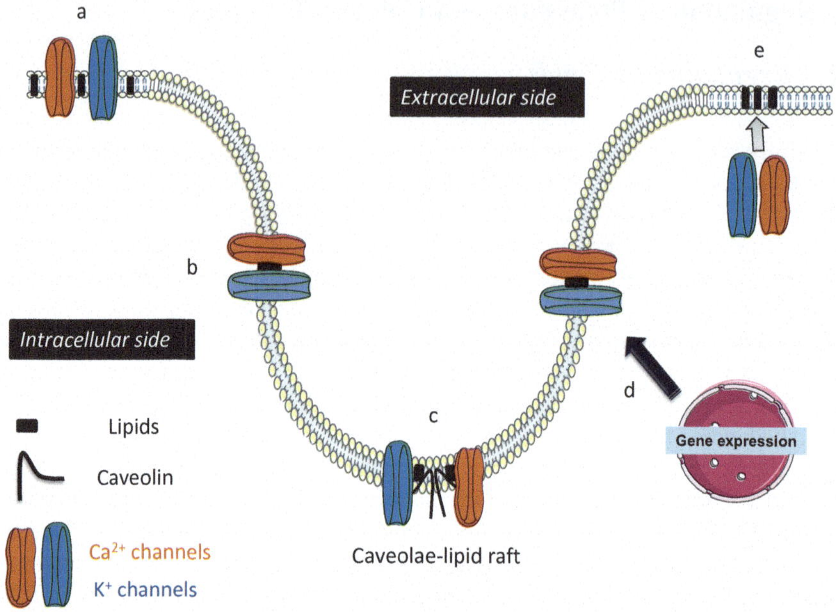

Fig. 2 Mechanisms by which lipids could favor the formation and activity of ion channel complexes. Lipids may act by (**a**) changing the biophysical properties of the membrane (Gueguinou et al. 2017; Herrera et al. 2017; Lundbaek et al. 1996; Maingret et al. 2000; Romero et al. 2019; Schagina et al. 1989, 1992), (**b**) inducing specific lipid-channel interactions (Elinder and Liin 2017; Hamilton et al. 2003; Jimenez-Garduno et al. 2014; Kacik et al. 2014; Singh et al. 2009; Tian et al. 2016; Zakany et al. 2018), (**c**) promoting interactions between ion channels and proteins such as caveolin (Alioua et al. 2008; Garg et al. 2009), (**d**) modulating ion channel expression (Lopes et al. 2018), and (**e**) modulating membrane insertion of ion channels (Pochynyuk et al. 2006)

acid, and omega-3 fatty acids (DHA) depends on the presence of electron double bonds and hydrophobicity of the ten carbons preceding the carboxyl head of the molecules (Kacik et al. 2014). Interestingly, Tian et al. detailed the atomic principles of the activation action of DHA on BKCa and found that the carboxylate group of DHA and the OH group of Y318 of BKCa form an ion dipole bond (Tian et al. 2016). Elinder and Liin reviewed the molecular sites of action and the molecular mechanism underlying the effects of PUFAs on voltage-gated ion channels (Elinder and Liin 2017). PUFAs were found to act on five different sites: the intracellular cavity, the extracellular entrance to the pore, the interface between the channel protein and the extracellular leaflet of the lipid bilayer, the voltage-sensor domain, and the interface between the extracellular leaflet of the lipid bilayer and the pore domain (Elinder and Liin 2017). $K_V10.1$ was found to be modulated by cholesterol in neurons (Jimenez-Garduno et al. 2014). Specifically, in plasma membranes isolated from mouse neuronal tissue, $K_V10.1$ was found to be split into a non-detergent-resistant membrane (DRM) fraction and a DRM fraction, associated with cholesterol and sphingolipid-rich domains, cytoskeleton integrity (actin), and

CaM/Ca^{2+} binding. Cytoskeleton integrity and cholesterol concentrations appear to act as stabilizing factors for $K_V10.1$ currents, which are increased when there are changes to the latter factors. In a well-conducted study using *Xenopus* oocytes, Zakany and colleagues aimed to investigate whether the effects of cholesterol on K^+ channels, using $K_V1.3$ as a reference channel, were mediated by the voltage sensor domains in the pore or whether it directly targeted the pore domain itself. They concluded that cholesterol modulated $K_V10.1$ in its pore domain (Zakany et al. 2018). The interaction between ion channels and proteins localized in nanodomains enriched with cholesterol and sphingolipids such as caveolin (Alioua et al. 2008; Garg et al. 2009) is another mechanism that may explain the regulation of ion channels by lipids. Finally, lipids can modulate membrane insertion of ion channels, as exemplified by PIP2, which promotes Epithelial Na channel (ENaC) insertion (Pochynyuk et al. 2006) in addition to ion channel expression, as demonstrated by the modulation of aquaporin AQP5 expression by DHA and EPA PUFAs (Lopes et al. 2018).

5.2 *Regulation by Peptides*

The integration of nonlipid molecules into the cell membrane may lead to an altered membrane structure, which may, in turn, indirectly change the activity of multiple signaling pathways. The multifunctional LL-37 peptide activates multiple membrane-associated proteins, transmembrane receptors of different classes and ion channels, triggering a variety of signal transduction pathways (Verjans et al. 2016).

The integration of peptides into the cellular membrane of target cells of both bacterial and eukaryotic origin, leading to disruption of the membrane structure, is a common characteristic of antimicrobial peptides, including LL-37. Several antimicrobial peptides (*AMPs*) form pores, leading to ion leakage. The resulting changes in membrane potential activate voltage-gated Ca^{2+} channels, leading to the induction of apoptosis (Sharma et al. 2016; Soletti et al. 2010). Such mechanisms have also been observed for LL-37 (Säll et al. 2013), leading to the hypothesis that it could be used in anticancer therapy. However, the mechanisms of action of this peptide are more complex, as it has been reported to both promote and suppress cancer depending on the cancer type (Piktel et al. 2016). In breast cancer, LL-37 promotes a metastatic phenotype (Weber et al. 2009). Consistent with this, LL-37 was also found to induce migration in breast cancer cell lines by activating the TRPV2 Ca^{2+} channel, which is recruited to pseudopodia through PI3K/AKT signaling. Entry of Ca^{2+} occurs through TRPV2, together with an efflux of K^+ through the BKCa channel (Gambade et al. 2016). Although there is currently no evidence of a physical complex between TRPV2 and BKCa, their colocalization in pseudopodia supports their functional association. Signaling by LL-37 does not appear to require its binding to a specific receptor in a conventional receptor-ligand interaction, as its all-D enantiomer shows identical activities. Instead, LL-37 was found to bind specifically to the surface of pseudopodia and caveolae, structures rich in cholesterol and known to harbor receptors activated by LL-37 (Gueguinou et al. 2015; Simons and Toomre 2000). Binding of LL-37 to these receptors results in a strong decrease in cell membrane fluidity, which can modify the kinetics and thus the activity of the transmembrane receptors (Yamamoto and Ando 2015). Taken together, these findings suggest that binding of LL-37 to the membrane interface may activate AKT signaling pathways and, consequently, Ca^{2+} signaling in an indirect manner. Binding studies of LL-37 model membranes have demonstrated that it adopts the conformation of an amphipathic helix, with the hydrophobic site inserted within the membrane interface (Sood et al. 2008). In these model studies, however, cholesterol was shown to attenuate its attachment (Sood and Kinnunen 2008). This apparent contradiction to findings in breast cancer cells would be resolved if LL-37 was found to be associated with nonlipid structures as well as the lipid bilayer. A recent investigation revealed that surface glycans, more specifically sulphated structures, were required to permit binding of LL-37 to the cell surface and, consequently, induce Ca^{2+} entry and cell migration. Syndecan-4, a proteoglycan associated with breast cancer and cell mobility, was identified to play critical roles in mediating cell surface binding and the activities of LL-37 (Habes et al. 2019). This suggests that syndecan-4 may serve as a "guide" for LL-37 to support its attachment to lipid raft domains. In conclusion, the activation of Ca^{2+} signaling and cell migration appears to involve the association of proteins, glycans and lipid structures, indicating a more complex cooperation of different classes of biomolecules than previously anticipated.

5.3 Regulation by the Sigma-1 Receptor Chaperone

The sigma-1 receptor (SIGMAR1, S1R) is a poorly characterized ER chaperone protein. In its resting state, S1R is coupled to binding immunoglobulin protein (BiP), another ER residing protein. Under conditions of ER stress, S1R dissociates from BiP and acts as an interorganelle signaling modulator (for review, Su et al. 2010; Tsai et al. 2009). In the central nervous system, S1R promotes cell survival in many diseases including stroke and other neurodegenerative diseases (e.g., amyotrophic lateral sclerosis and Alzheimer's disease) (Fukunaga et al. 2015; Kourrich et al. 2013; Penke et al. 2018). The S1R-dependent function can be mediated by protein-protein interactions with several protein superfamilies, including ion channels (Balasuriya et al. 2012, 2014). Emerging studies have revealed the presence of S1R in cancer cells. Interestingly, S1R plays a central role in the formation of ion channel complexes in cancer cells. In particular, S1R binds hERG α subunits and enhances hERG trafficking to the plasma membrane in K562 leukemia cells and transfected HEK293 cells, leading to increased current density (Balasuriya et al. 2014; Crottès et al. 2011, 2013). In myeloid leukemia and CRC, the rapid association between hERG and the $\beta 1$ subunit of integrin upon ECM stimulation requires S1R (see paragraph "Proteins associated with potassium and calcium channels"). Silencing of this chaperone abolishes both ECM-induced stimulation of hERG and the PI3/AKT pathway downstream of integrin stimulation. Consequently, S1R inhibition reduces migration, angiogenesis, and metastasis in vitro and in vivo in zebrafish and mouse models (Crottès et al. 2016). S1R also controls Ca^{2+} homeostasis by regulating the SK3/Orai1 association in CRC and breast cancer (see paragraph "Calcium and potassium channel complexes"). In fact, coimmunoprecipitation and FRET assays demonstrated that S1R binds SK3. S1R silencing was found to abrogate SK3-dependent SOCE and migration by forcing both SK3 and Orai1 out of caveolae lipid nanodomains. Interestingly, the sigma ligand igmesine mimicked these effects on Ca^{2+} influx and migration by dissociating Orai1 from SK3, the former being excluded from lipid caveolae nanodomains in MDA-MB-435s cancer cells. Notably, S1R is overexpressed in human CRC and breast cancer samples and is associated with higher-grade tumors in CRC and reduced overall survival in breast cancer patients (Gueguinou et al. 2017).

Together, these data suggest that S1R participates in the formation of ion channel complexes in cancer tissues and may represent a promising candidate to target ion channel-dependent signaling in cancers (Soriani and Rapetti-Mauss 2017).

5.4 *Regulation by Antibodies*

Monoclonal antibodies (mAbs) are currently recognized as a precision strategy to generate highly selective biologic inhibitors against cell surface-reachable antigens, which have been validated in numerous clinical trials. This is clearly highlighted by recent developments on the modulation of immune checkpoints, which rose to prominence as a means to treat a number of cancers. It is possible to distinguish a specific antigen from its nearest homologs. The therapeutic potential of ion channels and their modulators has been extensively reviewed elsewhere (Haustrate et al. 2019; Hutchings et al. 2019). To date, however, there are very few approved and/or marketed mAbs dedicated to ion channel blockade or activation. Due to its localization at the plasma membrane, therapeutic targeting of $K_V10.1$ with mAbs was first developed by Luis Pardo's team for several cancer models including breast, ovarian, and pancreatic cancers and glioma (Gomez-Varela et al. 2007; Napp et al. 2016; Pardo and Stuhmer 2008). mAb56, an IgGκ2b that targets $K_V10.1$, was the first mAb able to inhibit an ion channel current in cells. This antibody is highly selective and does not bind to human $K_v10.2$. mAb56 does not affect the $K_v11.1$ current that regulates cardiac repolarization. Another IgGκ2b, mAb62, was developed to visualize tumor cells without affecting their currents and to deliver therapeutics to the tumor. mAb62 labeled with a Cy5.5 maleimide monoreactive dye has been reported to accumulate at breast MDA-MB-435s engrafted tumor sites of immunodeficient mice, with a peak intensity observed 48 h after injection (excitation at 670 nm). mAb62 conjugated to a prodrug-activating enzyme β-D-galactosidase enabled the detection of activity in vivo at the tumor area.

It is worth noting that $K_V10.1$ is quickly internalized by endocytosis and recycled after its surface localization (Kohl et al. 2011); therefore, a classic fully human mAb could be inefficient in terms of pharmacokinetic optimization, so other formats such as antibody drug conjugates (Joubert et al. 2017) or bispecific antibodies should be developed and tested. Regarding bispecific antibodies, Hartung and colleagues developed a bispecific antibody comprising a single-chain antibody against an extracellular region of Kv10.1 (scFv62) and fused it to the human-soluble tumor necrosis factor-related *apoptosis*-inducing ligand (TRAIL), leading to a strategy that selectively induced apoptosis of Kv10.1-positive prostate tumor cells (Hartung et al. 2011). As previously mentioned, protein complexes of Orai1 and K^+ channels (Gueguinou et al. 2014, 2015; Mignen et al. 2017) contribute to Ca^{2+} influx, which could be deregulated in cancer cells. mAbs against human Orai1, inhibiting SOCE, have been developed by Amgen and Novo Nordisk and proposed as a novel therapeutic approach for the treatment of autoimmunity (Cox et al. 2013; Lin et al. 2013). Despite the number of studies suggesting changes in Orai1 expression in cancer cells and the involvement of this protein in oncogenic and metastatic processes (for review, Chalmers and Monteith 2018; Kappel et al. 2019), mAbs targeting Orai1 have not yet been evaluated as a potential therapeutic option to treat cancer. Recent work by Debant et al. clearly suggests that $STIM1_{PM}$ supports new innovative therapeutic perspectives, such as targeting $STIM1_{PM}$ for the treatment of CLL (Debant et al. 2019). mAbs to STIM1 may be associated with existing therapies that target BCR pathways, as combining an anti-STIM1 mAb with rituximab significantly reduces in vitro CLL B cell viability.

6 Conclusion

This review highlights new roles of Ca^{2+} and K^+ channel complexes in cancer and the potential use of modulators of these channels as a novel avenue for research in the treatment or prevention of cancer.

Acknowledgements This study was supported by grants from the University of Tours, the Région Centre-Val de Loire, INSERM, Canceropôle Grand Ouest, the Ligue Nationale Contre le Cancer, SATT Ouest Valorisation, the CANCEN Association, Tours Hospital Oncology Association (ACORT), the Institut National du Cancer (INCa PLBIO14-213 and PLBIO18-217) and the ministères de l'Europe et des affaires étrangères et de l'enseignement supérieur, de la recherche et de l'innovation. We thank Isabelle Domingo and Violetta Guérin for technical assistance and Catherine Leroy for secretarial support.

References

Alioua A, Lu R, Kumar Y, Eghbali M, Kundu P, Toro L, Stefani E (2008) Slo1 caveolin-binding motif, a mechanism of caveolin-1-Slo1 interaction regulating Slo1 surface expression. J Biol Chem 283:4808–4817

Azimi I, Monteith GR (2016) Plasma membrane ion channels and epithelial to mesenchymal transition in cancer cells. Endocr Relat Cancer 23:R517–R525. https://doi.org/10.1530/ERC-16-0334

Badaoui M, Mimsy-Julienne C, Saby C, Van Gulick L, Peretti M, Jeannesson P, Morjani H, Ouadid-Ahidouch H (2018) Collagen type 1 promotes survival of human breast cancer cells by overexpressing Kv10.1 potassium and Orai1 calcium channels through DDR1-dependent pathway. Oncotarget 9:24653–24671. https://doi.org/10.18632/oncotarget.19065

Balasuriya D, Stewart AP, Crottès D, Borgese F, Soriani O, Edwardson JM (2012) The sigma-1 receptor binds to the Nav1.5 voltage-gated Na+ channel with 4-fold symmetry. J Biol Chem 287:37021–37029. https://doi.org/10.1074/jbc.M112.382077

Balasuriya D, D'Sa L, Talker R, Dupuis E, Maurin F, Martin P, Borgese F, Soriani O, Edwardson JM (2014) A direct interaction between the sigma-1 receptor and the hERG voltage-gated K+ channel revealed by atomic force microscopy and homogeneous time-resolved fluorescence (HTRF®). J Biol Chem 289:32353–32363. https://doi.org/10.1074/jbc.M114.603506

Becchetti A, Petroni G, Arcangeli A (2019) Ion channel conformations regulate integrin-dependent signaling. Trends Cell Biol 29:298–307. https://doi.org/10.1016/j.tcb.2018.12.005

Bong AHL, Monteith GR (2018) Calcium signaling and the therapeutic targeting of cancer cells. Biochim Biophys Acta (BBA) Mol Cell Res 1865:1786–1794. https://doi.org/10.1016/j.bbamcr.2018.05.015

Buchanan PJ, McCloskey KD (2016) CaV channels and cancer: canonical functions indicate benefits of repurposed drugs as cancer therapeutics. Eur Biophys J 45:621–633

Chalmers SB, Monteith GR (2018) ORAI channels and cancer. Cell Calcium 74:160–167. https://doi.org/10.1016/j.ceca.2018.07.011

Chantome A, Potier-Cartereau M, Clarysse L, Fromont G, Marionneau-Lambot S, Gueguinou M, Pages JC, Collin C, Oullier T, Girault A, Arbion F, Haelters JP, Jaffres PA, Pinault M, Besson P, Joulin V, Bougnoux P, Vandier C (2013) Pivotal role of the lipid raft SK3-Orai1 complex in human cancer cell migration and bone metastases. Cancer Res 73:4852–4861

Cheng H, Wang S, Feng R (2016) STIM1 plays an important role in TGF-β-induced suppression of breast cancer cell proliferation. Oncotarget 7:16866–16878. https://doi.org/10.18632/oncotarget.7619

Cooper J, Giancotti FG (2019) Integrin signaling in cancer: Mechanotransduction, stemness, epithelial plasticity, and therapeutic resistance. Cancer Cell 35:347–367. https://doi.org/10.1016/j.ccell.2019.01.007

Cox JH, Hussell S, Søndergaard H, Roepstorff K, Bui J-V, Deer JR, Zhang J, Li Z-G, Lamberth K, Kvist PH, Padkjær S, Haase C, Zahn S, Odegard VH (2013) Antibody-mediated targeting of the Orai1 calcium channel inhibits T cell function. PLoS One 8:e82944. https://doi.org/10.1371/journal.pone.0082944

Crottès D, Martial S, Rapetti-Mauss R, Pisani DF, Loriol C, Pellissier B, Martin P, Chevet E, Borgese F, Soriani O (2011) Sig1R protein regulates hERG channel expression through a post-translational mechanism in leukemic cells. J Biol Chem 286:27947–27958. https://doi.org/10.1074/jbc.M111.226738

Crottès D, Guizouarn H, Martin P, Borgese F, Soriani O (2013) The sigma-1 receptor: a regulator of cancer cell electrical plasticity? Front Physiol 4:175. https://doi.org/10.3389/fphys.2013.00175

Crottès D, Rapetti-Mauss R, Alcaraz-Perez F, Tichet M, Gariano G, Martial S, Guizouarn H, Pellissier B, Loubat A, Popa A, Paquet A, Presta M, Tartare-Deckert S, Cayuela ML, Martin P, Borgese F, Soriani O (2016) SIGMAR1 regulates membrane electrical activity in response to extracellular matrix stimulation to drive cancer cell invasiveness. Cancer Res 76:607–618. https://doi.org/10.1158/0008-5472.CAN-15-1465

Davis FM, Azimi I, Faville RA, Peters AA, Jalink K, Putney JW, Goodhill GJ, Thompson EW, Roberts-Thomson SJ, Monteith GR (2014) Induction of epithelial-mesenchymal transition (EMT) in breast cancer cells is calcium signal dependent. Oncogene 33:2307–2316. https://doi.org/10.1038/onc.2013.187

Debant M, Burgos M, Hemon P, Buscaglia P, Fali T, Melayah S, Le Goux N, Vandier C, Potier-Cartereau M, Pers JO, Tempescul A, Berthou C, Bagacean C, Mignen O, Renaudineau Y (2019) STIM1 at the plasma membrane as a new target in progressive chronic lymphocytic leukemia. J Immunother Cancer 7:111

Déliot N, Constantin B (2015) Plasma membrane calcium channels in cancer: Alterations and consequences for cell proliferation and migration. Biochim Biophys Acta 1848:2512–2522. https://doi.org/10.1016/j.bbamem.2015.06.009

den Uil SH, Coupé VMH, Linnekamp JF, van den Broek E, Goos JACM, Delis-van Diemen PM, Belt EJT, van Grieken NCT, Scott PM, Vermeulen L, Medema JP, Bril H, Stockmann HBAC, Cormier RT, Meijer GA, Fijneman RJA (2016) Loss of KCNQ1 expression in stage II and stage III colon cancer is a strong prognostic factor for disease recurrence. Br J Cancer 115:1565–1574. https://doi.org/10.1038/bjc.2016.376

Ding XW, Luo HS, Jin X, Yan JJ, Ai YW (2007a) Aberrant expression of Eag1 potassium channels in gastric cancer patients and cell lines. Med Oncol 24:345–350

Ding XW, Luo HS, Yan JJ, An P, Lu P (2007b) Aberrant expression of EAG1 potassium channels in colorectal cancer. Chin J Pathol 36:410–411

Ding XW, Wang XG, Luo HS, Tan SY, Gao S, Luo B, Jiang H (2008) Expression and prognostic roles of Eag1 in resected esophageal squamous cell carcinomas. Dig Dis Sci 53:2039–2044. https://doi.org/10.1007/s10620-007-0116-7

Dubois C, Vanden Abeele F, Lehen'kyi V, Gkika D, Guarmit B, Lepage G, Slomianny C, Borowiec AS, Bidaux G, Benahmed M, Shuba Y, Prevarskaya N (2014) Remodeling of channel-forming ORAI proteins determines an oncogenic switch in prostate Cancer. Cancer Cell 26:19–32. https://doi.org/10.1016/j.ccr.2014.04.025

Elinder F, Liin SI (2017) Actions and mechanisms of polyunsaturated fatty acids on voltage-gated ion channels. Front Physiol 8:43

Fan H, Zhang M, Liu W (2018) Hypermethylated KCNQ1 acts as a tumor suppressor in hepatocellular carcinoma. Biochem Biophys Res Commun 503:3100–3107. https://doi.org/10.1016/j.bbrc.2018.08.099

Figiel S, Bery F, Chantôme A, Fontaine D, Pasqualin C, Maupoil V, Domingo I, Guibon R, Bruyère F, Potier-Cartereau M, Vandier C, Fromont G, Mahéo K (2019) A novel calcium-mediated EMT pathway controlled by lipids: an opportunity for prostate cancer adjuvant therapy. Cancers (Basel) 11. https://doi.org/10.3390/cancers11111814

Fiore A, Carraresi L, Morabito A, Polvani S, Fortunato A, Lastraioli E, Femia AP, De Lorenzo E, Caderni G, Arcangeli A (2013) Characterization of hERG1 channel role in mouse colorectal carcinogenesis. Cancer Med 2:583–594. https://doi.org/10.1002/cam4.72

Fukunaga K, Shinoda Y, Tagashira H (2015) The role of SIGMAR1 gene mutation and mitochondrial dysfunction in amyotrophic lateral sclerosis. J Pharmacol Sci 127:36–41. https://doi.org/10.1016/j.jphs.2014.12.012

Gambade A, Zreika S, Gueguinou M, Chourpa I, Fromont G, Bouchet AM, Burlaud-Gaillard J, Potier-Cartereau M, Roger S, Aucagne V, Chevalier S, Vandier C, Goupille C, Weber G (2016) Activation of TRPV2 and BKCa channels by the LL-37 enantiomers stimulates calcium entry and migration of cancer cells. Oncotarget 7:23785–23800

Garaud S, Taher TE, Debant M, Burgos M, Melayah S, Berthou C, Parikh K, Pers J-O, Luque-Paz D, Chiocchia G, Peppelenbosch M, Isenberg DA, Youinou P, Mignen O, Renaudineau Y, Mageed RA (2018) CD5 expression promotes IL-10 production through activation of the MAPK/Erk pathway and upregulation of TRPC1 channels in B lymphocytes. Cell Mol Immunol 15:158–170. https://doi.org/10.1038/cmi.2016.42

Garg V, Sun W, Hu K (2009) Caveolin-3 negatively regulates recombinant cardiac K(ATP) channels. Biochem Biophys Res Commun 385:472–477

Gomez-Varela D, Zwick-Wallasch E, Knotgen H, Sanchez A, Hettmann T, Ossipov D, Weseloh R, Contreras-Jurado C, Rothe M, Stuhmer W, Pardo LA (2007) Monoclonal antibody blockade of the human Eag1 potassium channel function exerts antitumor activity. Cancer Res 67:7343–7349. https://doi.org/10.1158/0008-5472.CAN-07-0107

Gueguinou M, Chantome A, Fromont G, Bougnoux P, Vandier C, Potier-Cartereau M (2014) KCa and Ca(2+) channels: the complex thought. Biochim Biophys Acta 1843:2322–2333

Gueguinou M, Gambade A, Felix R, Chantome A, Fourbon Y, Bougnoux P, Weber G, Potier-Cartereau M, Vandier C (2015) Lipid rafts, KCa/ClCa/Ca2+ channel complexes and EGFR signaling: novel targets to reduce tumor development by lipids? Biochim Biophys Acta 1848:2603–2620

Gueguinou M, Harnois T, Crottes D, Uguen A, Deliot N, Gambade A, Chantome A, Haelters JP, Jaffres PA, Jourdan ML, Weber G, Soriani O, Bougnoux P, Mignen O, Bourmeyster N, Constantin B, Lecomte T, Vandier C, Potier-Cartereau M (2016) SK3/TRPC1/Orai1 complex regulates SOCE-dependent colon cancer cell migration: a novel opportunity to modulate anti-EGFR mAb action by the alkyl-lipid Ohmline. Oncotarget 7:36168–36184

Gueguinou M, Crottes D, Chantome A, Rapetti-Mauss R, Potier-Cartereau M, Clarysse L, Girault A, Fourbon Y, Jezequel P, Guerin-Charbonnel C, Fromont G, Martin P, Pellissier B, Schiappa R, Chamorey E, Mignen O, Uguen A, Borgese F, Vandier C, Soriani O (2017) The SigmaR1 chaperone drives breast and colorectal cancer cell migration by tuning SK3-dependent Ca(2+) homeostasis. Oncogene 36:3640–3647

Habes C, Weber G, Goupille C (2019) Sulfated Glycoaminoglycans and proteoglycan Syndecan-4 are involved in membrane fixation of LL-37 and its pro-migratory effect in breast cancer cells. Biomolecules 9

Hamilton KL, Syme CA, Devor DC (2003) Molecular localization of the inhibitory arachidonic acid binding site to the pore of hIK1. J Biol Chem 278:16690–16697

Hartung F, Stuhmer W, Pardo LA (2011) Tumor cell-selective apoptosis induction through targeting of K(V)10.1 via bifunctional TRAIL antibody. Mol Cancer 10:109

Haustrate A, Hantute-Ghesquier A, Prevarskaya N, Lehen'kyi V (2019) Monoclonal antibodies targeting ion channels and their therapeutic potential. Front Pharmacol 10:606

Heitzmann D, Warth R (2008) Physiology and pathophysiology of potassium channels in gastrointestinal epithelia. Physiol Rev 88:1119–1182. https://doi.org/10.1152/physrev.00020.2007

Herrera FE, Sevrain CM, Jaffres PA, Couthon H, Grelard A, Dufourc EJ, Chantome A, Potier-Cartereau M, Vandier C, Bouchet AM (2017) Singular interaction between an Antimetastatic agent and the lipid bilayer: the Ohmline case. ACS Omega 2:6361–6370

Hu J, Qin K, Zhang Y, Gong J, Li N, Lv D, Xiang R, Tan X (2011) Downregulation of transcription factor Oct4 induces an epithelial-to-mesenchymal transition via enhancement of Ca2+ influx in breast cancer cells. Biochem Biophys Res Commun 411:786–791. https://doi.org/10.1016/j.bbrc.2011.07.025

Hutchings CJ, Colussi P, Clark TG (2019) Ion channels as therapeutic antibody targets. mAbs 11:265–296

Ibrahim S, Dakik H, Vandier C, Chautard R, Paintaud G, Mazurier F, Lecomte T, Gueguinou M, Raoul W (2019) Expression profiling of calcium channels and calcium-activated potassium channels in colorectal cancer. Cancers 11:561

Jespersen T, Grunnet M, Olesen S-P (2005) The KCNQ1 potassium channel: from gene to physiological function. Physiology (Bethesda) 20:408–416. https://doi.org/10.1152/physiol.00031.2005

Jimenez-Garduno AM, Mitkovski M, Alexopoulos IK, Sanchez A, Stuhmer W, Pardo LA, Ortega A (2014) KV10.1 K(+)-channel plasma membrane discrete domain partitioning and its functional correlation in neurons. Biochim Biophys Acta 1838:921–931. https://doi.org/10.1016/j.bbamem.2013.11.007

Joubert N, Denevault-Sabourin C, Bryden F, Viaud-Massuard MC (2017) Towards antibody-drug conjugates and prodrug strategies with extracellular stimuli-responsive drug delivery in the tumor microenvironment for cancer therapy. Eur J Med Chem 142:393–415

Kacik M, Olivan-Viguera A, Kohler R (2014) Modulation of K(Ca)3.1 channels by eicosanoids, omega-3 fatty acids, and molecular determinants. PloS One 9:e112081

Kappel S, Borgström A, Stokłosa P, Dörr K, Peinelt C (2019) Store-operated calcium entry in disease: beyond STIM/Orai expression levels. Semin Cell Dev Biol 94:66–73. https://doi.org/10.1016/j.semcdb.2019.01.003

Kohl T, Lorinczi E, Pardo LA, Stuhmer W (2011) Rapid internalization of the oncogenic K+ channel K(V)10.1. PloS One 6:e26329

Kourrich S, Hayashi T, Chuang J-Y, Tsai S-Y, Su T-P, Bonci A (2013) Dynamic interaction between sigma-1 receptor and Kv1.2 shapes neuronal and behavioral responses to cocaine. Cell 152:236–247. https://doi.org/10.1016/j.cell.2012.12.004

Lai W, Liu L, Zeng Y, Wu H, Xu H, Chen S, Chu Z (2013) KCNN4 channels participate in the EMT induced by PRL-3 in colorectal cancer. Med Oncol 30:566. https://doi.org/10.1007/s12032-013-0566-z

Lin F-F, Elliott R, Colombero A, Gaida K, Kelley L, Moksa A, Ho S-Y, Bykova E, Wong M, Rathanaswami P, Hu S, Sullivan JK, Nguyen HQ, McBride HJ (2013) Generation and characterization of fully human monoclonal antibodies against human Orai1 for autoimmune disease. J Pharmacol Exp Ther 345:225–238. https://doi.org/10.1124/jpet.112.202788

Lopes PA, Martins R, da Silva IV, Madeira MS, Prates JAM, Soveral G (2018) Modulation of aquaporin gene expression by n-3 long-chain PUFA lipid structures in white and brown adipose tissue from hamsters. Br J Nutr 120:1098–1106

Lundbaek JA, Birn P, Girshman J, Hansen AJ, Andersen OS (1996) Membrane stiffness and channel function. Biochemistry 35:3825–3830

Maingret F, Patel AJ, Lesage F, Lazdunski M, Honore E (2000) Lysophospholipids open the two-pore domain mechano-gated K(+) channels TREK-1 and TRAAK. J Biol Chem 275:10128–10133

Manji SS, Parker NJ, Williams RT, van Stekelenburg L, Pearson RB, Dziadek M, Smith PJ (2000) STIM1: a novel phosphoprotein located at the cell surface. Biochim Biophys Acta 1481:147–155. https://doi.org/10.1016/s0167-4838(00)00105-9

Marques-Carvalho MJ, Oppermann J, Munoz E, Fernandes AS, Gabant G, Cadene M, Heinemann SH, Schonherr R, Morais-Cabral JH (2016) Molecular insights into the mechanism of calmodulin inhibition of the EAG1 potassium channel. Structure 24:1742–1754. https://doi.org/10.1016/j.str.2016.07.020

Mignen O, Constantin B, Potier-Cartereau M, Penna A, Gautier M, Gueguinou M, Renaudineau Y, Shoji KF, Felix R, Bayet E, Buscaglia P, Debant M, Chantome A, Vandier C (2017) Constitutive calcium entry and cancer: updated views and insights. Eur Biophys J 46:395–413

Napp J, Pardo LA, Hartung F, Tietze LF, Stuhmer W, Alves F (2016) In vivo imaging of tumour xenografts with an antibody targeting the potassium channel Kv10.1. Eur Biophys J 45:721–733. https://doi.org/10.1007/s00249-016-1152-z

Nelson HA, Roe MW (2018) Molecular physiology and pathophysiology of stromal interaction molecules. Exp Biol Med (Maywood) 243:451–472. https://doi.org/10.1177/1535370218754524

Pardo LA, Stuhmer W (2008) Eag1: an emerging oncological target. Cancer Res 68:1611–1613. https://doi.org/10.1158/0008-5472.CAN-07-5710

Penke B, Fulop L, Szucs M, Frecska E (2018) The role of Sigma-1 receptor, an intracellular chaperone in neurodegenerative diseases. Curr Neuropharmacol 16:97–116. https://doi.org/10.2174/1570159X15666170529104323

Peretti M, Badaoui M, Girault A, Van Gulick L, Mabille MP, Tebbakha R, Sevestre H, Morjani H, Ouadid-Ahidouch H (2019) Original association of ion transporters mediates the ECM-induced breast cancer cell survival: Kv10.1-Orai1-SPCA2 partnership. Sci Rep 9:1175. https://doi.org/10.1038/s41598-018-37602-7

Piktel E, Niemirowicz K, Wnorowska U, Wątek M, Wollny T, Głuszek K, Góźdź S, Levental I, Bucki R (2016) The role of cathelicidin LL-37 in Cancer development. Arch Immunol Ther Exp 64:33–46. https://doi.org/10.1007/s00005-015-0359-5

Pillozzi S, Brizzi MF, Bernabei PA, Bartolozzi B, Caporale R, Basile V, Boddi V, Pegoraro L, Becchetti A, Arcangeli A (2007) VEGFR-1 (FLT-1), beta1 integrin, and hERG K+ channel for a

macromolecular signaling complex in acute myeloid leukemia: role in cell migration and clinical outcome. Blood 110:1238–1250. https://doi.org/10.1182/blood-2006-02-003772
Pochynyuk O, Medina J, Gamper N, Genth H, Stockand JD, Staruschenko A (2006) Rapid translocation and insertion of the epithelial Na+ channel in response to RhoA signaling. J Biol Chem 281:26520–26527
Qiu R, Lewis RS (2019) Structural features of STIM and Orai underlying store-operated calcium entry. Curr Opin Cell Biol 57:90–98. https://doi.org/10.1016/j.ceb.2018.12.012
Rapetti-Mauss R, Bustos V, Thomas W, McBryan J, Harvey H, Lajczak N, Madden SF, Pellissier B, Borgese F, Soriani O, Harvey BJ (2017) Bidirectional KCNQ1:β-catenin interaction drives colorectal cancer cell differentiation. Proc Natl Acad Sci U S A 114:4159–4164. https://doi.org/10.1073/pnas.1702913114
Romero LO, Massey AE, Mata-Daboin AD, Sierra-Valdez FJ, Chauhan SC, Cordero-Morales JF, Vasquez V (2019) Dietary fatty acids fine-tune Piezo1 mechanical response. Nat Commun 10:1200
Säll J, Carlsson M, Gidlöf O, Holm A, Humlén J, Ohman J, Svensson D, Nilsson B-O, Jönsson D (2013) The antimicrobial peptide LL-37 alters human osteoblast Ca2+ handling and induces Ca2+-independent apoptosis. J Innate Immun 5:290–300. https://doi.org/10.1159/000346587
Schaar A, Sukumaran P, Sun Y, Dhasarathy A, Singh BB (2016) TRPC1-STIM1 activation modulates transforming growth factor β-induced epithelial-to-mesenchymal transition. Oncotarget 7:80554–80567. https://doi.org/10.18632/oncotarget.12895
Schagina LV, Blasko K, Grinfeldt AE, Korchev YE, Lev AA (1989) Cholesterol-dependent gramicidin A channel inactivation in red blood cell membranes and lipid bilayer membranes. Biochim Biophys Acta 978:145–150
Schagina LV, Korchev YE, Grinfeldt AE, Lev AA, Blasto K (1992) Sterol specific inactivation of gramicidin A induced membrane cation permeability. Biochim Biophys Acta 1109:91–96
Schonherr R (2000) Inhibition of human ether a go-go potassium channels by Ca2+/calmodulin. EMBO J 19:3263–3271. https://doi.org/10.1093/emboj/19.13.3263
Sharma S, Sahoo N, Bhunia A (2016) Antimicrobial peptides and their pore/ion channel properties in neutralization of pathogenic microbes. Curr Top Med Chem 16:46–53. https://doi.org/10.2174/1568026615666150703115454
Shuttleworth TJ, Thompson JL, Mignen O (2007) STIM1 and the noncapacitative ARC channels. Cell Calcium 42:183–191. https://doi.org/10.1016/j.ceca.2007.01.012
Simons K, Toomre D (2000) Lipid rafts and signal transduction. Nat Rev 1:31–39
Singh DK, Rosenhouse-Dantsker A, Nichols CG, Enkvetchakul D, Levitan I (2009) Direct regulation of prokaryotic Kir channel by cholesterol. J Biol Chem 284:30727–30736
Soletti RC, del Barrio L, Daffre S, Miranda A, Borges HL, Moura-Neto V, Lopez MG, Gabilan NH (2010) Peptide gomesin triggers cell death through L-type channel calcium influx, MAPK/ERK, PKC and PI3K signaling and generation of reactive oxygen species. Chem Biol Interact 186:135–143. https://doi.org/10.1016/j.cbi.2010.04.012
Sood R, Kinnunen PK (2008) Cholesterol, lanosterol, and ergosterol attenuate the membrane association of LL-37(W27F) and temporin L. Biochim Biophys Acta 1778:1460–1466
Sood R, Domanov Y, Pietiainen M, Kontinen VP, Kinnunen PK (2008) Binding of LL-37 to model biomembranes: insight into target vs host cell recognition. Biochim Biophys Acta 1778:983–996
Soriani O, Rapetti-Mauss R (2017) Sigma 1 receptor and ion channel dynamics in Cancer. Adv Exp Med Biol 964:63–77. https://doi.org/10.1007/978-3-319-50174-1_6
Su T-P, Hayashi T, Maurice T, Buch S, Ruoho AE (2010) The sigma-1 receptor chaperone as an inter-organelle signaling modulator. Trends Pharmacol Sci 31:557–566. https://doi.org/10.1016/j.tips.2010.08.007
Than BLN, Goos JACM, Sarver AL, O'Sullivan MG, Rod A, Starr TK, Fijneman RJA, Meijer GA, Zhao L, Zhang Y, Largaespada DA, Scott PM, Cormier RT (2014) The role of KCNQ1 in mouse and human gastrointestinal cancers. Oncogene 33:3861–3868. https://doi.org/10.1038/onc.2013.350

Thompson JL, Mignen O, Shuttleworth TJ (2013) The ARC channel – an endogenous store-independent Orai Channel. In: Current topics in membranes. Elsevier, Amsterdam, pp 125–148. https://doi.org/10.1016/B978-0-12-407870-3.00006-8

Tian Y, Aursnes M, Hansen TV, Tungen JE, Galpin JD, Leisle L, Ahern CA, Xu R, Heinemann SH, Hoshi T (2016) Atomic determinants of BK channel activation by polyunsaturated fatty acids. Proc Natl Acad Sci U S A 113:13905–13910

Tsai S-Y, Hayashi T, Mori T, Su T-P (2009) Sigma-1 receptor chaperones and diseases. Cent Nerv Syst Agents Med Chem 9:184–189. https://doi.org/10.2174/1871524910909030184

Uhlen M, Fagerberg L, Hallstrom BM, Lindskog C, Oksvold P, Mardinoglu A, Sivertsson A, Kampf C, Sjostedt E, Asplund A, Olsson I, Edlund K, Lundberg E, Navani S, Szigyarto CA, Odeberg J, Djureinovic D, Takanen JO, Hober S, Alm T, Edqvist PH, Berling H, Tegel H, Mulder J, Rockberg J, Nilsson P, Schwenk JM, Hamsten M, von Feilitzen K, Forsberg M, Persson L, Johansson F, Zwahlen M, von Heijne G, Nielsen J, Ponten F (2015) Proteomics Tissue-based map of the human proteome. Science 347:1260419. https://doi.org/10.1126/science.1260419

Verjans ET, Zels S, Luyten W, Landuyt B, Schoofs L (2016) Molecular mechanisms of LL-37-induced receptor activation: an overview. Peptides 85:16–26

Weber G, Chamorro CI, Granath F, Liljegren A, Zreika S, Saidak Z, Sandstedt B, Rotstein S, Mentaverri R, Sánchez F, Pivarcsi A, Ståhle M (2009) Human antimicrobial protein hCAP18/LL-37 promotes a metastatic phenotype in breast cancer. Breast Cancer Res 11:R6. https://doi.org/10.1186/bcr2221

Williams RT, Senior PV, Van Stekelenburg L, Layton JE, Smith PJ, Dziadek MA (2002) Stromal interaction molecule 1 (STIM1), a transmembrane protein with growth suppressor activity, contains an extracellular SAM domain modified by N-linked glycosylation. Biochim Biophys Acta 1596:131–137. https://doi.org/10.1016/s0167-4838(02)00211-x

Yamamoto K, Ando J (2015) Vascular endothelial cell membranes differentiate between stretch and shear stress through transitions in their lipid phases. Am J Physiol 309:H1178–H1185

Zakany F, Pap P, Papp F, Kovacs T, Nagy P, Peter M, Szente L, Panyi G, Varga Z (2018) Determining the target of membrane sterols on voltage-gated potassium channels. Biochim Biophys Acta Mol Cell Biol Lipids. https://doi.org/10.1016/j.bbalip.2018.12.006

Rev Physiol Biochem Pharmacol (2022) 183: 177–216
https://doi.org/10.1007/112_2020_30

Published online: 7 August 2020

Solute Carrier Transportome in Chemotherapy-Induced Adverse Drug Reactions

Jason T. Anderson, Kevin M. Huang, Maryam B. Lustberg, Alex Sparreboom, and Shuiying Hu

Contents

Abstract Members of the solute carrier (SLC) family of transporters are responsible for the cellular influx of a broad range of endogenous compounds and xenobiotics. These proteins are highly expressed in the gastrointestinal tract and eliminating organs such as the liver and kidney, and are considered to be of particular importance in governing drug absorption and elimination. Many of the same transporters are also expressed in a wide variety of organs targeted by clinically important anticancer

A.S. and S.H. originated the concept, J.A. wrote the paper, all authors contributed to manuscript review and editing, and all authors approved the final version.

J. T. Anderson, K. M. Huang, A. Sparreboom, and S. Hu (✉)
Division of Pharmaceutics and Pharmacology, College of Pharmacy and Comprehensive Cancer Center, The Ohio State University, Columbus, OH, USA
e-mail: hu.1333@osu.edu

M. B. Lustberg
Department of Medical Oncology, The Ohio State University, Comprehensive Cancer Center, Columbus, OH, USA

drugs, directly affect cellular sensitivity to these agents, and indirectly influence treatment-related side effects. Furthermore, targeted intervention strategies involving the use of transport inhibitors have been recently developed, and have provided promising lead candidates for combinatorial therapies associated with decreased toxicity. Gaining a better understanding of the complex interplay between transporter-mediated on-target and off-target drug disposition will help guide the further development of these novel treatment strategies to prevent drug accumulation in toxicity-associated organs, and improve the safety of currently available treatment modalities. In this report, we provide an update on this rapidly emerging field with particular emphasis on anticancer drugs belonging to the classes of taxanes, platinum derivatives, nucleoside analogs, and anthracyclines.

Keywords Adverse drug reactions · Anticancer · Solute carrier · Toxicity

1 Introduction

The expression and localization of transporters within specific tissues contributes to a dynamic interplay between intracellular substrate concentrations and the extracellular environment of various cell types. Disruption of this sensitive balance has the potential to modify intracellular accumulation of xenobiotics and may contribute to increases in the incidence and severity of tissue-specific organ damage. In this article, we will focus on the contribution of solute carrier (SLC) family members to the initial cellular influx of substrates and how these proteins can directly contribute to toxicities associated with small-molecule oncology drugs. Of note, transporters involved in cellular efflux, such as members of the ATP-binding cassette (ABC) family, might also play a role in toxicities associated with the same drugs. However, these proteins are predicted to mitigate the risk of toxicity rather than precipitating toxicity, and this area is beyond the scope of the present review. Rather than presenting a comprehensive overview of the field, we aimed in this article to highlight examples of well-established toxicities associated with commonly used anticancer drugs that are dependent on specific solute carriers in order to stimulate discussion within the transporter community and further advance the field.

1.1 Transporter Function

A selectively permeable plasma membrane is a ubiquitous feature of all life forms (Grecco et al. 2011; Singer and Nicolson 1972), and membrane transporters are the key regulators of this selective cellular permeability (Kaback et al. 2001). These proteins mediate the uptake and efflux of many endogenous metabolites such as amino acids, nucleosides, sugars, as well as many dietary compounds and therapeutic agents (Borst and Elferink 2002). Therefore, along with their essential contribution to normal physiology and pathophysiology, membrane transporters are also key determinants of therapeutic responses to drugs, including unwanted adverse events.

The human genome encodes more than 400 membrane-transporter genes belonging to two major super-families: ABC transporters and SLC transporters, which are involved in most essential biological processes (Borst and Elferink 2002; He et al. 2009; Hediger et al. 2013; Nigam 2015). Among these, about 20 "multi-specific" transporters belonging to either super-family have been widely implicated in drug transport (DeGorter et al. 2012; Giacomini et al. 2012; Nigam 2015). Tissue types that are involved in the absorption, distribution, metabolism, and excretion (ADME), such as the kidney, liver, intestine, and endothelial barriers, have high expression of transporters that accumulate substrates within these organ types (International Transporter Consortium et al. 2010). At the cellular level, transporter-mediated uptake or efflux can involve drug sensitive or resistant (Borst et al. 2000) phenotypes in target cells, thereby affecting therapeutic efficacy. Conversely, transporter-mediated drug accumulation in non-target cells can contribute to drug toxicity profiles (Sprowl and Sparreboom 2014). Consequently, drug transporters, in addition to drug-metabolizing enzymes, have emerged in recent years to be critical determinants of drug disposition, therapeutic efficacy, toxicity profiles, and drug–drug.

1.2 Role of SLCs in Toxicity

Increasing evidence has confirmed that SLC expression and localization at non-target tissues can play an important role in drug distribution and subsequent toxicity profiles (Sprowl and Sparreboom 2014; Yang and Han 2019). In addition, the ability of drugs to compete for the natural substrates of these transporters can potentially lead to altered cellular function and trigger unwanted adverse reactions. Since virtually all currently used oncology drugs can cause severe dose-limiting toxic side effects and, in some cases, cause life-threatening toxicities associated with organ damage, knowledge of specific SLCs that recognize such drugs can theoretically contribute to the development of improved and safer treatment strategies. Furthermore, the contribution of SLCs to observed differences in chemotherapeutic clinical response rates and toxicity profiles between genders remains poorly understood. The identification of pharmacodynamic biomarkers of SLC function that can potentially guide treatment decisions has led to the design of strategies involving the administration of concurrent medications that ameliorate the incidence and/or severity of these side effects. Besides a direct contribution of SLCs to the tissue-specific uptake of anticancer drugs, these proteins can also indirectly contribute to altered drug distribution patterns due to their involvement in clearance mechanisms in organs of elimination such as the liver and kidney. In the last few decades, technological advances in cloning have resulted in the identification of several important

SLC families mediating the transport of organic cations and organic anions, the so-called organic cation transporters (OCTs), organic anion transporters (OATs), and organic anion transporting polypeptides (OATPs).

1.2.1 Organic Cation Transport

About 40% of approved prescription drugs are positively charged at neutral pH ("organic cations"), and the membrane transport of these agents depends on facilitated carriers. In recent years, considerable progress has been made in the study of transporters belonging to the OCT family. Subsequent studies in heterologous expression models have confirmed that members of this transporter family mediate the cellular uptake of many structurally diverse endogenous compounds and an increasingly large number of cationic anticancer drugs. The organic cation transporters OCT1 (*SLC22A1*) and OCT2 (*SLC22A2*) have particular relevance in this connection, since they are highly expressed at the basolateral membranes of hepatocytes and renal tubular cells, respectively, and these proteins are considered major transporters in the secretion of organic cations from the circulation into the liver and kidney. Consequently, OCT1 and OCT2 facilitate the hepatocellular and renal excretion of organic cationic compounds and play an important role in governing systemic elimination of many drugs. The contribution of OCTs to hepatic and renal organic cation secretion was first conclusively demonstrated from the clearance of the prototypic organic cation, tetraethylammonium (TEA), in mice from which Oct1, Oct2, or both were eliminated (Jonker et al. 2001, 2003). These studies demonstrated that the renal TEA clearance in the Oct1-null mouse [Oct1(−/−)] was actually increased, reflecting (1) reduced hepatic clearance, (2) elevated plasma concentrations of TEA, and (3) sufficient expression of Oct2 in the kidney to handle the increased plasma load. Renal TEA clearance in the Oct2-null mouse [Oct2(−/−)] was unchanged compared to control, reflecting sufficient expression of Oct1 in the murine kidney to efficiently clear TEA, even in the absence of Oct2. Most importantly, the elimination of both Oct1 and Oct2 in mice [Oct1/2(−/−)] completely eliminated active secretion of TEA. Although mice express substantial levels of both Oct1 and Oct2 in the kidney, in humans there is strong agreement that OCT2 dominates renal organic cation transport, whereas OCT1 dominates hepatic organic cation transport.

In the liver, a transporter belonging to the class of multidrug and toxin extrusion (MATE) proteins called MATE1 (*SLC47A1*) is localized at the bile-canalicular membrane of hepatocytes, and forms a functional unit with the basolaterally expressed OCT1 to mediate the secretion of many cationic drugs from the circulation across the hepatocyte into the bile (Fig. 1). In the luminal membrane of the proximal tubular epithelium, MATE1 cooperates with the basolaterally expressed OCT2 in the vectorial renal secretion of organic cations (Fig. 1). Previously reported experimental data indicate that the tubular secretion of several dual OCT2/MATE1 substrates, including cisplatin (Franke et al. 2010b) and oxaliplatin (Sprowl et al. 2013a), is decreased in Oct1/2(−/−) mice, but not in Oct1(−/−) and Oct2(−/−) mice (Filipski

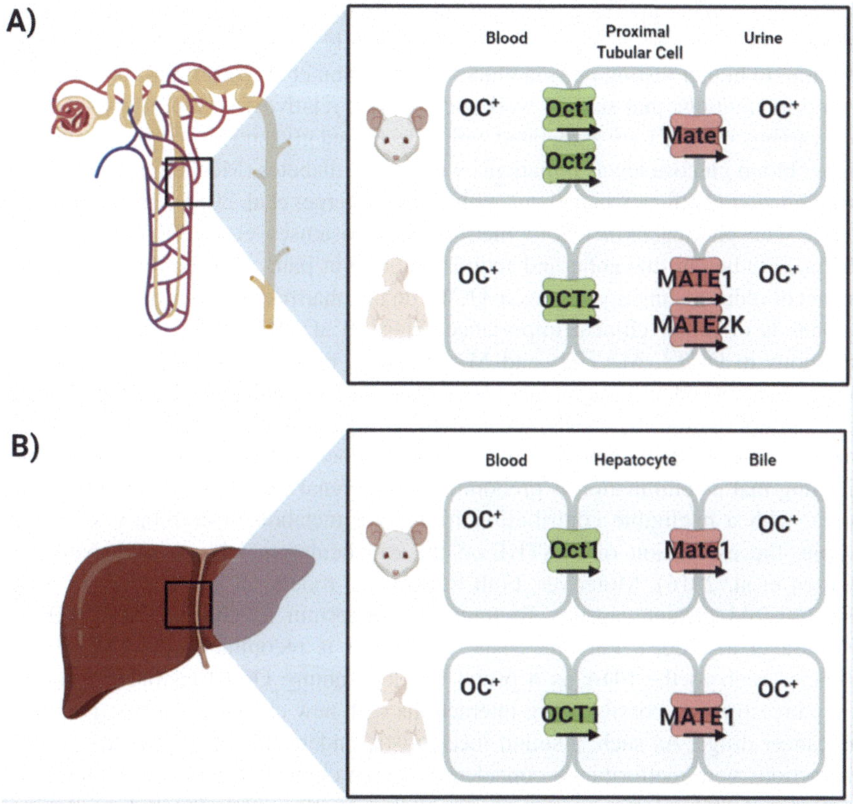

Fig. 1 Schematic depiction of the vectorial transport of organic cations (OC+) in the kidney (**a**) and liver (**b**) of rodents and humans

et al. 2009), as well as in Mate1 knockout [Mate1(–/–)] mice (Li et al. 2013; Nakamura et al. 2010). The contribution of these transporters to the hepatic and renal handling of xenobiotics in mammals has been most extensively studied for the biguanide analog metformin, a first-line medication for the treatment of type 2 diabetes that has intrinsic anticancer and chemo-preventative properties as well (Chae et al. 2016; Heckman-Stoddard et al. 2016). These studies have demonstrated that OCT1 mediates >75% of metformin uptake into hepatocytes, and that OCT2 (Oct1/2 in mice) mediates >60% of metformin uptake into renal tubular cells (Higgins et al. 2012). Consequently, the hepatic exposure of metformin is decreased ~4–8 fold in Oct1(–/–) mice (Shu et al. 2007; Wang et al. 2002), decreased 4.2-fold in Oct1/2 (–/–) mice (Higgins et al. 2012), and increased 8.4-fold in Mate1(–/–) mice (Li et al. 2013). In the case of Oct1/2- or Mate1-deficiency, this resulted in significantly altered systemic exposure to metformin (Higgins et al. 2012; Tsuda et al. 2009).

Because hepatic uptake is required for inhibition of gluconeogenesis, the pharmacodynamic properties of metformin are markedly attenuated in Oct1(−/−) mice resulting in higher fasting plasma glucose levels (Shu et al. 2007). This is consistent with prior findings that several well-documented, relatively common loss-of-function variants in OCT1 are associated with a decreased glycemic response resulting in higher blood glucose levels in patients with type 2 diabetes (McCreight et al. 2016), and with altered efficacy in patients with cancer (Joerger et al. 2015), but not with the steady-state pharmacokinetics of metformin (Christensen et al. 2015; Stage et al. 2015a). Studies in dizygotic and monozygotic twin pairs have confirmed that the impact of reduced-function alleles in OCT1 on the pharmacokinetics of metformin in humans is of minor clinical importance (Stage et al. 2015b). In contrast, genetic variations in OCT2, MATE1, and MATE2-K, a related transporter expressed in human but not rodent kidney, have been shown to significantly affect the pharmacokinetics of metformin (Pawlyk et al. 2014; Wang and Weinshilboum 2014). These studies are consistent with the known physicochemical properties of metformin dictating that its elimination is predominantly by renal excretion (up to 90% of the dose), with a negligible contribution from liver metabolism or biliary secretion, despite the expression of MATE1 on the canalicular membrane of hepatocytes (Jensen et al. 2016). Moreover, both human and rodent OCT2 have about a 10- and 100-fold greater capacity to transport metformin as compared with OCT1 (Kimura et al. 2005). On this basis, metformin is recommended by regulatory agencies such as the FDA as a probe for determining OCT2-mediated transport when investigating possible drug interactions with new chemical entities, including anticancer drugs. As such, a sound mechanistic understanding of pharmacokinetic interactions with metformin is considered of high clinical importance. It should be pointed out that interactions involving altered levels within the kidney are not necessarily reflected in an equivalent change in systemic exposure (Sprowl and Sparreboom 2014). Nonetheless, published studies have demonstrated that many known OCT2 inhibitors, including cimetidine, dolutegravir (Song et al. 2016; Zong et al. 2014), ranitidine (Cho et al. 2014), and verapamil (Cho and Chung 2016) can significantly increase the area under the curve (AUC) of metformin (Koepsell 2015). Furthermore, it has been recommended that dose adjustments of metformin be considered to maintain optimal glycemic control when patients are starting/stopping these agents while taking metformin (Song et al. 2016). The clinical impact of these interactions is further supported by the recent finding in a population of >400,000 incidental metformin users that the risk of early therapy discontinuation, as a proxy for intolerance, is associated with the concomitant use of drugs that are known to inhibit OCT2 and/or MATE1 (Stage et al. 2016).

1.2.2 Organic Anion Transport

In the liver, two transporters belonging to the class of OATPs called OATP1B1 (*SLCO1B1*) and OATP1B3 (*SLCO1B3*; Fig. 2) are localized at basolateral membrane of hepatocytes and mediate the uptake of a remarkably broad range of

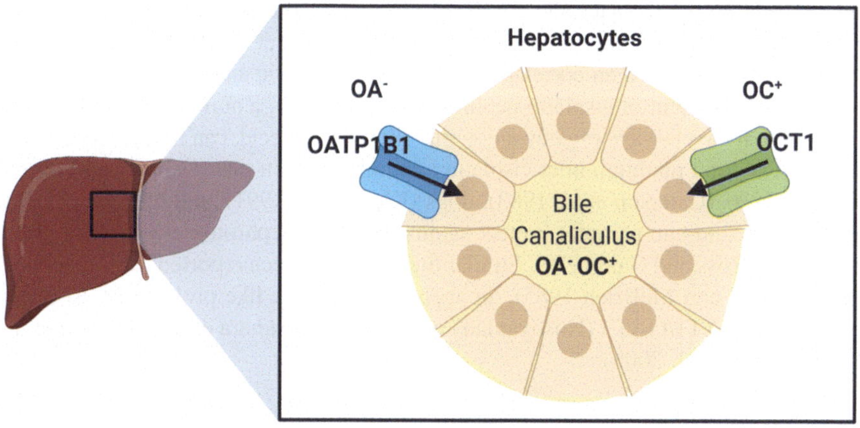

Fig. 2 Schematic depiction of the transport of organic anions (OA-) and organic cations (OC+) in human hepatocytes by OCT1 and OATP1B1

substrates from the circulation across the outer membrane of hepatocytes. These agents include charged organic anions (e.g., methotrexate and statins), endogenous and xenobiotic glucuronides (e.g., of bilirubin and sorafenib), charged organic cations (e.g., tyrosine kinase inhibitors such as imatinib), polar zwitterions (e.g., fexofenadine), uncharged hydrophobic agents (e.g., taxanes such as paclitaxel), as well as chemically diverse platinum-based therapeutics (e.g., cisplatin) (Lancaster et al. 2013). The in vivo pharmacological characterization of this transport mechanism was first performed in mice knockout for the orthologue transporter Oatp1b2 [Oatp1b2(–/–)] (Zaher et al. 2008) with the prototypical substrate pravastatin, a member of the class of statins used for the treatment of dyslipidemia and the prevention of cardiovascular disease. This initial work demonstrated that the liver-to-plasma ratios of pravastatin were reduced by four to fivefold in Oatp1b2(–/–) mice, which was independently verified in another Oatp1b2-deficient mouse strain (Salphati et al. 2014). Subsequent investigation demonstrated that 95–99% of pravastatin uptake into the liver is mediated by OATP1B1 (Izumi et al. 2018), and that the altered liver-to-plasma ratios of pravastatin, as well as other substrates (Zimmerman et al. 2013), observed in Oatp1b2(–/–) mice can be at least partially restored in humanized transgenic animals with liver-specific expression of OATP1B1 (Salphati et al. 2014).

Historically, the two major transport families contributing the renal secretion of organic compounds were divided into two different groups named "organic cation system" and "organic anion system"; classical substrate for the former included TEA, which was inhibited by cimetidine, while the latter was shown to efficiently transport p-aminohippurate (PAH) and inhibited by probenecid (Hagenbuch 2010). It was not until the late 1990s that the transport systems responsible for the organic anion systems in the kidney were identified as OAT1 (SLC22A3) and OAT3 (SLC22A5) (VanWert et al. 2010). Even though early studies showed that organic anions such as pyrazionate and PAH did not impact the accumulation of organic cations such as cisplatin in renal cortex slices (Safirstein et al. 1984), recent in vivo investigation suggested an correlation between the organic anion system disruption and the development of cisplatin nephrotoxicity. Of note, a number of studies have reported that the classical organic anion inhibitor, probenecid, can reduce the tubular secretion of total platinum after cisplatin administration in rats (Osman and Litterst 1983), rabbits (Caterson et al. 1983), dogs (Klein et al. 1991), and humans (Jacobs et al. 1984), and can partially protect against cisplatin nephrotoxicity in mice (Ban et al. 1994; Ross and Gale 1979). Similar findings have been reported for furosemide (Daley-Yates and McBrien 1985), an agent now known, like probenecid, to be an inhibitor of OAT1/OAT3 (Hosoyamada et al. 1999; Kusuhara et al. 1999; Lu et al. 1999; Vallon et al. 2008).

2 Toxicity Induced by Chemotherapeutics

2.1 OATPs and Paclitaxel-Induced Peripheral Neuropathy

Paclitaxel is a taxane antineoplastic agent that elicits its antitumor effects by disrupting the microtubule dynamics and causing mitotic arrest and cell death in a variety of tumor types. Paclitaxel remains among the most widely used drugs in the treatment of a variety of solid tumors, including early-stage breast cancer, but its clinical use is associated with debilitating damage to peripheral nerves (neuropathy). This damage is a tremendous health problem worldwide and remains one of the most important complications of contemporary oncology regimens as it may limit further use of curative-intent treatment and/or may cause long-term quality of life concerns. Since the majority of patients with cancer receiving paclitaxel-based chemotherapy are at high risk of experiencing peripheral neuropathy, a large community of cancer survivors could potentially benefit from the SLC contributing to this chemotherapy-induced toxicity.

Several tubulin poisons, including paclitaxel, induce a chronic, dose-dependent sensory peripheral neuropathy that is characterized by tingling, numbness, increased sensitivity to cold and touch, and burning pain of the distal extremities. The incidence of this side effect is particularly high in the case of paclitaxel, as it occurs in up to 70–80% in patients with breast cancer (De Iuliis et al. 2015). With continued dosing, the painful symptoms increase in severity and can persist for years (Peters et al. 2007), or even cause a lifelong functional impairment that impacts quality of life (Mielke et al. 2006). The mechanistic basis of this side effect has remained uncertain until relatively recently (Carozzi et al. 2015). Previously, studies have reported that paclitaxel is able to induce injury to sensory neurons resulting in morphological and biochemical changes in the dorsal root ganglion (DRG) satellite glial cells, proliferation of macrophages within the peripheral nervous system, activation and increases within the microglial and astrocyte populations within the spinal cord (Brewer et al. 2016; Marmiroli and Cavaletti 2016). Additionally, paclitaxel also causes acute pain syndrome in patients that precipitates within 1–3 days of paclitaxel administration and resides within 1 week. This acute pain syndrome has been postulated to be caused by the activation of toll-like receptor 4 (TLR4) in the DRG and spinal dorsal horn (Yan et al. 2015), which has been shown to be the main site of paclitaxel accumulation within the nervous system (Cavaletti et al. 2000).

The notion that the DRG neurons play a central role in the elicit side effects associated with paclitaxel chemotherapy is supported by the rationale that paclitaxel has easy accessibility to the DRG and subsequent accumulation which is supported by reports of detectable levels of paclitaxel in the spinal cord and sciatic nerve, presumably mediated by the transport along the centripetal and centrifugal branches of the axon in DRGs (Cavaletti et al. 2000). Furthermore, previous cellular uptake studies have demonstrated that paclitaxel, along with other structurally related taxanes such as docetaxel, accumulates via a facilitated transport mechanism (Smith et al. 2005). Consequently, after administration of paclitaxel, tissue disposition patterns and resulting pathological changes are restricted to cell types that are capable of transporting the movement of paclitaxel from the extracellular environment. This is consistent with accumulating evidence that the transmembrane transport of paclitaxel is mediated by specific OATPs. In particular, it has been reported that both paclitaxel and docetaxel are transported substrates of human OATP1B1 and OATP1B3 (Baker et al. 2009; de Graan et al. 2012; Smith et al. 2005, 2007), as well as the single functional homologue Oatp1b2 in mice (Nieuweboer et al. 2014) and rats (Franke et al. 2010a; Nieuweboer et al. 2014). These findings have been independently verified (Iusuf et al. 2015; Marada et al. 2015; Sun et al. 2016; van de Steeg et al. 2011, 2013), and are consistent with in vitro studies that have identified paclitaxel as a potent inhibitor of OATP1B1- (Gui et al. 2008, 2009) and OATP1B3-mediated transport (Gui et al. 2008; Letschert et al. 2006; Yamaguchi et al. 2008). However, none of the other known nine human OATPs are capable of transporting paclitaxel (Svoboda et al. 2011), and similar results have been obtained with docetaxel (Lee et al. 2015). Thus, the reported differing affinities highlight the notion that OATPs capable of transporting paclitaxel need to be expressed in tissues such that the drug can cross the plasma membrane and then exert cellular injury. In this context, it is noteworthy that the Oatp1b2 protein is expressed in mouse DRG (Sprowl et al. 2013a), and an independent study evaluating all 15 mouse OATPs confirmed expression of Oatp1b2 in murine neurons (Feurstein et al. 2010).

Comparative pharmacokinetic analysis after a clinically relevant dose of paclitaxel in wild-type mice and Oatp1b2(−/−) produced a modest increase in systemic exposure (Durmus et al. 2015). The minor change in systemic exposure of Oatp1b2-deficient animals to paclitaxel suggests rodents recapitulate pharmacokinetic observations in human patients and serves as an appropriate preclinical model (Nieuweboer et al. 2014). The absence of significant changes in systemic exposure also demonstrates that the genotype-dependent differences are unlikely to influence the extent of paclitaxel-induced peripheral neuropathy. In line with docetaxel studies (de Graan et al. 2012), the genetic deficiency of Oatp1b2 contributed to significant decreases in liver and DRG uptake of paclitaxel (Nieuweboer et al. 2014). The diminished accumulation of paclitaxel in DRGs from Oatp1b2-deficient mice demonstrates the preclinical utility of evaluating peripheral neuropathy, namely the Von Frey Hairs test, to assess mechanical allodynia (Boehmerle et al. 2014; Peters et al. 2007; Yan et al. 2015). After paclitaxel treatment, wild-type mice experience a 50% decrease in sensitivity to mechanical stimulation. Although several reports have suggested that metabolites of paclitaxel may also contribute to peripheral neuropathy (Sparreboom et al. 1995), this phenotype is casually related to paclitaxel itself as direct administration of clinically relevant concentrations of three major paclitaxel liver metabolites: 6α-hydroxy-paclitaxel, 3′-p-hydroxy-paclitaxel, and 6α,3′-*p*-dihydroxy-paclitaxel did not produce neuropathic pain, when compared to the parent drug. In contrast to wild-type mice, Oatp1b2 deficiency recapitulated mechanical sensitivity that resembled baseline values or vehicle-treated group, and is protected from acute paclitaxel-induced peripheral neuropathy (Leblanc et al. 2018). Thermal sensitivity and electrical nerve conductance were also preserved in Oatp1b2(−/−) after chronic treatment. In support of Oatp1b2's functional involvement in accumulation of paclitaxel in DRGs, cabazitaxel, a taxane derivative approved for the treatment of prostate cancer, rarely causes peripheral neuropathy (Omlin et al. 2015) and is not transported by OATP1B-type transporters (Nieuweboer et al. 2014).

Various approaches have been proposed to predict, prevent, and/or treat paclitaxel-induced peripheral neuropathy (Scripture et al. 2006). The predictive strategies have predominantly focused on the search for hereditary biomarkers that could identify patients at increased risk of toxicity through candidate gene (Boora et al. 2016; Green et al. 2009; Hertz et al. 2012, 2013; Leskela et al. 2011; Sissung et al. 2006; Tanabe et al. 2017; Abraham et al. 2014; de Graan et al. 2013) or genome-wide association studies (Baldwin et al. 2012; Bergmann et al. 2013; Lam et al. 2016; Sucheston et al. 2011). However, the findings from these studies done to date have identified non-overlapping single or pathway biomarker associations that preclude immediate clinical implementation (Brewer et al. 2016; Schneider et al. 2015a, b; Frederiks et al. 2015; Hertz 2013). In addition, the decision to act on a toxicity biomarker is hampered in many diseases by the lack of available alternative treatments to replace paclitaxel and/or the need for a patient-tailored reduction in the paclitaxel dose to prevent toxicity, which will have negative effects on the disease management.

To date there have been more than 40 randomized controlled clinical trials of agents to prevent or treat peripheral neuropathy associated with paclitaxel, and these trials have not provided convincing evidence for a clinically beneficial agent (Hershman et al. 2014). One area of research currently being pursued is based on the hypothesis that agents with inhibitory properties toward OATP1B-type transporters could be exploited as neuro-protectants in conjunction with paclitaxel-based chemotherapy (Fig. 3). One of the candidate inhibitors is nilotinib, an inhibitor of the Bcr-Abl kinase used in the treatment of certain leukemias. Pretreatment with this agent protected against acute and chronic forms of paclitaxel-induced peripheral neuropathy in mice, without affecting the disposition properties of paclitaxel or its antitumor properties (Leblanc et al. 2018). The optimal pharmaceutical and pharmacological properties of nilotinib provide rationale for an excellent modulator of

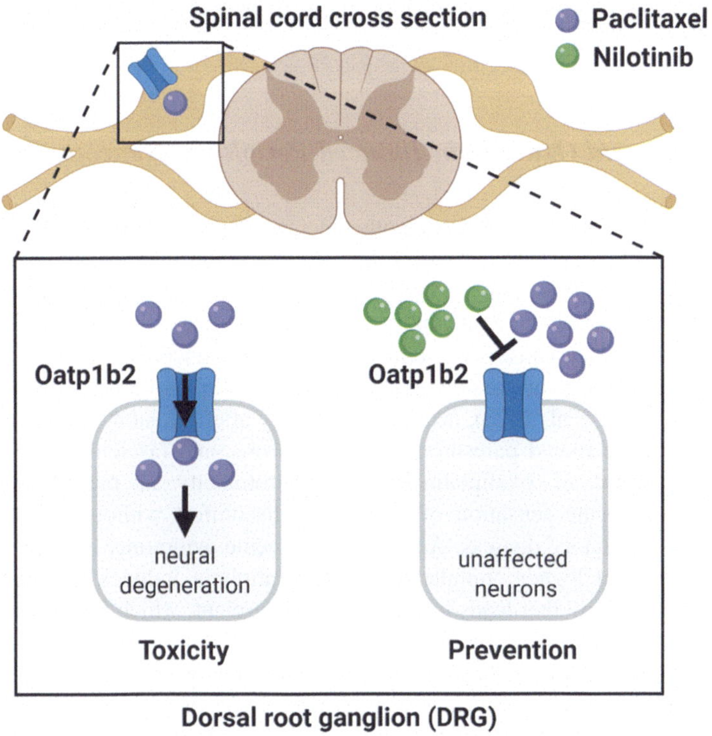

Fig. 3 Proposed model of paclitaxel-induced injury to the peripheral nervous system. Murine Oatp1b2 mediates intracellular accumulation of paclitaxel, leading to peripheral neuropathy and neural degeneration (left). These effects can be blocked by the Oatp1b2 inhibitor nilotinib (right)

the off-target cellular disposition of paclitaxel and subsequent toxicities. Nilotinib is orally bioavailable and has a long half-life resulting from its relatively slow systemic clearance (Xia et al. 2012); this slow clearance will ensure sufficient nilotinib levels to modulate the cellular disposition of paclitaxel. Interestingly, high-dose TKI pulse-exposure, in contrast to a chronic low-dose daily exposure, is becoming a more well-received concept in the treatment of various cancer (Lipka et al. 2012). The increase of familiarity of clinicians with this high-dose pulse strategy of nilotinib will ultimately help facilitate the translation of our propose concept of utilizing nilotinib, as a transporter inhibitor, as adjunct therapy in paclitaxel-based chemotherapies to reduce toxicities. Interestingly, recent preclinical study of nilotinib-paclitaxel combination showed exquisite activity (Holbeck et al. 2017), resulting in tumor regressions in models of breast cancer xenografts with no tumor regrowth observed for more than 80 days following the end of therapy. Although confirmation of these initial findings is required, in patient-derived tumor models that more faithfully represent the characteristics of primary human breast cancer compared with xenografted cell lines (Hidalgo et al. 2014), these combined initial observations indicate that combining paclitaxel with inhibitors of OATP1B-type transporters such as nilotinib has the potential to simultaneously reduce toxicities and increase anticancer effects.

2.2 OCT2 and Oxaliplatin-Induced Peripheral Neurotoxicity

Oxaliplatin is a platinum-based chemotherapeutic that is widely used in the treatment of colorectal and gastric cancers, but its clinical use is associated with debilitating damage to peripheral nerves. This side effect remains one of the most important complications of contemporary oncology regimens as it may limit further treatment and/or may cause long-term quality of life concerns. The characteristic pattern of peripheral neurotoxicity associated with oxaliplatin affects up to 92% of patients (Argyriou et al. 2008), occurs immediately after infusion, and is characterized by cold-exacerbated paresthesia, muscle spasms, and fasciculations. Another distinctive feature of oxaliplatin-induced neurotoxicity is pharyngolaryngeal dysesthesia, an acute sensation of respiratory discomfort without any objective evidence of respiratory distress. Although these acute symptoms typically resolve within 1 week, at higher cumulative doses, oxaliplatin induces a dose-limiting sensory neurotoxicity that leads to functional impairment, which could last several months to even years following discontinuation of treatment or in more severe cases may display permanent, incomplete recovery (McWhinney et al. 2009).

The cellular and molecular mechanisms underlying oxaliplatin-induced neurotoxicity remain incompletely understood to this day. However, studies performed during the last decade have provided crucial insights into the pathophysiological events that contribute to the development of oxaliplatin-induced neurotoxicity. Unlike the central nervous system, the peripheral nervous system, consisting of nerves and ganglia outside of the brain and spinal cord, is not protected by the blood–brain barrier, and is particularly sensitive to oxaliplatin-induced injury. Within the peripheral nervous system, oxaliplatin accumulates and causes toxicity only in the sensory neurons, while the motor neurons are spared (Thompson et al. 1984). As with paclitaxel, the DRG has emerged as the major target of oxaliplatin-induced neurotoxicity (Cavaletti et al. 2001; McKeage et al. 2001), and both structural and functional alterations in the DRG are thought to be the major reason for the development of the observed toxicity (Jamieson et al. 2005). Morphological and anatomical examinations have determined key structural changes in the DRG during oxaliplatin treatment, including alterations in the nuclear/nucleolar morphology, selective atrophy of a subpopulation of cells, platinum-DNA adduct formation, and apoptosis (Cavaletti et al. 2001). The functional changes include sodium channel dysfunction, leading to altered peripheral-nerve excitability, as well as oxaliplatin-dependent changes in the function and expression of several ion channels in the DRG leading to increased nociception (Trevisan et al. 2013; Zhao et al. 2012). The ion channels are activated by mechanical, thermal, chemical, and noxious stimuli, and oxaliplatin-induced changes in their function and expression contribute to the sensory neuropathic symptoms observed in patients treated with oxaliplatin.

Notably, oxaliplatin-induced neurotoxicity is not associated with axonal degeneration as observed in diabetic neuropathy, and damage to the DRG is believed to be causative in the development of neurotoxicity (Gregg et al. 1992).

The key event that triggers the development of pathological changes in the structure and function of DRG is the initial accumulation of oxaliplatin. Multiple studies have shown that oxaliplatin preferentially accumulates in DRG (McKeage et al. 2001), accounting for its selective toxicity to the peripheral sensory system. Unlike sensory neurons, the motor neurons are spared due to lack of oxaliplatin accumulation, highlighting the role of oxaliplatin uptake in the development of neuropathic symptoms. Importantly, in patients, the severity of neuropathy correlates with the platinum concentration in peripheral nerves (Gregg et al. 1992), confirming that oxaliplatin uptake is the major determinant of neurotoxicity. Interestingly, oxaliplatin levels in DRG remain high for prolonged time periods, even after discontinuation of treatment (Holmes et al. 1998). Although efflux transporters (or lack thereof) may be playing a role in the highly prolonged levels of oxaliplatin in DRGs, the transporter-mediated initial influx of oxaliplatin is one of the earliest steps driving cellular accumulation and believed to be the major determinant for subsequent neurotoxicity.

Identification of the biological processes involved in oxaliplatin accumulation in DRG is critical to understanding the mechanisms responsible for oxaliplatin-induced neurotoxicity. Several recent studies have proposed candidate transporters of oxaliplatin and related platinum complexes in heterologous cell culture models. In particular, OCTs belonging to the *SLC22A* family that includes OCT2 (*SLC22A2*) have been implicated in the transport of oxaliplatin and other platinum-based drugs (Sprowl et al. 2013c). Using genetic and pharmacological strategies, in vivo evidence has been accumulating for the direct role of OCT2 in the development of oxaliplatin associated neurotoxicity (Sprowl et al. 2013a). More specifically, recent studies involving genetically engineered mouse models have documented that the acute and chronic forms of oxaliplatin-induced peripheral neurotoxicity are dependent on OCT2, a transporter that regulates the transfer of drug from the circulation into satellite glial cells (SGC) in the DRG. These studies also demonstrate that tyrosine phosphorylation of OCT2 by the protein kinase YES1 is essential for function and targeting this post-translational modification can be exploited to modulate transport activity (Hu and Sprowl 2018; Sprowl et al. 2013a, c, 2016).

In addition to OCT2, several other uptake transporters in rodents have been postulated to be of potential relevance to the pharmacokinetics and toxicity of oxaliplatin, including Oct1 (Li et al. 2011), Oct3 (Yokoo et al. 2008), Octn1 (Nishida et al. 2018), Mate1 (Jong et al. 2011), Ctr1 (Ip et al. 2013), and Oatp1b2 (Lancaster et al. 2013). Although these transporters are all expressed in isolated DRGs from wild-type mice, uptake studies in engineered HEK293 cells overexpressing individual transporters have suggested that oxaliplatin may not be a transported substrate of these OCTs and OATPs, and that transport by OCT2 was more efficient than that observed for MATE1 and OCT3 (Sprowl et al. 2013a; Zhang et al. 2006). Additional studies are required to define the individual and collective contributions of these alternative neuronal uptake mechanisms for oxaliplatin.

Preliminary studies performed to resolve this question experimentally with the use of primary SGCs cultures from mouse DRGs have indicated that about 80% of oxaliplatin uptake in these cells is mediated by OCT2, suggesting that the collective contribution of alternative mechanisms, including passive diffusion, may be minimal.

Targeting OCT2 is particularly appealing as an approach in ameliorating oxaliplatin-induced toxicities because potent and specific inhibitors are unlikely to sacrifice treatment efficacy in view of the general lack of OCT2 expression in tumor cells (Franke et al. 2010b), including colorectal cancers (Sprowl et al. 2013a). Importantly, since OCT2 inhibition may reduce the initial accumulation of platinum-based agents, the neuroprotective effects of OCT2 inhibition observed in the acute toxicity models may also lead to reduction in chronic pain and neuropathic symptoms frequently observed in patients. This possibility is strongly supported by the recent finding that patients receiving oxaliplatin-based chemotherapy who manifest symptoms of the acute neurotoxicity syndrome are those who will also eventually develop the chronic toxicity (Argyriou et al. 2013).

2.3 OCT2 and Cisplatin-Induced Nephrotoxicity

Cisplatin is another DNA-crosslinking, platinum-based chemotherapeutic agent that is among the most widely used anticancer drugs in both adult and pediatric populations (Dasari and Bernard Tchounwou 2014). In the conventional treatment regimens in which the drug is administered once every 3 weeks, dose-limiting side effects include renal tubular dysfunction (nephrotoxicity), and to a lesser extent, hearing loss (ototoxicity) and damage to peripheral nerves. Severe and irreversible damage to the kidney remains the single most important complication of cisplatin treatment as it may limit further treatment or even threaten life. This side effect primarily affects the S3 segment of the renal proximal tubules and occurs in up to 40% of patients despite intensive prophylactic measures, including extensive pre- and post-hydration regimens with hypertonic saline (Arany and Safirstein 2003; de Jongh et al. 2003). Furthermore, about 20% of all acute renal failure cases among hospitalized patients are due to cisplatin-containing chemotherapy (Berns and Ford 1997). As previously observed with toxicities associated with paclitaxel and oxaliplatin, the exact pathogenesis of cisplatin-related chronic toxicities and identity of SLCs involved in these processes, in which quiescent cells are selectively damaged, has remained unclear until relatively recently (Waissbluth and Daniel 2013; Yao et al. 2007).

Using transfected HEK293 cells, it was previously reported that cisplatin is a substrate for OCT2, as indicated by saturable uptake with an estimated Michaelis-Menten constant of 11 μM (Filipski et al. 2008). The localization of OCT2 in the S3 segment of the renal proximal tubules (Leibbrandt et al. 1995) suggests that OCT2 may be a key regulator in the renal elimination of cisplatin and may indirectly regulate the extent to which the drug causes kidney damage. This hypothesis has been verified in several studies with the use of Oct1/2(−/−) mice (Filipski et al. 2009), which are partially protected against cisplatin nephrotoxicity (Ciarimboli et al. 2010). Additionally, higher expression of OCT2 in male rodents correlated with a greater propensity and susceptibility to proximal tubular injury compared to female rodents. Subsequent investigation demonstrated that OCT2 is also highly expressed in the cochlea, and that Oct1/2(−/−) mice are completely protected from platinum-induced ototoxicity (Ciarimboli et al. 2010; Lanvers-Kaminsky et al. 2015) as well as from cisplatin-mediated peripheral neurotoxicity (Sprowl et al. 2013a; Hucke et al. 2019).

The demonstration that this solute carrier plays an important role in all clinically relevant platinum-related toxicities provides a rationale for the development of therapeutic interventions for cisplatin-containing regimens in patients involving the use of specific inhibitors of OCT2. Targeting OCT2 is particularly appealing as an approach in ameliorating cisplatin-induced toxicities because potent and specific inhibitors are unlikely to compromise treatment efficacy in view of the general lack of OCT2 expression in tumor cells (Franke et al. 2010b). However, the incomplete protection against cisplatin-associated renal tubular damage in Oct1/2(−/−) mice suggests the existence of a secondary pathway that contributes, independently of Oct1/Oct2-mediated renal tubular drug uptake, to cisplatin-induced renal damage (Sprowl et al. 2014). Recent rodent studies have suggested that the OCT2-independent pathway is regulated by the transporters OAT1 and OAT3, which mediate the accumulation of a nephrotoxic, mercapturic acid metabolite of cisplatin formed in the γ-glutamyltranspeptidase pathway (Hu et al. 2017).

Over the last three decades, various approaches have been proposed to afford tissue protection during cisplatin treatment, although most of these interventions have not been evaluated in animal models or humans. Indeed, there is still no known preventative treatment for cisplatin-induced renal dysfunction (dos Santos et al. 2012), ototoxicity (Langer et al. 2013; Travis et al. 2014), or neurotoxicity (Albers et al. 2014). Agents that have advanced to clinical testing, such as amifostine, are associated with intrinsic toxicity and, more importantly, do not appear to have a major impact on ameliorating the risk of developing severe nephrotoxicity (Gallegos-Castorena et al. 2007; Sastry and Kellie 2005), ototoxicity (Duval and Daniel 2012), or neurotoxicity (Hensley et al. 2009). Furthermore, the clinical application of many alternate strategies has been hampered by the recognition that (1) cisplatin has multiple intracellular targets and hence blocking a single injurious event will only have partial protective effects and (2) the protective approach may diminish the antitumor effects of cisplatin given the overlap in cell death signaling pathways between normal cells and tumor cells. Therefore, an ideal approach is to simultaneously protect the kidneys and other afflicted tissues such as cochlea and peripheral nerves without affecting the therapeutic effects in tumors. The development of such an approach would rely on the identification of the critical differences between normal and malignant cells during cisplatin treatment.

One of these agents, cimetidine, has shown some promise in ameliorating cisplatin-induced nephrotoxicity (Franke et al. 2010b) and ototoxicity (Ciarimboli et al. 2010), as well as oxaliplatin-induced neurotoxicity in experimental mouse models (Sprowl et al. 2013a). To obtain preliminary evidence for the usefulness of adding cimetidine to cisplatin-based therapy in cancer patients, randomized cross-over trials (Sprowl et al. 2013b) have been performed and demonstrated that inhibition of OCT2 function by cimetidine did not affect the antitumor or disposition properties of cisplatin. However, cimetidine did not completely eradicate renal tubular damage, in line with another recent clinical trial indicating that the renoprotective effects associated with cimetidine are only partial (Zhang and Zhou 2012). This confirms in vitro studies suggesting that cimetidine is an inefficient and non-specific inhibitor of OCT2 (Ito et al. 2012; Motohashi et al. 2004), and that identification of alternate OCT2 inhibitors is urgently needed. Such agents would ideally have (1) high potency, (2) high specificity, (3) low drug–drug interaction potential, (4) intrinsic antitumor properties, (5) favorable pharmaceutical properties, (6) non-overlapping toxicity profiles, and (7) potentially other renoprotective features, including inhibition of OAT1 and OAT3. Among possible candidates for further exploration, palbociclib, an FDA-approved inhibitor of CDK4/6 used in the treatment of breast cancer, is of particular interest because it has been previously shown that other inhibitors of cyclin dependent kinase are able to afford protection against cisplatin-induced kidney injury in experimental models (Price et al. 2009). Preliminary studies performed in mice have suggested that cisplatin-induced nephrotoxicity can be mitigated by pretreatment with palbociclib through a mechanism that is partially dependent on OCT2 (Pabla et al. 2015). Due to significant overlap of OCT/MATE inhibitors, it is important that the design of these intervention strategies aimed at selectively targeting OCT2-mediated uptake does not increase the residence time in proximal tubular cells due to unintended inhibition of MATE1-mediated efflux.

2.4 *ENT1/OCTN1 and Cytarabine-Related Toxicities*

Cytarabine is a nucleoside analog belonging to the family of antimetabolites due to its similarity in chemical structure to that of endogenous nucleosides. Cytarabine is utilized in a variety of leukemia subtypes but is a mainstay in the treatment of acute myeloid leukemia (AML) where it is an integral component of first-line therapy. All of the endogenous and xenobiotic nucleoside analogs are polar hydrophilic compounds that are poorly membrane permeable and require functional nucleoside transporter proteins to enter cells.

Nucleoside transporters facilitate the accumulation of both endogenous nucleosides and nucleoside-derived drugs that are utilized as anticancer and antiviral agents (Baldwin et al. 1999). Some cell types, including brain, enterocytes, and bone marrow cells, rely heavily on the nucleoside salvage pathway due to their inability to synthesis nucleosides de novo and thus rely heavily on the extracellular milieu for their primary source of nucleosides for use in RNA and DNA synthesis (Murray 1971). Nucleoside transporters play an integral role in the maintenance of extracellular concentrations of nucleosides, which are available to bind to receptors and modulate a variety of physiological processes. Transporter-mediated transport of nucleosides is thus a critical determinant in the salvage and consequently, nucleoside-mediated toxicity in many cell types (Griffith and Jarvis 1996).

The two major classes of nucleoside transporters in mammalian cells and tissues consist of equilibrative nucleoside transporters (ENTs) and concentrative nucleoside transporters (CNTs). Two proteins of the former class, ENT1 (*SLC29A1*) and ENT2 (*SLC29A2*), are known to mediate the transport of purine and pyrimidine nucleosides across biological membranes down their concentration gradients. These transporters exhibit broad substrate selectivity and are subdivided based on their sensitivity (ENT1) or resistance (ENT2) to inhibition by nanomolar concentrations of nitrobenzylmercaptopurine ribonucleoside (NBMPR) (Damaraju et al. 2003). In addition to ENT1, several non-canonical putative nucleoside transporters, including OCTN1, can potentially transport nucleoside analogs in a manner that is sensitive to nanomolar concentrations of NBMPR (Drenberg et al. 2017).

The clinical use of cytarabine is associated with dose-limiting damage to normal bone marrow (myelosuppression), which occurs in the majority of patients, as well as with damage to the skin (toxic erythema), and these complications may require dose-modification, limit further treatment, or even threaten life (Hwang et al. 2012; Zhang et al. 2014). Interestingly, OCTN1 is highly expressed in several organs of particular relevance to cytarabine-based chemotherapy regimens, including myeloid progenitor cells in the bone marrow, proximal tubular cells in the kidney (Kobayashi et al. 2004), and epidermal keratinocytes in the skin (Wu et al. 2000). Preliminary evidence pointing to potential causality of this connection has come from population-based genetic studies indicating that patients with a functional polymorphic germline variant of OCTN1 (L503F; rs1050152) experience an increased frequency of febrile neutropenia (Drenberg et al. 2015). Furthermore, overexpression of this genetic variant (L503F) in HEK293 cells is associated with increased transport function (Urban et al. 2007) and increased formation of Ara-CTP, the active triphosphorylated form of cytarabine (Drenberg et al. 2017).

Consistent with the thesis that OCTN1 may be contributing to cytarabine-related toxicities, it has been reported that this transport system may also be operational for related cytotoxic nucleoside analogs such as gemcitabine, an agent used in the treatment of pancreatic cancer with activity in certain subtypes of AML (Drenberg et al. 2019), that causes dose-limiting anemia and neutropenia. Although this site of toxicity directly aligns with the expression profile of OCTN1, further study is warranted to determine the contribution of individual SLCs to gemcitabine-related side effects. In this context, it is worth pointing out that the intracellular accumulation of nucleoside analogs such as cytarabine was recently reported to occur independently of OCTN1 (Tschirka et al. 2018). In this work, the authors used an LC-MS/MS-based method to measure the intracellular levels of unchanged cytarabine, whereas prior studies involved the use of an analysis based on the measurement of total radioactivity [i.e., the total of parent drug and metabolite(s)]. This is a potentially important methodological difference as cytarabine can undergo rapid enzyme-mediated metabolism once inside cells to form mono-, di-, and tri-phosphorylated forms (Owens et al. 1992), which may easily escape detection and result in underestimating the actual extent of uptake. This possibility is supported by the finding that cytarabine is rapidly and extensively phosphorylated in HEK293 cells (Drenberg et al. 2017), and by the demonstration that in a comparative analysis intracellular levels of total radioactivity originating from cytarabine in cells overexpressing OCTN1 are high, while levels of the unchanged parent drug, as measured by LC-MS/MS, remain undetectable (Anderson et al. 2019).

2.5 OATPs and Irinotecan-Mediated Neutropenia and Diarrhea

Irinotecan is a prodrug of the topoisomerase I inhibitor, SN-38, and is utilized in a variety of chemotherapy containing regimens that are used to treat solid tumors such as colorectal cancers. Irinotecan is known to cause several debilitating adverse events such as myelosuppression and diarrhea. In contrast to the metabolism of irinotecan, which has been well documented and characterized (de Man et al. 2018), the pharmacokinetic processes of relevance to irinotecan disposition are less well characterized and are likely dependent on the interplay of drug transporters residing in organs such as the liver. In this context, the contribution of hepatocellular uptake transporters in the disposition of irinotecan remains poorly understood. Preclinical reports have shown that after irinotecan administration, the systemic exposure of SN-38, the active metabolite of irinotecan, is highly impacted by the deficiency of Oatp1a- and Oatp1b-type carriers in murine models and expression of human OATP1B1 or OATP1B3 (Iusuf et al. 2014). Furthermore, only a small fraction of the administered dose of irinotecan is excreted in the bile (0.9% for SN-38 and 3% for SN-38-glucuronide) (de Jong et al. 2006), which supports the notion that a large fraction of these metabolites formed within hepatocytes and being transported back into the system circulation, presumably by the hepatic efflux transporter ABCC3 (Kitamura et al. 2010). This increased SN-38 efflux can then be taken up again by adjacent hepatocytes in an OATP1B-mediated manner for further glucuronidation and/or biliary excretion termed "hepatocyte hopping" (Iusuf et al. 2012). This unusual mechanism is hypothesized as a physiological mechanism to overcome hepatocellular saturation and to efficiently facilitate detoxification through utilizing other elimination pathways such as Phase II conjugation and transporter-mediated excretion into the bile. This unusual mechanism for irinotecan elimination is supported by the observation of patients with functional variants of OATP1B1 that are associated with altered exposure to SN-38 and at risk for severe toxicity following irinotecan-based chemotherapies (Di Paolo et al. 2011).

Interestingly, in contrast to SN-38 and other known glucuronide metabolites (Ni et al. 2010; Zimmerman et al. 2013), SN-38-glucuronide accumulation is not mediated by OATP1B1 (Nozawa et al. 2005). Clinical observation of excessive SN-38-glucuronide buildup in the systemic circulation relative to unconjugated

SN-38 could be explained by the lack of an efficient hepatocellular uptake mechanism for SN-38-glucuronide (Innocenti et al. 2014).

Since formation of SN-38 from irinotecan is essential to the therapy-related diarrhea (Fujita et al. 2016), recent studies have focused on connecting functional expression of intestinal transporters to the occurrence of this side effect. This work has resulted in the identification of OATP2B1 as a putative carrier of SN-38 on the basis of uptake studies performed in *Xenopus* oocytes (Fujita et al. 2016). This transporter is highly expressed in the small intestine (Tamai et al. 2000), and sensitive to inhibition by cyclosporine (Chen et al. 2020), an agent that has been exploited as a therapeutic to prevent the dose-limiting diarrhea associated with irinotecan treatment (Chester et al. 2003). The availability of a recently developed Oatp2b1-deficient mouse model will allow the unequivocal demonstration of a causal connection of OATP2B1-mediated transport of SN-38 with irinotecan-related diarrhea (Medwid et al. 2019).

2.6 SLCs and Doxorubicin-Related Cardiotoxicity

Doxorubicin, an anthracycline-derived DNA-intercalator, is widely used in the treatment of multiple tumor types, including breast cancers and soft tissue sarcomas. Common side effects associated with the use of doxorubicin include acute nausea and vomiting, mucositis, alopecia, and tissue extravasation. More serious, dose-limiting toxicities associated with doxorubicin include myelosuppression and cardiotoxicity, and these side effects are dependent on the cumulative dose administered. In particular, the risk for patients developing congestive heart failure is estimated at 5% (Von Hoff et al. 1979; Kremer et al. 2001), 18% (Kremer et al. 2001), and 36% (Swain et al. 2003) for cumulative doses of doxorubicin of <500 mg/m^2, 500–600 mg/m^2, and >600 mg/m^2, respectively. Manifestation of acute cardiotoxicity presents as arrhythmias or ventricular dysfunction. However, since the myocardium has limited regenerative capacity (Lionetti and Ventura 2013; Yamada et al. 2015), chronic cardiotoxicity induced by anthracyclines culminates into dilated cardiomyopathy and congestive heart failure (Boucek Jr. et al. 1997; Lipshultz et al. 2013), which can occur months or even years after cessation of therapy. The risk of developing chronic cardiotoxicity is particularly high in young adult and adolescent cancer survivors (Lipshultz et al. 2013). Clinical risk factors associated with doxorubicin-induced cardiotoxicity also include pre-existing cardiac dysfunction and age (Doyle et al. 2005) as well as prior therapy involving radiation or chemotherapy (Singal and Iliskovic 1998).

The mechanisms by which anthracyclines such as doxorubicin accumulate into cardiomyocytes remain largely unknown. At physiological pH 7.4, the hydrophobic weak base doxorubicin is slightly cationic (Raghunand et al. 1999), and this recognition has resulted in recent efforts to connect uptake of anthracyclines to OCTs that can explain cell-type specific toxicity profiles. In particular, studies have demonstrated that overexpression of OCTN1 (Okabe et al. 2008), OCT1 (Andreev et al.

2016), or OCT6 (Okabe et al. 2005; Ota et al. 2007) is associated with significantly increased drug uptake and sensitivity of leukemic and ovarian cancer cells following exposure to doxorubicin. Furthermore, studies in *C. elegans* and *Danio rerio* corroborate involvement of OCT1 and possibly OCT2 (Brosseau et al. 2015) in the cellular uptake of doxorubicin. It is likely that cardiac expression of OCTs capable of transporting doxorubicin in the myocardium contributes as an initiating event that ultimately leads to treatment-related cardiotoxicity.

Among the class of OCTs, studies have confirmed the presence of several members at both the mRNA and protein levels in the human heart, with OCTN2 having the highest expression, followed by OCT3, OCTN1, and OCT1, while OCT2 is undetectable (Grube et al. 2011). In view of the predominant expression of OCT2 in the kidney, it is possible that this transport mechanism contributes to urinary excretion of anthracyclines and to doxorubicin-induced nephrotoxicity (Filipski et al. 2009; Ayla et al. 2011). Additional investigation has indicated that expression of these transporters is confined to either the vasculature (OCTN1, OCTN2, and OCT3) or to cardiomyocytes (OCTN1 and OCT1) (Grube et al. 2011; Iwata et al. 2008) of the myocardium, although other studies have demonstrated ubiquitous localization in the heart (Nishimura and Naito 2008). It has been suggested that OCTN2 is involved in the uptake of certain cardiovascular drugs, and is spatially regulated in a rat cardiomyopathy model (Iwata et al. 2008). Indeed, supplementation with the high-affinity OCTN2 substrate, L-carnitine, can reduce doxorubicin-induced upregulation of heart fatty acid binding protein (H-FABP) (Sayed-Ahmed et al. 2010), suggesting the potential for an OCTN2-mediated mechanism for doxorubicin uptake. In addition, expression of the structurally related transporter OCTN1 is correlated with augmenting the blockage of HERG potassium channels and potentiating *torsade de pointes* (McBride et al. 2009), which can lead to serious cardiac arrhythmias.

OCT1 and OCT3 have also been linked to cardiovascular drug response, in particular to certain beta-blockers (Bachmakov et al. 2009; Kubo et al. 2013; Hassan et al. 2016), which are used principally as prophylactic treatment in managing cardiovascular function after anthracycline therapy. Additionally, OCT3 has been identified as an important regulator of neurological and cardiovascular response to endogenous substrates such as epinephrine and norepinephrine (Zwart et al. 2001; Hanafy et al. 2012; Ayala-Lopez et al. 2015; Zhu et al. 2012). Furthermore, case-control studies of the OCT3 gene locus have suggested its involvement in coronary vascular development, and multiples variants, including rs9381439, rs2048327, rs18190126, and rs9349379, were previously associated with decreased risk for coronary artery disease (Tregouet et al. 2009; Wang et al. 2016). Definitive demonstration of a direct contribution of any of the candidate transporters to doxorubicin-induced cardiotoxicity, for example, in mice with individual SLC deficiencies, remains warranted, and may provide opportunities for the future design of preventative intervention strategies with the use of selective transport inhibitors.

Since hepatobiliary excretion is the main elimination pathway of doxorubicin (Legha et al. 1982; Bronchud et al. 1990; Robert et al. 1985, 1987; Maniez-Devos et al. 1986), recent studies have attempted to connect the function of hepatic OATPs with the disposition of doxorubicin (Gong and Kim 2013). These investigations have shown that deficiency of Oatp1a- and Oatp1b-type transporters in mice is associated with increases in systemic exposure to doxorubicin. In addition, introduction of the human transporters OATP1B1, OATP1B3, or OATP1A2 in the livers of these knockout mice can partially recapitulate the pharmacokinetic profile of doxorubicin observed in wild-type mice (Durmus et al. 2014). Drug uptake studies in cell-based models engineered to overexpress human OATP1A2 variants have corroborated the results in mice, and are consistent with previous reports involving other substrates (Lee et al. 2017). Although several related transporters, including OATP2B1, OATP3A1, and OATP4A1, are known to be expressed in the heart (Hanafy et al. 2012; Grube et al. 2006; Atilano-Roque and Joy 2017), there is currently an apparent lack of documentation that establishes a correlation of these OATPs with the cardiac uptake of anthracyclines, and thus their direct relevance to cardiovascular function after doxorubicin treatment remains uncertain.

3 Toxicity Induced by Targeted Therapeutics

In oncology, the last two decades have seen a dramatic transition from the use of traditional cytotoxic chemotherapy to the emergence of a new paradigm in rational drug design coupled with an uprising in the development of targeted agents, including the tyrosine kinase inhibitors (TKIs). To date, >45 different TKIs have received approval by the FDA for the treatment of a variety of diseases that were previously essentially resistant to standard chemotherapy, and many more can be expected to become available in the future (Drenberg et al. 2013). However, while TKIs offer possibly a number of important theoretical advantages over conventional cytotoxic chemotherapy, they are still afflicted by some of the same problems, including an extensive inter-individual pharmacokinetic variability, the existence of a rather narrow therapeutic window, and the occurrence of multiple, debilitating adverse events (Drenberg et al. 2013).

Previous investigations on drug transporters and their contribution to pharmacological effects of TKIs have often exclusively focused on the effects on measures of systemic exposure, ignoring effects on local drug uptake and cellular retention in healthy tissues. Other investigations have tended to focus solely on the cellular target regulating pharmacological effects while ignoring the effects on systemic and/or local drug exposure. However, recently developed conceptual frameworks for an integrated approach have started to address questions related to the relevance of specific SLCs to the local tissue exposure of TKIs.

3.1 OAT6 and Sorafenib-Mediated Skin Toxicity

Sorafenib is a multi-kinase inhibitor (MKI) utilized in unresectable hepatocellular carcinoma, advanced renal cell carcinoma, and thyroid carcinoma. Common debilitating adverse events of sorafenib include fatigue, infection, alopecia, hand-foot skin reaction, and rash. Of the listed adverse events, cutaneous adverse effects are among the most frequently observed toxicities with many TKIs, and their intensity can significantly affect both quality of life and health care economics (Macdonald et al. 2015). In one study, 40% of renal cell carcinoma patients taking sorafenib had a dermatologic reaction (Kane et al. 2006).

A particularly painful complication seen most frequently during the early weeks of use with MKIs such as sorafenib and regorafenib is known as hand-foot skin reaction (HFSR), in which hyperkeratotic plaques develop predominantly over sites of pressure or friction (Inaba et al. 2011; Lipworth et al. 2009). These plaques may have significant inflammation and xerotic hyperkeratosis, often in a bilateral symmetric distribution, causing pain and debilitation that interfere with activities of daily living (Macdonald et al. 2015). Sequential biopsy specimens from patients receiving MKIs have revealed progressive accumulation of hyperkeratosis with focal parakeratosis. The clinical incidence of HFSR varies among MKIs with a particularly high incidence being observed with sorafenib (>60%) (Zimmerman et al. 2016), and this appears to be unrelated to increased excretion of MKIs through sweat (Jain et al. 2010).

The hair follicle is a specialized mini-organ that is critically dependent on programmed keratinocyte differentiation (Botchkarev and Paus 2003; Cotsarelis 2006), and disruption of hair-follicle cycling or morphology is indicative of keratinocyte toxicity. Recent studies have indicated that sorafenib accumulates extensively in primary human epidermal keratinocytes compared to a panel of other TKIs, can decrease cell viability, and increase apoptosis (Zimmerman et al. 2016). The mechanism by which sorafenib is taken up into keratinocytes is concentration-, time-, and temperature-dependent. Since sorafenib is a poorly permeable compound, it is plausible that its uptake into keratinocytes is predominantly a transporter-mediated process. This hypothesis was verified by the recent demonstration that uptake of sorafenib in keratinocytes is dependent on OAT6, identified from a transportome-wide gene silencing screen (Tian et al. 2018), and that sorafenib-induced injury to keratinocytes in a mouse model of HFSR can be reversed by pretreatment with the OAT6 inhibitor, probenecid (Fig. 4) (Zimmerman et al. 2016). The translational significance of intervention strategies derived from the combinatorial use of probenecid and sorafenib or regorafenib (Belum et al. 2013) requires further investigation.

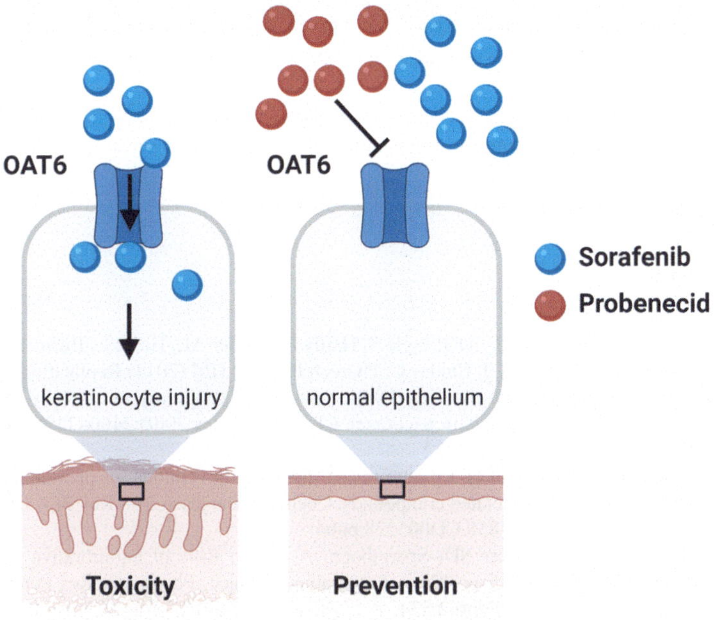

Fig. 4 Proposed model of sorafenib-induced keratinocyte injury. Organic anionic transport 6 (OAT6) mediates intracellular concentrations of sorafenib, leading to cytotoxicity and keratinocyte injury (left). These effects can be blocked by the OAT6 inhibitor probenecid (right)

4 Conclusions

In order to better understand the contribution of SLCs to debilitating side effects associated with anticancer drugs, there is an urgent need to further characterize the role of these proteins in the transport of drugs in both target and off-target tissues. This research would require the (1) expansion of the SLC proteomics field in various tissue types associated with toxicities of importance, (2) metabolomics approaches to further understand the trafficking of endogenous substrates of each transporter, and (3) the availability of agnostic screening platforms that allow for more rapid identification of drug-transporter and inhibitor-transporter pairs. Understanding the delicate balance of on-target and off-target tissue accumulation could then be exploited as a basis for the development of predictive tools as well as for the discovery of novel intervention strategies with application in the clinic in order to improve the safety of currently available treatment modalities.

Acknowledgement This work was supported in part by the National Institutes of Health Grants R01CA215802 (to A.S.), R01CA238946 (to S.H., M.B.L), and by the OSU Comprehensive Cancer Center using Pelotonia funds. The content is solely the responsibility of the authors and does not necessarily represent the official views of the funding agencies.

Conflicts of Interest The authors have no conflicts of interests.

References

Abraham JE, Guo Q, Dorling L, Tyrer J, Ingle S, Hardy R, Vallier AL, Hiller L, Burns R, Jones L, Bowden SJ, Dunn JA, Poole CJ, Caldas C, Pharoah PP, Earl HM (2014) Replication of genetic polymorphisms reported to be associated with taxane-related sensory neuropathy in patients with early breast cancer treated with paclitaxel. Clin Cancer Res 20(9):2466–2475. https://doi.org/10.1158/1078-0432.CCR-13-3232

Albers JW, Chaudhry V, Cavaletti G, Donehower RC (2014) Interventions for preventing neuropathy caused by cisplatin and related compounds. Cochrane Database Syst Rev 3(3):CD005228. https://doi.org/10.1002/14651858.CD005228.pub4

Anderson JT, Hu S, Fu Q, Baker SD, Sparreboom A (2019) Role of equilibrative nucleoside transporter 1 (ENT1) in the disposition of cytarabine in mice. Pharmacol Res Perspect 7(6): e00534. https://doi.org/10.1002/prp2.534

Andreev E, Brosseau N, Carmona E, Mes-Masson AM, Ramotar D (2016) The human organic cation transporter OCT1 mediates high affinity uptake of the anticancer drug daunorubicin. Sci Rep 6:20508. https://doi.org/10.1038/srep20508

Arany I, Safirstein RL (2003) Cisplatin nephrotoxicity. Semin Nephrol 23(5):460–464

Argyriou AA, Polychronopoulos P, Iconomou G, Chroni E, Kalofonos HP (2008) A review on oxaliplatin-induced peripheral nerve damage. Cancer Treat Rev 34(4):368–377. https://doi.org/10.1016/j.ctrv.2008.01.003

Argyriou AA, Cavaletti G, Briani C, Velasco R, Bruna J, Campagnolo M, Alberti P, Bergamo F, Cortinovis D, Cazzaniga M, Santos C, Papadimitriou K, Kalofonos HP (2013) Clinical pattern and associations of oxaliplatin acute neurotoxicity: a prospective study in 170 patients with colorectal cancer. Cancer 119(2):438–444. https://doi.org/10.1002/cncr.27732

Atilano-Roque A, Joy MS (2017) Characterization of simvastatin acid uptake by organic anion transporting polypeptide 3A1 (OATP3A1) and influence of drug-drug interaction. Toxicol In Vitro 45(Pt 1):158–165. https://doi.org/10.1016/j.tiv.2017.09.002

Ayala-Lopez N, Jackson WF, Burnett R, Wilson JN, Thompson JM, Watts SW (2015) Organic cation transporter 3 contributes to norepinephrine uptake into perivascular adipose tissue. Am J Physiol Heart Circ Physiol 309(11):H1904–H1914. https://doi.org/10.1152/ajpheart.00308.2015

Ayla S, Seckin I, Tanriverdi G, Cengiz M, Eser M, Soner BC, Oktem G (2011) Doxorubicin induced nephrotoxicity: protective effect of nicotinamide. Int J Cell Biol 2011:390238. https://doi.org/10.1155/2011/390238

Bachmakov I, Glaeser H, Endress B, Morl F, Konig J, Fromm MF (2009) Interaction of beta-blockers with the renal uptake transporter OCT2. Diabetes Obes Metab 11(11):1080–1083. https://doi.org/10.1111/j.1463-1326.2009.01076.x

Baker SD, Verweij J, Cusatis GA, van Schaik RH, Marsh S, Orwick SJ, Franke RM, Hu S, Schuetz EG, Lamba V, Messersmith WA, Wolff AC, Carducci MA, Sparreboom A (2009) Pharmacogenetic pathway analysis of docetaxel elimination. Clin Pharmacol Ther 85 (2):155–163. https://doi.org/10.1038/clpt.2008.95

Baldwin SA, Mackey JR, Cass CE, Young JD (1999) Nucleoside transporters: molecular biology and implications for therapeutic development. Mol Med Today 5(5):216–224. https://doi.org/10.1016/S1357-4310(99)01459-8
Baldwin RM, Owzar K, Zembutsu H, Chhibber A, Kubo M, Jiang C, Watson D, Eclov RJ, Mefford J, McLeod HL, Friedman PN, Hudis CA, Winer EP, Jorgenson EM, Witte JS, Shulman LN, Nakamura Y, Ratain MJ, Kroetz DL (2012) A genome-wide association study identifies novel loci for paclitaxel-induced sensory peripheral neuropathy in CALGB 40101. Clin Cancer Res 18(18):5099–5109. https://doi.org/10.1158/1078-0432.CCR-12-1590
Ban M, Hettich D, Huguet N (1994) Nephrotoxicity mechanism of cis-platinum (II) diamine dichloride in mice. Toxicol Lett 71(2):161–168
Belum VR, Wu S, Lacouture ME (2013) Risk of hand-foot skin reaction with the novel multikinase inhibitor regorafenib: a meta-analysis. Investig New Drugs 31(4):1078–1086. https://doi.org/10.1007/s10637-013-9977-0
Bergmann TK, Vach W, Feddersen S, Eckhoff L, Green H, Herrstedt J, Brosen K (2013) GWAS-based association between RWDD3 and TECTA variants and paclitaxel induced neuropathy could not be confirmed in Scandinavian ovarian cancer patients. Acta Oncol 52(4):871–874. https://doi.org/10.3109/0284186X.2012.707787
Berns JS, Ford PA (1997) Renal toxicities of antineoplastic drugs and bone marrow transplantation. Semin Nephrol 17(1):54–66
Boehmerle W, Huehnchen P, Peruzzaro S, Balkaya M, Endres M (2014) Electrophysiological, behavioral and histological characterization of paclitaxel, cisplatin, vincristine and bortezomib-induced neuropathy in C57Bl/6 mice. Sci Rep 4:6370. https://doi.org/10.1038/srep06370
Boora GK, Kanwar R, Kulkarni AA, Abyzov A, Sloan J, Ruddy KJ, Banck MS, Loprinzi CL, Beutler AS (2016) Testing of candidate single nucleotide variants associated with paclitaxel neuropathy in the trial NCCTG N08C1 (Alliance). Cancer Med 5(4):631–639. https://doi.org/10.1002/cam4.625
Borst P, Elferink RO (2002) Mammalian ABC transporters in health and disease. Annu Rev Biochem 71:537–592. https://doi.org/10.1146/annurev.biochem.71.102301.093055
Borst P, Evers R, Kool M, Wijnholds J (2000) A family of drug transporters: the multidrug resistance-associated proteins. J Natl Cancer Inst 92(16):1295–1302
Botchkarev VA, Paus R (2003) Molecular biology of hair morphogenesis: development and cycling. J Exp Zool B Mol Dev Evol 298(1):164–180. https://doi.org/10.1002/jez.b.33
Boucek RJ Jr, Dodd DA, Atkinson JB, Oquist N, Olson RD (1997) Contractile failure in chronic doxorubicin-induced cardiomyopathy. J Mol Cell Cardiol 29(10):2631–2640
Brewer JR, Morrison G, Dolan ME, Fleming GF (2016) Chemotherapy-induced peripheral neuropathy: current status and progress. Gynecol Oncol 140(1):176–183. https://doi.org/10.1016/j.ygyno.2015.11.011
Bronchud MH, Margison JM, Howell A, Lind M, Lucas SB, Wilkinson PM (1990) Comparative pharmacokinetics of escalating doses of doxorubicin in patients with metastatic breast cancer. Cancer Chemother Pharmacol 25(6):435–439. https://doi.org/10.1007/bf00686055
Brosseau N, Andreev E, Ramotar D (2015) Complementation of the yeast model system reveals that Caenorhabditis elegans OCT-1 is a functional transporter of anthracyclines. PLoS One 10(7): e0133182. https://doi.org/10.1371/journal.pone.0133182
Carozzi VA, Canta A, Chiorazzi A (2015) Chemotherapy-induced peripheral neuropathy: what do we know about mechanisms? Neurosci Lett 596:90–107. https://doi.org/10.1016/j.neulet.2014.10.014
Caterson R, Etheredge S, Snitch P, Duggin G (1983) Mechanisms of renal excretion of cisdichlorodiamine platinum. Res Commun Chem Pathol Pharmacol 41(2):255–264
Cavaletti G, Cavalletti E, Oggioni N, Sottani C, Minoia C, D'Incalci M, Zucchetti M, Marmiroli P, Tredici G (2000) Distribution of paclitaxel within the nervous system of the rat after repeated intravenous administration. Neurotoxicology 21(3):389–393
Cavaletti G, Tredici G, Petruccioli MG, Donde E, Tredici P, Marmiroli P, Minoia C, Ronchi A, Bayssas M, Etienne GG (2001) Effects of different schedules of oxaliplatin treatment on the

peripheral nervous system of the rat. Eur J Cancer 37(18):2457–2463. https://doi.org/10.1016/s0959-8049(01)00300-8
Chae YK, Arya A, Malecek MK, Shin DS, Carneiro B, Chandra S, Kaplan J, Kalyan A, Altman JK, Platanias L, Giles F (2016) Repurposing metformin for cancer treatment: current clinical studies. Oncotarget 7(26):40767–40780. https://doi.org/10.18632/oncotarget.8194
Chen M, Neul C, Schaeffeler E, Frisch F, Winter S, Schwab M, Koepsell H, Hu S, Laufer S, Baker SD, Sparreboom A, Nies AT (2020) Sorafenib activity and disposition in liver cancer does not depend on organic cation transporter 1. Clin Pharmacol Ther 107(1):227–237. https://doi.org/10.1002/cpt.1588
Chester JD, Joel SP, Cheeseman SL, Hall GD, Braun MS, Perry J, Davis T, Button CJ, Seymour MT (2003) Phase I and pharmacokinetic study of intravenous irinotecan plus oral ciclosporin in patients with fuorouracil-refractory metastatic colon cancer. J Clin Oncol 21(6):1125–1132. https://doi.org/10.1200/JCO.2003.08.049
Cho SK, Chung JY (2016) The MATE1 rs2289669 polymorphism affects the renal clearance of metformin following ranitidine treatment. Int J Clin Pharmacol Ther 54(4):253–262. https://doi.org/10.5414/CP202473
Cho SK, Kim CO, Park ES, Chung JY (2014) Verapamil decreases the glucose-lowering effect of metformin in healthy volunteers. Br J Clin Pharmacol 78(6):1426–1432. https://doi.org/10.1111/bcp.12476
Christensen MM, Hojlund K, Hother-Nielsen O, Stage TB, Damkier P, Beck-Nielsen H, Brosen K (2015) Steady-state pharmacokinetics of metformin is independent of the OCT1 genotype in healthy volunteers. Eur J Clin Pharmacol 71(6):691–697. https://doi.org/10.1007/s00228-015-1853-8
Ciarimboli G, Deuster D, Knief A, Sperling M, Holtkamp M, Edemir B, Pavenstadt H, Lanvers-Kaminsky C, am Zehnhoff-Dinnesen A, Schinkel AH, Koepsell H, Jurgens H, Schlatter E (2010) Organic cation transporter 2 mediates cisplatin-induced oto- and nephrotoxicity and is a target for protective interventions. Am J Pathol 176(3):1169–1180. https://doi.org/10.2353/ajpath.2010.090610
Cotsarelis G (2006) Epithelial stem cells: a folliculocentric view. J Invest Dermatol 126(7):1459–1468. https://doi.org/10.1038/sj.jid.5700376
Daley-Yates PT, McBrien DC (1985) The renal fractional clearance of platinum antitumour compounds in relation to nephrotoxicity. Biochem Pharmacol 34(9):1423–1428
Damaraju VL, Damaraju S, Young JD, Baldwin SA, Mackey J, Sawyer MB, Cass CE (2003) Nucleoside anticancer drugs: the role of nucleoside transporters in resistance to cancer chemotherapy. Oncogene 22(47):7524–7536. https://doi.org/10.1038/sj.onc.1206952
Dasari S, Bernard Tchounwou P (2014) Cisplatin in cancer therapy: molecular mechanisms of action. Eur J Pharmacol 740C:364–378. https://doi.org/10.1016/j.ejphar.2014.07.025
de Graan AJ, Lancaster CS, Obaidat A, Hagenbuch B, Elens L, Friberg LE, de Bruijn P, Hu S, Gibson AA, Bruun GH, Corydon TJ, Mikkelsen TS, Walker AL, Du G, Loos WJ, van Schaik RH, Baker SD, Mathijssen RH, Sparreboom A (2012) Influence of polymorphic OATP1B-type carriers on the disposition of docetaxel. Clin Cancer Res 18(16):4433–4440. https://doi.org/10.1158/1078-0432.CCR-12-0761
de Graan AJ, Elens L, Sprowl JA, Sparreboom A, Friberg LE, van der Holt B, de Raaf PJ, de Bruijn P, Engels FK, Eskens FA, Wiemer EA, Verweij J, Mathijssen RH, van Schaik RH (2013) CYP3A4*22 genotype and systemic exposure affect paclitaxel-induced neurotoxicity. Clin Cancer Res 19(12):3316–3324. https://doi.org/10.1158/1078-0432.CCR-12-3786
De Iuliis F, Taglieri L, Salerno G, Lanza R, Scarpa S (2015) Taxane induced neuropathy in patients affected by breast cancer: literature review. Crit Rev Oncol Hematol 96(1):34–45. https://doi.org/10.1016/j.critrevonc.2015.04.011
de Jong FA, Kitzen JJ, de Bruijn P, Verweij J, Loos WJ (2006) Hepatic transport, metabolism and biliary excretion of irinotecan in a cancer patient with an external bile drain. Cancer Biol Ther 5(9):1105–1110

de Jongh FE, van Veen RN, Veltman SJ, de Wit R, van der Burg ME, van den Bent MJ, Planting AS, Graveland WJ, Stoter G, Verweij J (2003) Weekly high-dose cisplatin is a feasible treatment option: analysis on prognostic factors for toxicity in 400 patients. Br J Cancer 88 (8):1199–1206. https://doi.org/10.1038/sj.bjc.6600884

de Man FM, Goey AKL, van Schaik RHN, Mathijssen RHJ, Bins S (2018) Individualization of irinotecan treatment: a review of pharmacokinetics, pharmacodynamics, and pharmacogenetics. Clin Pharmacokinet 57(10):1229–1254. https://doi.org/10.1007/s40262-018-0644-7

DeGorter MK, Xia CQ, Yang JJ, Kim RB (2012) Drug transporters in drug efficacy and toxicity. Annu Rev Pharmacol Toxicol 52:249–273. https://doi.org/10.1146/annurev-pharmtox-010611-134529

Di Paolo A, Bocci G, Polillo M, Del Re M, Di Desidero T, Lastella M, Danesi R (2011) Pharmacokinetic and pharmacogenetic predictive markers of irinotecan activity and toxicity. Curr Drug Metab 12(10):932–943. https://doi.org/10.2174/138920011798062283

dos Santos NA, Carvalho Rodrigues MA, Martins NM, dos Santos AC (2012) Cisplatin-induced nephrotoxicity and targets of nephroprotection: an update. Arch Toxicol 86(8):1233–1250. https://doi.org/10.1007/s00204-012-0821-7

Doyle JJ, Neugut AI, Jacobson JS, Grann VR, Hershman DL (2005) Chemotherapy and cardiotoxicity in older breast cancer patients: a population-based study. J Clin Oncol 23 (34):8597–8605. https://doi.org/10.1200/JCO.2005.02.5841

Drenberg CD, Baker SD, Sparreboom A (2013) Integrating clinical pharmacology concepts in individualized therapy with tyrosine kinase inhibitors. Clin Pharmacol Ther 93(3):215–219. https://doi.org/10.1038/clpt.2012.247

Drenberg CD, Paugh SW, Pounds SB, Shi L, Orwick SJ, Li L, Hu S, Gibson AA, Ribeiro RC, Rubnitz JE, Evans WE, Sparreboom A, Baker SD (2015) Inherited variation in OATP1B1 is associated with treatment outcome in acute myeloid leukemia. Clin Pharmacol Ther. https://doi.org/10.1002/cpt.315

Drenberg CD, Gibson AA, Pounds SB, Shi L, Rhinehart DP, Li L, Hu S, Du G, Nies AT, Schwab M, Pabla N, Blum W, Gruber TA, Baker SD, Sparreboom A (2017) OCTN1 is a high-affinity carrier of nucleoside analogues. Cancer Res 77(8):2102–2111. https://doi.org/10.1158/0008-5472.CAN-16-2548

Drenberg CD, Shelat A, Dang J, Cotton A, Orwick SJ, Li M, Jeon JY, Fu Q, Buelow DR, Pioso M, Hu S, Inaba H, Ribeiro RC, Rubnitz JE, Gruber TA, Guy RK, Baker SD (2019) A high-throughput screen indicates gemcitabine and JAK inhibitors may be useful for treating pediatric AML. Nat Commun 10(1):2189. https://doi.org/10.1038/s41467-019-09917-0

Durmus S, Naik J, Buil L, Wagenaar E, van Tellingen O, Schinkel AH (2014) In vivo disposition of doxorubicin is affected by mouse Oatp1a/1b and human OATP1A/1B transporters. Int J Cancer 135(7):1700–1710. https://doi.org/10.1002/ijc.28797

Durmus S, Lozano-Mena G, van Esch A, Wagenaar E, van Tellingen O, Schinkel AH (2015) Preclinical mouse models to study human OATP1B1- and OATP1B3-mediated drug-drug interactions in vivo. Mol Pharm 12(12):4259–4269. https://doi.org/10.1021/acs.molpharmaceut.5b00453

Duval M, Daniel SJ (2012) Meta-analysis of the efficacy of amifostine in the prevention of cisplatin ototoxicity. J Otolaryngol Head Neck Surg 41(5):309–315

Feurstein D, Kleinteich J, Heussner AH, Stemmer K, Dietrich DR (2010) Investigation of microcystin congener-dependent uptake into primary murine neurons. Environ Health Perspect 118(10):1370–1375. https://doi.org/10.1289/ehp.0901289

Filipski KK, Loos WJ, Verweij J, Sparreboom A (2008) Interaction of cisplatin with the human organic cation transporter 2. Clin Cancer Res 14(12):3875–3880. https://doi.org/10.1158/1078-0432.CCR-07-4793

Filipski KK, Mathijssen RH, Mikkelsen TS, Schinkel AH, Sparreboom A (2009) Contribution of organic cation transporter 2 (OCT2) to cisplatin-induced nephrotoxicity. Clin Pharmacol Ther 86(4):396–402. https://doi.org/10.1038/clpt.2009.139

Franke RM, Carducci MA, Rudek MA, Baker SD, Sparreboom A (2010a) Castration-dependent pharmacokinetics of docetaxel in patients with prostate cancer. J Clin Oncol 28(30):4562–4567. https://doi.org/10.1200/JCO.2010.30.7025

Franke RM, Kosloske AM, Lancaster CS, Filipski KK, Hu C, Zolk O, Mathijssen RH, Sparreboom A (2010b) Influence of Oct1/Oct2-deficiency on cisplatin-induced changes in urinary N-acetyl-beta-D-glucosaminidase. Clin Cancer Res 16(16):4198–4206. https://doi.org/10.1158/1078-0432.CCR-10-0949

Frederiks CN, Lam SW, Guchelaar HJ, Boven E (2015) Genetic polymorphisms and paclitaxel- or docetaxel-induced toxicities: a systematic review. Cancer Treat Rev 41(10):935–950. https://doi.org/10.1016/j.ctrv.2015.10.010

Fujita D, Saito Y, Nakanishi T, Tamai I (2016) Organic anion transporting polypeptide (OATP)2B1 contributes to gastrointestinal toxicity of anticancer drug SN-38, active metabolite of Irinotecan hydrochloride. Drug Metab Dispos 44(1):1–7. https://doi.org/10.1124/dmd.115.066712

Gallegos-Castorena S, Martinez-Avalos A, Mohar-Betancourt A, Guerrero-Avendano G, Zapata-Tarres M, Medina-Sanson A (2007) Toxicity prevention with amifostine in pediatric osteosarcoma patients treated with cisplatin and doxorubicin. Pediatr Hematol Oncol 24(6):403–408. https://doi.org/10.1080/08880010701451244

Giacomini KM, Yee SW, Ratain MJ, Weinshilboum RM, Kamatani N, Nakamura Y (2012) Pharmacogenomics and patient care: one size does not fit all. Sci Transl Med 4 (153):153ps118. https://doi.org/10.1126/scitranslmed.3003471

Gong IY, Kim RB (2013) Impact of genetic variation in OATP transporters to drug disposition and response. Drug Metab Pharmacokinet 28(1):4–18. https://doi.org/10.2133/dmpk.dmpk-12-rv-099

Grecco HE, Schmick M, Bastiaens PI (2011) Signaling from the living plasma membrane. Cell 144 (6):897–909. https://doi.org/10.1016/j.cell.2011.01.029

Green H, Soderkvist P, Rosenberg P, Mirghani RA, Rymark P, Lundqvist EA, Peterson C (2009) Pharmacogenetic studies of paclitaxel in the treatment of ovarian cancer. Basic Clin Pharmacol Toxicol 104(2):130–137. https://doi.org/10.1111/j.1742-7843.2008.00351.x

Gregg RW, Molepo JM, Monpetit VJ, Mikael NZ, Redmond D, Gadia M, Stewart DJ (1992) Cisplatin neurotoxicity: the relationship between dosage, time, and platinum concentration in neurologic tissues, and morphologic evidence of toxicity. J Clin Oncol 10(5):795–803. https://doi.org/10.1200/JCO.1992.10.5.795

Griffith DA, Jarvis SM (1996) Nucleoside and nucleobase transport systems of mammalian cells. Biochim Biophys Acta 1286(3):153–181

Grube M, Kock K, Oswald S, Draber K, Meissner K, Eckel L, Bohm M, Felix SB, Vogelgesang S, Jedlitschky G, Siegmund W, Warzok R, Kroemer HK (2006) Organic anion transporting polypeptide 2B1 is a high-affinity transporter for atorvastatin and is expressed in the human heart. Clin Pharmacol Ther 80(6):607–620. https://doi.org/10.1016/j.clpt.2006.09.010

Grube M, Ameling S, Noutsias M, Kock K, Triebel I, Bonitz K, Meissner K, Jedlitschky G, Herda LR, Reinthaler M, Rohde M, Hoffmann W, Kuhl U, Schultheiss HP, Volker U, Felix SB, Klingel K, Kandolf R, Kroemer HK (2011) Selective regulation of cardiac organic cation transporter novel type 2 (OCTN2) in dilated cardiomyopathy. Am J Pathol 178 (6):2547–2559. https://doi.org/10.1016/j.ajpath.2011.02.020

Gui C, Miao Y, Thompson L, Wahlgren B, Mock M, Stieger B, Hagenbuch B (2008) Effect of pregnane X receptor ligands on transport mediated by human OATP1B1 and OATP1B3. Eur J Pharmacol 584(1):57–65. https://doi.org/10.1016/j.ejphar.2008.01.042

Gui C, Wahlgren B, Lushington GH, Hagenbuch B (2009) Identification, Ki determination and CoMFA analysis of nuclear receptor ligands as competitive inhibitors of OATP1B1-mediated estradiol-17beta-glucuronide transport. Pharmacol Res 60(1):50–56. https://doi.org/10.1016/j.phrs.2009.03.004

Hagenbuch B (2010) Drug uptake systems in liver and kidney: a historic perspective. Clin Pharmacol Ther 87(1):39–47. https://doi.org/10.1038/clpt.2009.235

Hanafy S, El-Kadi AO, Jamali F (2012) Effect of inflammation on molecular targets and drug transporters. J Pharm Pharm Sci 15(3):361–375

Hassan OT, Hassan RT, Arora RR (2016) Organic cation transporter-mediated clearance of cardiovascular drugs: a pharmacological perspective. Am J Ther 23(3):e855–e861. https://doi.org/10.1097/MJT.0000000000000148

He L, Vasiliou K, Nebert DW (2009) Analysis and update of the human solute carrier (SLC) gene superfamily. Hum Genomics 3(2):195–206

Heckman-Stoddard BM, Gandini S, Puntoni M, Dunn BK, DeCensi A, Szabo E (2016) Repurposing old drugs to chemoprevention: the case of metformin. Semin Oncol 43 (1):123–133. https://doi.org/10.1053/j.seminoncol.2015.09.009

Hediger MA, Clemencon B, Burrier RE, Bruford EA (2013) The ABCs of membrane transporters in health and disease (SLC series): introduction. Mol Asp Med 34(2–3):95–107. https://doi.org/10.1016/j.mam.2012.12.009

Hensley ML, Hagerty KL, Kewalramani T, Green DM, Meropol NJ, Wasserman TH, Cohen GI, Emami B, Gradishar WJ, Mitchell RB, Thigpen JT, Trotti A 3rd, von Hoff D, Schuchter LM (2009) American Society of Clinical Oncology 2008 clinical practice guideline update: use of chemotherapy and radiation therapy protectants. J Clin Oncol 27(1):127–145. https://doi.org/10.1200/JCO.2008.17.2627

Hershman DL, Lacchetti C, Dworkin RH, Lavoie Smith EM, Bleeker J, Cavaletti G, Chauhan C, Gavin P, Lavino A, Lustberg MB, Paice J, Schneider B, Smith ML, Smith T, Terstriep S, Wagner-Johnston N, Bak K, Loprinzi CL, American Society of Clinical Oncology (2014) Prevention and management of chemotherapy-induced peripheral neuropathy in survivors of adult cancers: American Society of Clinical Oncology clinical practice guideline. J Clin Oncol 32(18):1941–1967. https://doi.org/10.1200/JCO.2013.54.0914

Hertz DL (2013) Germline pharmacogenetics of paclitaxel for cancer treatment. Pharmacogenomics 14(9):1065–1084. https://doi.org/10.2217/pgs.13.90

Hertz DL, Motsinger-Reif AA, Drobish A, Winham SJ, McLeod HL, Carey LA, Dees EC (2012) CYP2C8*3 predicts benefit/risk profile in breast cancer patients receiving neoadjuvant paclitaxel. Breast Cancer Res Treat 134(1):401–410. https://doi.org/10.1007/s10549-012-2054-0

Hertz DL, Roy S, Motsinger-Reif AA, Drobish A, Clark LS, McLeod HL, Carey LA, Dees EC (2013) CYP2C8*3 increases risk of neuropathy in breast cancer patients treated with paclitaxel. Ann Oncol 24(6):1472–1478. https://doi.org/10.1093/annonc/mdt018

Hidalgo M, Amant F, Biankin AV, Budinska E, Byrne AT, Caldas C, Clarke RB, de Jong S, Jonkers J, Maelandsmo GM, Roman-Roman S, Seoane J, Trusolino L, Villanueva A (2014) Patient-derived xenograft models: an emerging platform for translational cancer research. Cancer Discov 4(9):998–1013. https://doi.org/10.1158/2159-8290.CD-14-0001

Higgins JW, Bedwell DW, Zamek-Gliszczynski MJ (2012) Ablation of both organic cation transporter (OCT)1 and OCT2 alters metformin pharmacokinetics but has no effect on tissue drug exposure and pharmacodynamics. Drug Metab Dispos 40(6):1170–1177. https://doi.org/10.1124/dmd.112.044875

Holbeck SL, Camalier R, Crowell JA, Govindharajulu JP, Hollingshead MG, Anderson LW, Polley EC, Rubinstein L, Srivastava AK, Wilsker DF, Collins JM, Doroshow JH (2017) The National Cancer Institute ALMANAC: a comprehensive screening resource for the detection of anticancer drug pairs with enhanced therapeutic activity. Cancer Res. https://doi.org/10.1158/0008-5472.CAN-17-0489

Holmes J, Stanko J, Varchenko M, Ding H, Madden VJ, Bagnell CR, Wyrick SD, Chaney SG (1998) Comparative neurotoxicity of oxaliplatin, cisplatin, and ormaplatin in a Wistar rat model. Toxicol Sci 46(2):342–351. https://doi.org/10.1006/toxs.1998.2558

Hosoyamada M, Sekine T, Kanai Y, Endou H (1999) Molecular cloning and functional expression of a multispecific organic anion transporter from human kidney. Am J Phys 276(1):F122–F128. https://doi.org/10.1152/ajprenal.1999.276.1.F122

Hu S, Sprowl JA (2018) Strategies to reduce solute carrier-mediated toxicity. Clin Pharmacol Ther. https://doi.org/10.1002/cpt.1185

Hu S, Leblanc AF, Gibson AA, Hong KW, Kim JY, Janke LJ, Li L, Vasilyeva A, Finkelstein DB, Sprowl JA, Sweet DH, Schlatter E, Ciarimboli G, Schellens J, Baker SD, Pabla N, Sparreboom A (2017) Identification of OAT1/OAT3 as contributors to cisplatin toxicity. Clin Transl Sci 10 (5):412–420. https://doi.org/10.1111/cts.12480
Hucke A, Rinschen MM, Bauer OB, Sperling M, Karst U, Koppen C, Sommer K, Schroter R, Ceresa C, Chiorazzi A, Canta A, Semperboni S, Marmiroli P, Cavaletti G, Schlatt S, Schlatter E, Pavenstadt H, Heitplatz B, Van Marck V, Sparreboom A, Barz V, Knief A, Deuster D, Zehnhoff-Dinnesen AA, Ciarimboli G (2019) An integrative approach to cisplatin chronic toxicities in mice reveals importance of organic cation-transporter-dependent protein networks for renoprotection. Arch Toxicol 93(10):2835–2848. https://doi.org/10.1007/s00204-019-02557-9
Hwang YY, Trendell-Smith NJ, Yeung CK, Kwong YL (2012) Fatal palmar-plantar erythrodysesthesia after clofarabine and cytarabine chemotherapy. Acta Haematol 128 (3):151–153. https://doi.org/10.1159/000338828
Inaba H, Rubnitz JE, Coustan-Smith E, Li L, Furmanski BD, Mascara GP, Heym KM, Christensen R, Onciu M, Shurtleff SA, Pounds SB, Pui CH, Ribeiro RC, Campana D, Baker SD (2011) Phase I pharmacokinetic and pharmacodynamic study of the multikinase inhibitor sorafenib in combination with clofarabine and cytarabine in pediatric relapsed/refractory leukemia. J Clin Oncol 29(24):3293–3300. https://doi.org/10.1200/JCO.2011.34.7427
Innocenti F, Schilsky RL, Ramirez J, Janisch L, Undevia S, House LK, Das S, Wu K, Turcich M, Marsh R, Karrison T, Maitland ML, Salgia R, Ratain MJ (2014) Dose-finding and pharmacokinetic study to optimize the dosing of irinotecan according to the UGT1A1 genotype of patients with cancer. J Clin Oncol 32(22):2328–2334. https://doi.org/10.1200/JCO.2014.55.2307
International Transporter Consortium, Giacomini KM, Huang SM, Tweedie DJ, Benet LZ, Brouwer KL, Chu X, Dahlin A, Evers R, Fischer V, Hillgren KM, Hoffmaster KA, Ishikawa T, Keppler D, Kim RB, Lee CA, Niemi M, Polli JW, Sugiyama Y, Swaan PW, Ware JA, Wright SH, Yee SW, Zamek-Gliszczynski MJ, Zhang L (2010) Membrane transporters in drug development. Nat Rev Drug Discov 9(3):215–236. https://doi.org/10.1038/nrd3028
Ip V, Liu JJ, McKeage MJ (2013) Evaluation of effects of copper histidine on copper transporter 1-mediated accumulation of platinum and oxaliplatin-induced neurotoxicity in vitro and in vivo. Clin Exp Pharmacol Physiol 40(6):371–378. https://doi.org/10.1111/1440-1681.12088
Ito S, Kusuhara H, Yokochi M, Toyoshima J, Inoue K, Yuasa H, Sugiyama Y (2012) Competitive inhibition of the luminal efflux by multidrug and toxin extrusions, but not basolateral uptake by organic cation transporter 2, is the likely mechanism underlying the pharmacokinetic drug-drug interactions caused by cimetidine in the kidney. J Pharmacol Exp Ther 340(2):393–403. https://doi.org/10.1124/jpet.111.184986
Iusuf D, van de Steeg E, Schinkel AH (2012) Hepatocyte hopping of OATP1B substrates contributes to efficient hepatic detoxification. Clin Pharmacol Ther 92(5):559–562. https://doi.org/10.1038/clpt.2012.143
Iusuf D, Ludwig M, Elbatsh A, van Esch A, van de Steeg E, Wagenaar E, van der Valk M, Lin F, van Tellingen O, Schinkel AH (2014) OATP1A/1B transporters affect irinotecan and SN-38 pharmacokinetics and carboxylesterase expression in knockout and humanized transgenic mice. Mol Cancer Ther 13(2):492–503. https://doi.org/10.1158/1535-7163.Mct-13-0541
Iusuf D, Hendrikx JJ, van Esch A, van de Steeg E, Wagenaar E, Rosing H, Beijnen JH, Schinkel AH (2015) Human OATP1B1, OATP1B3 and OATP1A2 can mediate the in vivo uptake and clearance of docetaxel. Int J Cancer 136(1):225–233. https://doi.org/10.1002/ijc.28970
Iwata D, Kato Y, Wakayama T, Sai Y, Kubo Y, Iseki S, Tsuji A (2008) Involvement of carnitine/organic cation transporter OCTN2 (SLC22A5) in distribution of its substrate carnitine to the heart. Drug Metab Pharmacokinet 23(3):207–215
Izumi S, Nozaki Y, Kusuhara H, Hotta K, Mochizuki T, Komori T, Maeda K, Sugiyama Y (2018) Relative activity factor (RAF)-based scaling of uptake clearance mediated by organic anion

transporting polypeptide (OATP) 1B1 and OATP1B3 in human hepatocytes. Mol Pharm. https://doi.org/10.1021/acs.molpharmaceut.8b00138

Jacobs C, Coleman CN, Rich L, Hirst K, Weiner MW (1984) Inhibition of cis-diamminedichloroplatinum secretion by the human kidney with probenecid. Cancer Res 44(8):3632–3635

Jain L, Gardner ER, Figg WD, Chernick MS, Kong HH (2010) Lack of association between excretion of sorafenib in sweat and hand-foot skin reaction. Pharmacotherapy 30(1):52–56. https://doi.org/10.1592/phco.30.1.52

Jamieson SM, Liu J, Connor B, McKeage MJ (2005) Oxaliplatin causes selective atrophy of a subpopulation of dorsal root ganglion neurons without inducing cell loss. Cancer Chemother Pharmacol 56(4):391–399. https://doi.org/10.1007/s00280-004-0953-4

Jensen JB, Sundelin EI, Jakobsen S, Gormsen LC, Munk OL, Frokiaer J, Jessen N (2016) [11C]-metformin distribution in the liver and small intestine using dynamic PET in mice demonstrates tissue-specific transporter dependency. Diabetes. https://doi.org/10.2337/db16-0032

Joerger M, van Schaik RH, Becker ML, Hayoz S, Pollak M, Cathomas R, Winterhalder R, Gillessen S, Rothermundt C (2015) Multidrug and toxin extrusion 1 and human organic cation transporter 1 polymorphisms in patients with castration-resistant prostate cancer receiving metformin (SAKK 08/09). Prostate Cancer Prostatic Dis 18(2):167–172. https://doi.org/10.1038/pcan.2015.8

Jong NN, Nakanishi T, Liu JJ, Tamai I, McKeage MJ (2011) Oxaliplatin transport mediated by organic cation/carnitine transporters OCTN1 and OCTN2 in overexpressing human embryonic kidney 293 cells and rat dorsal root ganglion neurons. J Pharmacol Exp Ther 338(2):537–547. https://doi.org/10.1124/jpet.111.181297

Jonker JW, Wagenaar E, Mol CA, Buitelaar M, Koepsell H, Smit JW, Schinkel AH (2001) Reduced hepatic uptake and intestinal excretion of organic cations in mice with a targeted disruption of the organic cation transporter 1 (Oct1 [Slc22a1]) gene. Mol Cell Biol 21(16):5471–5477. https://doi.org/10.1128/MCB.21.16.5471-5477.2001

Jonker JW, Wagenaar E, Van Eijl S, Schinkel AH (2003) Deficiency in the organic cation transporters 1 and 2 (Oct1/Oct2 [Slc22a1/Slc22a2]) in mice abolishes renal secretion of organic cations. Mol Cell Biol 23(21):7902–7908

Kaback HR, Sahin-Toth M, Weinglass AB (2001) The kamikaze approach to membrane transport. Nat Rev Mol Cell Biol 2(8):610–620. https://doi.org/10.1038/35085077

Kane RC, Farrell AT, Saber H, Tang S, Williams G, Jee JM, Liang C, Booth B, Chidambaram N, Morse D, Sridhara R, Garvey P, Justice R, Pazdur R (2006) Sorafenib for the treatment of advanced renal cell carcinoma. Clin Cancer Res 12(24):7271–7278. https://doi.org/10.1158/1078-0432.CCR-06-1249

Kimura N, Masuda S, Tanihara Y, Ueo H, Okuda M, Katsura T, Inui K (2005) Metformin is a superior substrate for renal organic cation transporter OCT2 rather than hepatic OCT1. Drug Metab Pharmacokinet 20(5):379–386

Kitamura Y, Kusuhara H, Sugiyama Y (2010) Functional characterization of multidrug resistance-associated protein 3 (mrp3/abcc3) in the basolateral efflux of glucuronide conjugates in the mouse small intestine. J Pharmacol Exp Ther 332(2):659–666. https://doi.org/10.1124/jpet.109.156943

Klein J, Bentur Y, Cheung D, Moselhy G, Koren G (1991) Renal handling of cisplatin: interactions with organic anions and cations in the dog. Clin Invest Med 14(5):388–394

Kobayashi D, Aizawa S, Maeda T, Tsuboi I, Yabuuchi H, Nezu J, Tsuji A, Tamai I (2004) Expression of organic cation transporter OCTN1 in hematopoietic cells during erythroid differentiation. Exp Hematol 32(12):1156–1162. https://doi.org/10.1016/j.exphem.2004.08.009

Koepsell H (2015) Role of organic cation transporters in drug-drug interaction. Expert Opin Drug Metab Toxicol 11(10):1619–1633. https://doi.org/10.1517/17425255.2015.1069274

Kremer LC, van Dalen EC, Offringa M, Ottenkamp J, Voute PA (2001) Anthracycline-induced clinical heart failure in a cohort of 607 children: long-term follow-up study. J Clin Oncol 19(1):191–196. https://doi.org/10.1200/JCO.2001.19.1.191

Kubo Y, Shimizu Y, Kusagawa Y, Akanuma S, Hosoya K (2013) Propranolol transport across the inner blood-retinal barrier: potential involvement of a novel organic cation transporter. J Pharm Sci 102(9):3332–3342. https://doi.org/10.1002/jps.23535

Kusuhara H, Sekine T, Utsunomiya-Tate N, Tsuda M, Kojima R, Cha SH, Sugiyama Y, Kanai Y, Endou H (1999) Molecular cloning and characterization of a new multispecific organic anion transporter from rat brain. J Biol Chem 274(19):13675–13680

Lam SW, Frederiks CN, van der Straaten T, Honkoop AH, Guchelaar HJ, Boven E (2016) Genotypes of CYP2C8 and FGD4 and their association with peripheral neuropathy or early dose reduction in paclitaxel-treated breast cancer patients. Br J Cancer 115(11):1335–1342. https://doi.org/10.1038/bjc.2016.326

Lancaster CS, Sprowl JA, Walker AL, Hu S, Gibson AA, Sparreboom A (2013) Modulation of OATP1B-type transporter function alters cellular uptake and disposition of platinum chemotherapeutics. Mol Cancer Ther 12(8):1537–1544. https://doi.org/10.1158/1535-7163.MCT-12-0926

Langer T, am Zehnhoff-Dinnesen A, Radtke S, Meitert J, Zolk O (2013) Understanding platinum-induced ototoxicity. Trends Pharmacol Sci 34(8):458–469. https://doi.org/10.1016/j.tips.2013.05.006

Lanvers-Kaminsky C, Sprowl JA, Malath I, Deuster D, Eveslage M, Schlatter E, Mathijssen RH, Boos J, Jurgens H, Am Zehnhoff-Dinnesen AG, Sparreboom A, Ciarimboli G (2015) Human OCT2 variant c.808G>T confers protection effect against cisplatin-induced ototoxicity. Pharmacogenomics 16(4):323–332. https://doi.org/10.2217/pgs.14.182

Leblanc AF, Sprowl JA, Alberti P, Chiorazzi A, Arnold WD, Gibson AA, Hong KW, Pioso MS, Chen M, Huang KM, Chodisetty V, Costa O, Florea T, de Bruijn P, Mathijssen RH, Reinbolt RE, Lustberg MB, Sucheston-Campbell LE, Cavaletti G, Sparreboom A, Hu S (2018) OATP1B2 deficiency protects against paclitaxel-induced neurotoxicity. J Clin Invest 128 (2):816–825. https://doi.org/10.1172/JCI96160

Lee HH, Leake BF, Teft W, Tirona RG, Kim RB, Ho RH (2015) Contribution of hepatic organic anion-transporting polypeptides to docetaxel uptake and clearance. Mol Cancer Ther 14 (4):994–1003. https://doi.org/10.1158/1535-7163.MCT-14-0547

Lee HH, Leake BF, Kim RB, Ho RH (2017) Contribution of organic anion-transporting polypeptides 1A/1B to doxorubicin uptake and clearance. Mol Pharmacol 91(1):14–24. https://doi.org/10.1124/mol.116.105544

Legha SS, Benjamin RS, Mackay B, Ewer M, Wallace S, Valdivieso M, Rasmussen SL, Blumenschein GR, Freireich EJ (1982) Reduction of doxorubicin cardiotoxicity by prolonged continuous intravenous infusion. Ann Intern Med 96(2):133–139

Leibbrandt ME, Wolfgang GH, Metz AL, Ozobia AA, Haskins JR (1995) Critical subcellular targets of cisplatin and related platinum analogs in rat renal proximal tubule cells. Kidney Int 48 (3):761–770

Leskela S, Jara C, Leandro-Garcia LJ, Martinez A, Garcia-Donas J, Hernando S, Hurtado A, Vicario JC, Montero-Conde C, Landa I, Lopez-Jimenez E, Cascon A, Milne RL, Robledo M, Rodriguez-Antona C (2011) Polymorphisms in cytochromes P450 2C8 and 3A5 are associated with paclitaxel neurotoxicity. Pharmacogenomics J 11(2):121–129. https://doi.org/10.1038/tpj.2010.13

Letschert K, Faulstich H, Keller D, Keppler D (2006) Molecular characterization and inhibition of amanitin uptake into human hepatocytes. Toxicol Sci 91(1):140–149. https://doi.org/10.1093/toxsci/kfj141

Li S, Chen Y, Zhang S, More SS, Huang X, Giacomini KM (2011) Role of organic cation transporter 1, OCT1 in the pharmacokinetics and toxicity of cis-diammine(pyridine) chloroplatinum(II) and oxaliplatin in mice. Pharm Res 28(3):610–625. https://doi.org/10.1007/s11095-010-0312-6

Li Q, Guo D, Dong Z, Zhang W, Zhang L, Huang SM, Polli JE, Shu Y (2013) Ondansetron can enhance cisplatin-induced nephrotoxicity via inhibition of multiple toxin and extrusion proteins

(MATEs). Toxicol Appl Pharmacol 273(1):100–109. https://doi.org/10.1016/j.taap.2013.08.024

Lionetti V, Ventura C (2013) Regenerative medicine approach to repair the failing heart. Vasc Pharmacol 58(3):159–163. https://doi.org/10.1016/j.vph.2013.01.002

Lipka DB, Wagner MC, Dziadosz M, Schnoder T, Heidel F, Schemionek M, Melo JV, Kindler T, Muller-Tidow C, Koschmieder S, Fischer T (2012) Intracellular retention of ABL kinase inhibitors determines commitment to apoptosis in CML cells. PLoS One 7(7):e40853. https://doi.org/10.1371/journal.pone.0040853

Lipshultz SE, Adams MJ, Colan SD, Constine LS, Herman EH, Hsu DT, Hudson MM, Kremer LC, Landy DC, Miller TL, Oeffinger KC, Rosenthal DN, Sable CA, Sallan SE, Singh GK, Steinberger J, Cochran TR, Wilkinson JD (2013) Long-term cardiovascular toxicity in children, adolescents, and young adults who receive cancer therapy: pathophysiology, course, monitoring, management, prevention, and research directions: a scientific statement from the American Heart Association. Circulation 128(17):1927–1995. https://doi.org/10.1161/CIR.0b013e3182a88099

Lipworth AD, Robert C, Zhu AX (2009) Hand-foot syndrome (hand-foot skin reaction, palmar-plantar erythrodysesthesia): focus on sorafenib and sunitinib. Oncology 77(5):257–271. https://doi.org/10.1159/000258880

Lu R, Chan BS, Schuster VL (1999) Cloning of the human kidney PAH transporter: narrow substrate specificity and regulation by protein kinase C. Am J Phys 276(2):F295–F303. https://doi.org/10.1152/ajprenal.1999.276.2.F295

Macdonald JB, Macdonald B, Golitz LE, LoRusso P, Sekulic A (2015) Cutaneous adverse effects of targeted therapies: Part II: Inhibitors of intracellular molecular signaling pathways. J Am Acad Dermatol 72(2):221–236. https://doi.org/10.1016/j.jaad.2014.07.033

Maniez-Devos DM, Baurain R, Lesne M, Trouet A (1986) Doxorubicin and daunorubicin plasmatic, hepatic and renal disposition in the rabbit with or without enterohepatic circulation. J Pharmacol 17(1):1–13

Marada VV, Florl S, Kuhne A, Burckhardt G, Hagos Y (2015) Interaction of human organic anion transporter polypeptides 1B1 and 1B3 with antineoplastic compounds. Eur J Med Chem 92:723–731. https://doi.org/10.1016/j.ejmech.2015.01.011

Marmiroli P, Cavaletti G (2016) Drugs for the treatment of peripheral neuropathies. Expert Opin Pharmacother 17(3):381–394. https://doi.org/10.1517/14656566.2016.1120719

McBride BF, Yang T, Liu K, Urban TJ, Giacomini KM, Kim RB, Roden DM (2009) The organic cation transporter, OCTN1, expressed in the human heart, potentiates antagonism of the HERG potassium channel. J Cardiovasc Pharmacol 54(1):63–71. https://doi.org/10.1097/FJC.0b013e3181abc288

McCreight LJ, Bailey CJ, Pearson ER (2016) Metformin and the gastrointestinal tract. Diabetologia 59(3):426–435. https://doi.org/10.1007/s00125-015-3844-9

McKeage MJ, Hsu T, Screnci D, Haddad G, Baguley BC (2001) Nucleolar damage correlates with neurotoxicity induced by different platinum drugs. Br J Cancer 85(8):1219–1225. https://doi.org/10.1054/bjoc.2001.2024

McWhinney SR, Goldberg RM, McLeod HL (2009) Platinum neurotoxicity pharmacogenetics. Mol Cancer Ther 8(1):10–16. https://doi.org/10.1158/1535-7163.MCT-08-0840

Medwid S, Li MMJ, Knauer MJ, Lin K, Mansell SE, Schmerk CL, Zhu C, Griffin KE, Yousif MD, Dresser GK, Schwarz UI, Kim RB, Tirona RG (2019) Fexofenadine and rosuvastatin pharmacokinetics in mice with targeted disruption of organic anion transporting polypeptide 2B1. Drug Metab Dispos 47(8):832–842. https://doi.org/10.1124/dmd.119.087619

Mielke S, Sparreboom A, Mross K (2006) Peripheral neuropathy: a persisting challenge in paclitaxel-based regimes. Eur J Cancer 42(1):24–30. https://doi.org/10.1016/j.ejca.2005.06.030

Motohashi H, Uwai Y, Hiramoto K, Okuda M, Inui K (2004) Different transport properties between famotidine and cimetidine by human renal organic ion transporters (SLC22A). Eur J Pharmacol 503(1–3):25–30. https://doi.org/10.1016/j.ejphar.2004.09.032

Murray AW (1971) The biological significance of purine salvage. Annu Rev Biochem 40:811–826. https://doi.org/10.1146/annurev.bi.40.070171.004115
Nakamura T, Yonezawa A, Hashimoto S, Katsura T, Inui K (2010) Disruption of multidrug and toxin extrusion MATE1 potentiates cisplatin-induced nephrotoxicity. Biochem Pharmacol 80 (11):1762–1767. https://doi.org/10.1016/j.bcp.2010.08.019
Ni W, Ji J, Dai Z, Papp A, Johnson AJ, Ahn S, Farley KL, Lin TS, Dalton JT, Li X, Jarjoura D, Byrd JC, Sadee W, Grever MR, Phelps MA (2010) Flavopiridol pharmacogenetics: clinical and functional evidence for the role of SLCO1B1/OATP1B1 in flavopiridol disposition. PLoS One 5(11):e13792. https://doi.org/10.1371/journal.pone.0013792
Nieuweboer AJ, Hu S, Gui C, Hagenbuch B, Ghobadi Moghaddam-Helmantel IM, Gibson AA, de Bruijn P, Mathijssen RH, Sparreboom A (2014) Influence of drug formulation on OATP1B-mediated transport of paclitaxel. Cancer Res 74(11):3137–3145. https://doi.org/10.1158/0008-5472.CAN-13-3634
Nigam SK (2015) What do drug transporters really do? Nat Rev Drug Discov 14(1):29–44. https://doi.org/10.1038/nrd4461
Nishida K, Takeuchi K, Hosoda A, Sugano S, Morisaki E, Ohishi A, Nagasawa K (2018) Ergothioneine ameliorates oxaliplatin-induced peripheral neuropathy in rats. Life Sci 207:516–524. https://doi.org/10.1016/j.lfs.2018.07.006
Nishimura M, Naito S (2008) Tissue-specific mRNA expression profiles of human solute carrier transporter superfamilies. Drug Metab Pharmacokinet 23(1):22–44
Nozawa T, Minami H, Sugiura S, Tsuji A, Tamai I (2005) Role of organic anion transporter OATP1B1 (OATP-C) in hepatic uptake of irinotecan and its active metabolite, 7-ethyl-10-hydroxycamptothecin: in vitro evidence and effect of single nucleotide polymorphisms. Drug Metab Dispos 33(3):434–439. https://doi.org/10.1124/dmd.104.001909
Okabe M, Unno M, Harigae H, Kaku M, Okitsu Y, Sasaki T, Mizoi T, Shiiba K, Takanaga H, Terasaki T, Matsuno S, Sasaki I, Ito S, Abe T (2005) Characterization of the organic cation transporter SLC22A16: a doxorubicin importer. Biochem Biophys Res Commun 333 (3):754–762. https://doi.org/10.1016/j.bbrc.2005.05.174
Okabe M, Szakacs G, Reimers MA, Suzuki T, Hall MD, Abe T, Weinstein JN, Gottesman MM (2008) Profiling SLCO and SLC22 genes in the NCI-60 cancer cell lines to identify drug uptake transporters. Mol Cancer Ther 7(9):3081–3091. https://doi.org/10.1158/1535-7163.MCT-08-0539
Omlin A, Sartor O, Rothermundt C, Cathomas R, De Bono JS, Shen L, Su Z, Gillessen S (2015) Analysis of side effect profile of alopecia, nail changes, peripheral neuropathy, and dysgeusia in prostate cancer patients treated with docetaxel and cabazitaxel. Clin Genitourin Cancer 13(4): e205–e208. https://doi.org/10.1016/j.clgc.2015.01.010
Osman NM, Litterst CL (1983) Effect of probenecid and N'-methylnicotinamide on renal handling of cis-dichlorodiammineplatinum-II in rats. Cancer Lett 19(1):107–111
Ota K, Ito K, Akahira J, Sato N, Onogawa T, Moriya T, Unno M, Abe T, Niikura H, Takano T, Yaegashi N (2007) Expression of organic cation transporter SLC22A16 in human epithelial ovarian cancer: a possible role of the adriamycin importer. Int J Gynecol Pathol 26(3):334–340. https://doi.org/10.1097/01.pgp.0000236951.33914.1b
Owens JK, Shewach DS, Ullman B, Mitchell BS (1992) Resistance to 1-β-d-arabinofuranosylcytosine in human T-lymphoblasts mediated by mutations within the deoxycytidine kinase gene. Cancer Res 52(9):2389–2393
Pabla N, Gibson AA, Buege M, Ong SS, Li L, Hu S, Du G, Sprowl JA, Vasilyeva A, Janke LJ, Schlatter E, Chen T, Ciarimboli G, Sparreboom A (2015) Mitigation of acute kidney injury by cell-cycle inhibitors that suppress both CDK4/6 and OCT2 functions. Proc Natl Acad Sci U S A 112(16):5231–5236. https://doi.org/10.1073/pnas.1424313112
Pawlyk AC, Giacomini KM, McKeon C, Shuldiner AR, Florez JC (2014) Metformin pharmacogenomics: current status and future directions. Diabetes 63(8):2590–2599. https://doi.org/10.2337/db13-1367

Peters CM, Jimenez-Andrade JM, Jonas BM, Sevcik MA, Koewler NJ, Ghilardi JR, Wong GY, Mantyh PW (2007) Intravenous paclitaxel administration in the rat induces a peripheral sensory neuropathy characterized by macrophage infiltration and injury to sensory neurons and their supporting cells. Exp Neurol 203(1):42–54. https://doi.org/10.1016/j.expneurol.2006.07.022

Price PM, Safirstein RL, Megyesi J (2009) The cell cycle and acute kidney injury. Kidney Int 76 (6):604–613. https://doi.org/10.1038/ki.2009.224

Raghunand N, He X, van Sluis R, Mahoney B, Baggett B, Taylor CW, Paine-Murrieta G, Roe D, Bhujwalla ZM, Gillies RJ (1999) Enhancement of chemotherapy by manipulation of tumour pH. Br J Cancer 80(7):1005–1011. https://doi.org/10.1038/sj.bjc.6690455

Robert J, Vrignaud P, Nguyen-Ngoc T, Iliadis A, Mauriac L, Hurteloup P (1985) Comparative pharmacokinetics and metabolism of doxorubicin and epirubicin in patients with metastatic breast cancer. Cancer Treat Rep 69(6):633–640

Robert J, Bui NB, Vrignaud P (1987) Pharmacokinetics of doxorubicin in sarcoma patients. Eur J Clin Pharmacol 31(6):695–699

Ross DA, Gale GR (1979) Reduction of the renal toxicity of cis-dichlorodiammineplatinum(II) by probenecid. Cancer Treat Rep 63(5):781–787

Safirstein R, Miller P, Guttenplan JB (1984) Uptake and metabolism of cisplatin by rat kidney. Kidney Int 25(5):753–758

Salphati L, Chu X, Chen L, Prasad B, Dallas S, Evers R, Mamaril-Fishman D, Geier EG, Kehler J, Kunta J, Mezler M, Laplanche L, Pang J, Rode A, Soars MG, Unadkat JD, van Waterschoot RA, Yabut J, Schinkel AH, Scheer N (2014) Evaluation of organic anion transporting polypeptide 1B1 and 1B3 humanized mice as a translational model to study the pharmacokinetics of statins. Drug Metab Dispos 42(8):1301–1313. https://doi.org/10.1124/dmd.114.057976

Sastry J, Kellie SJ (2005) Severe neurotoxicity, ototoxicity and nephrotoxicity following high-dose cisplatin and amifostine. Pediatr Hematol Oncol 22(5):441–445. https://doi.org/10.1080/08880010590964381

Sayed-Ahmed MM, Al-Shabanah OA, Hafez MM, Aleisa AM, Al-Rejaie SS (2010) Inhibition of gene expression of heart fatty acid binding protein and organic cation/carnitine transporter in doxorubicin cardiomyopathic rat model. Eur J Pharmacol 640(1–3):143–149. https://doi.org/10.1016/j.ejphar.2010.05.002

Schneider BP, Hershman DL, Loprinzi C (2015a) Symptoms: chemotherapy-induced peripheral neuropathy. Adv Exp Med Biol 862:77–87. https://doi.org/10.1007/978-3-319-16366-6_6

Schneider BP, Li L, Radovich M, Shen F, Miller KD, Flockhart DA, Jiang G, Vance G, Gardner L, Vatta M, Bai S, Lai D, Koller D, Zhao F, O'Neill A, Smith ML, Railey E, White C, Partridge A, Sparano J, Davidson NE, Foroud T, Sledge GW Jr (2015b) Genome-wide association studies for taxane-induced peripheral neuropathy in ECOG-5103 and ECOG-1199. Clin Cancer Res 21 (22):5082–5091. https://doi.org/10.1158/1078-0432.CCR-15-0586

Scripture CD, Figg WD, Sparreboom A (2006) Peripheral neuropathy induced by paclitaxel: recent insights and future perspectives. Curr Neuropharmacol 4(2):165–172

Shu Y, Sheardown SA, Brown C, Owen RP, Zhang S, Castro RA, Ianculescu AG, Yue L, Lo JC, Burchard EG, Brett CM, Giacomini KM (2007) Effect of genetic variation in the organic cation transporter 1 (OCT1) on metformin action. J Clin Invest 117(5):1422–1431. https://doi.org/10.1172/JCI30558

Singal PK, Iliskovic N (1998) Doxorubicin-induced cardiomyopathy. N Engl J Med 339 (13):900–905. https://doi.org/10.1056/NEJM199809243391307

Singer SJ, Nicolson GL (1972) The fluid mosaic model of the structure of cell membranes. Science 175(4023):720–731

Sissung TM, Mross K, Steinberg SM, Behringer D, Figg WD, Sparreboom A, Mielke S (2006) Association of ABCB1 genotypes with paclitaxel-mediated peripheral neuropathy and neutropenia. Eur J Cancer 42(17):2893–2896. https://doi.org/10.1016/j.ejca.2006.06.017

Smith NF, Acharya MR, Desai N, Figg WD, Sparreboom A (2005) Identification of OATP1B3 as a high-affinity hepatocellular transporter of paclitaxel. Cancer Biol Ther 4(8):815–818

Smith NF, Marsh S, Scott-Horton TJ, Hamada A, Mielke S, Mross K, Figg WD, Verweij J, McLeod HL, Sparreboom A (2007) Variants in the SLCO1B3 gene: interethnic distribution and association with paclitaxel pharmacokinetics. Clin Pharmacol Ther 81(1):76–82. https://doi.org/10.1038/sj.clpt.6100011

Song IH, Zong J, Borland J, Jerva F, Wynne B, Zamek-Gliszczynski MJ, Humphreys JE, Bowers GD, Choukour M (2016) The effect of Dolutegravir on the pharmacokinetics of metformin in healthy subjects. J Acquir Immune Defic Syndr. https://doi.org/10.1097/QAI.0000000000000983

Sparreboom A, Huizing MT, Boesen JJ, Nooijen WJ, van Tellingen O, Beijnen JH (1995) Isolation, purification, and biological activity of mono- and dihydroxylated paclitaxel metabolites from human feces. Cancer Chemother Pharmacol 36(4):299–304. https://doi.org/10.1007/BF00689047

Sprowl JA, Sparreboom A (2014) Uptake carriers and oncology drug safety. Drug Metab Dispos 42(4):611–622. https://doi.org/10.1124/dmd.113.055806

Sprowl JA, Ciarimboli G, Lancaster CS, Giovinazzo H, Gibson AA, Du G, Janke LJ, Cavaletti G, Shields AF, Sparreboom A (2013a) Oxaliplatin-induced neurotoxicity is dependent on the organic cation transporter OCT2. Proc Natl Acad Sci U S A 110(27):11199–11204. https://doi.org/10.1073/pnas.1305321110

Sprowl JA, Mathijssen RH, Sparreboom A (2013b) Can erlotinib ameliorate cisplatin-induced toxicities? J Clin Oncol 31(27):3442–3443. https://doi.org/10.1200/JCO.2013.50.8184

Sprowl JA, Ness RA, Sparreboom A (2013c) Polymorphic transporters and platinum pharmacodynamics. Drug Metab Pharmacokinet 28(1):19–27. https://doi.org/10.2133/dmpk.dmpk-12-rv-073

Sprowl JA, Lancaster CS, Pabla N, Hermann E, Kosloske AM, Gibson AA, Li L, Zeeh D, Schlatter E, Janke LJ, Ciarimboli G, Sparreboom A (2014) Cisplatin-induced renal injury is independently mediated by OCT2 and p53. Clin Cancer Res 20(15):4026–4035. https://doi.org/10.1158/1078-0432.CCR-14-0319

Sprowl JA, Ong SS, Gibson AA, Hu S, Du G, Lin W, Li L, Bharill S, Ness RA, Stecula A, Offer SM, Diasio RB, Nies AT, Schwab M, Cavaletti G, Schlatter E, Ciarimboli G, Schellens JH, Isacoff EY, Sali A, Chen T, Baker SD, Sparreboom A, Pabla N (2016) A phosphotyrosine switch regulates organic cation transporters. Nat Commun 7:10880. https://doi.org/10.1038/ncomms10880

Stage TB, Brosen K, Christensen MM (2015a) A comprehensive review of drug-drug interactions with metformin. Clin Pharmacokinet 54(8):811–824. https://doi.org/10.1007/s40262-015-0270-6

Stage TB, Damkier P, Pedersen RS, Christensen MM, Christiansen L, Christensen K, Brosen K (2015b) A twin study of the trough plasma steady-state concentration of metformin. Pharmacogenet Genomics 25(5):259–262. https://doi.org/10.1097/FPC.0000000000000133

Stage TB, Lee MP, Hallas J, Christensen MM, Brosen K, Christensen K, Gagne JJ, Pottegard A (2016) Early discontinuation of metformin in individuals treated with inhibitors of transporters of metformin. Basic Clin Pharmacol Toxicol 118(6):487–495. https://doi.org/10.1111/bcpt.12579

Sucheston LE, Zhao H, Yao S, Zirpoli G, Liu S, Barlow WE, Moore HC, Thomas Budd G, Hershman DL, Davis W, Ciupak GL, Stewart JA, Isaacs C, Hobday TJ, Salim M, Hortobagyi GN, Gralow JR, Livingston RB, Albain KS, Hayes DF, Ambrosone CB (2011) Genetic predictors of taxane-induced neurotoxicity in a SWOG phase III intergroup adjuvant breast cancer treatment trial (S0221). Breast Cancer Res Treat 130(3):993–1002. https://doi.org/10.1007/s10549-011-1671-3

Sun X, Li J, Guo C, Xing H, Xu J, Wen Y, Qiu Z, Zhang Q, Zheng Y, Chen X, Zhao D (2016) Pharmacokinetic effects of curcumin on docetaxel mediated by OATP1B1, OATP1B3 and CYP450s. Drug Metab Pharmacokinet 31(4):269–275. https://doi.org/10.1016/j.dmpk.2016.02.005

Svoboda M, Wlcek K, Taferner B, Hering S, Stieger B, Tong D, Zeillinger R, Thalhammer T, Jager W (2011) Expression of organic anion-transporting polypeptides 1B1 and 1B3 in ovarian cancer cells: relevance for paclitaxel transport. Biomed Pharmacother 65(6):417–426. https://doi.org/10.1016/j.biopha.2011.04.031

Swain SM, Whaley FS, Ewer MS (2003) Congestive heart failure in patients treated with doxorubicin: a retrospective analysis of three trials. Cancer 97(11):2869–2879. https://doi.org/10.1002/cncr.11407

Tamai I, Nezu J, Uchino H, Sai Y, Oku A, Shimane M, Tsuji A (2000) Molecular identification and characterization of novel members of the human organic anion transporter (OATP) family. Biochem Biophys Res Commun 273(1):251–260. https://doi.org/10.1006/bbrc.2000.2922

Tanabe Y, Shimizu C, Hamada A, Hashimoto K, Ikeda K, Nishizawa D, Hasegawa J, Shimomura A, Ozaki Y, Tamura N, Yamamoto H, Yunokawa M, Yonemori K, Takano T, Kawabata H, Tamura K, Fujiwara Y (2017) Paclitaxel-induced sensory peripheral neuropathy is associated with an ABCB1 single nucleotide polymorphism and older age in Japanese. Cancer Chemother Pharmacol 79(6):1179–1186. https://doi.org/10.1007/s00280-017-3314-9

Thompson SW, Davis LE, Kornfeld M, Hilgers RD, Standefer JC (1984) Cisplatin neuropathy. Clinical, electrophysiologic, morphologic, and toxicologic studies. Cancer 54(7):1269–1275

Tian A, Lu H, Zhang J, Fu S, Jiang Z, Lam W, Guan F, Chen L, Feng L, Cheng Y (2018) Multikinase inhibitor sorafenib induces skin toxicities in tumor-bearing mice. Cancer Chemother Pharmacol 81(6):1025–1033. https://doi.org/10.1007/s00280-018-3575-y

Travis LB, Fossa SD, Sesso HD, Frisina RD, Herrmann DN, Beard CJ, Feldman DR, Pagliaro LC, Miller RC, Vaughn DJ, Einhorn LH, Cox NJ, Dolan ME (2014) Chemotherapy-induced peripheral neurotoxicity and ototoxicity: new paradigms for translational genomics. J Natl Cancer Inst 106(5). https://doi.org/10.1093/jnci/dju044

Tregouet DA, Konig IR, Erdmann J, Munteanu A, Braund PS, Hall AS, Grosshennig A, Linsel-Nitschke P, Perret C, DeSuremain M, Meitinger T, Wright BJ, Preuss M, Balmforth AJ, Ball SG, Meisinger C, Germain C, Evans A, Arveiler D, Luc G, Ruidavets JB, Morrison C, van der Harst P, Schreiber S, Neureuther K, Schafer A, Bugert P, El Mokhtari NE, Schrezenmeir J, Stark K, Rubin D, Wichmann HE, Hengstenberg C, Ouwehand W, Wellcome Trust Case Control Consortium, Cardiogenics Consortium, Ziegler A, Tiret L, Thompson JR, Cambien F, Schunkert H, Samani NJ (2009) Genome-wide haplotype association study identifies the SLC22A3-LPAL2-LPA gene cluster as a risk locus for coronary artery disease. Nat Genet 41 (3):283–285. https://doi.org/10.1038/ng.314

Trevisan G, Materazzi S, Fusi C, Altomare A, Aldini G, Lodovici M, Patacchini R, Geppetti P, Nassini R (2013) Novel therapeutic strategy to prevent chemotherapy-induced persistent sensory neuropathy by TRPA1 blockade. Cancer Res 73(10):3120–3131. https://doi.org/10.1158/0008-5472.CAN-12-4370

Tschirka J, Kreisor M, Betz J, Grundemann D (2018) Substrate selectivity check of the ergothioneine transporter. Drug Metab Dispos 46(6):779–785. https://doi.org/10.1124/dmd.118.080440

Tsuda M, Terada T, Mizuno T, Katsura T, Shimakura J, Inui K (2009) Targeted disruption of the multidrug and toxin extrusion 1 (MATE1) gene in mice reduces renal secretion of metformin. Mol Pharmacol 75(6):1280–1286. https://doi.org/10.1124/mol.109.056242

Urban TJ, Yang C, Lagpacan LL, Brown C, Castro RA, Taylor TR, Huang CC, Stryke D, Johns SJ, Kawamoto M, Carlson EJ, Ferrin TE, Burchard EG, Giacomini KM (2007) Functional effects of protein sequence polymorphisms in the organic cation/ergothioneine transporter OCTN1 (SLC22A4). Pharmacogenet Genomics 17(9):773–782. https://doi.org/10.1097/FPC.0b013e3281c6d08e

Vallon V, Rieg T, Ahn SY, Wu W, Eraly SA, Nigam SK (2008) Overlapping in vitro and in vivo specificities of the organic anion transporters OAT1 and OAT3 for loop and thiazide diuretics. Am J Physiol Renal Physiol 294(4):F867–F873. https://doi.org/10.1152/ajprenal.00528.2007

van de Steeg E, van Esch A, Wagenaar E, van der Kruijssen CM, van Tellingen O, Kenworthy KE, Schinkel AH (2011) High impact of Oatp1a/1b transporters on in vivo disposition of the hydrophobic anticancer drug paclitaxel. Clin Cancer Res 17(2):294–301. https://doi.org/10.1158/1078-0432.CCR-10-1980

van de Steeg E, van Esch A, Wagenaar E, Kenworthy KE, Schinkel AH (2013) Influence of human OATP1B1, OATP1B3, and OATP1A2 on the pharmacokinetics of methotrexate and paclitaxel in humanized transgenic mice. Clin Cancer Res 19(4):821–832. https://doi.org/10.1158/1078-0432.CCR-12-2080

VanWert AL, Gionfriddo MR, Sweet DH (2010) Organic anion transporters: discovery, pharmacology, regulation and roles in pathophysiology. Biopharm Drug Dispos 31(1):1–71. https://doi.org/10.1002/bdd.693

Von Hoff DD, Layard MW, Basa P, Davis HL Jr, Von Hoff AL, Rozencweig M, Muggia FM (1979) Risk factors for doxorubicin-induced congestive heart failure. Ann Intern Med 91 (5):710–717

Waissbluth S, Daniel SJ (2013) Cisplatin-induced ototoxicity: transporters playing a role in cisplatin toxicity. Hear Res 299:37–45. https://doi.org/10.1016/j.heares.2013.02.002

Wang L, Weinshilboum R (2014) Metformin pharmacogenomics: biomarkers to mechanisms. Diabetes 63(8):2609–2610. https://doi.org/10.2337/db14-0609

Wang DS, Jonker JW, Kato Y, Kusuhara H, Schinkel AH, Sugiyama Y (2002) Involvement of organic cation transporter 1 in hepatic and intestinal distribution of metformin. J Pharmacol Exp Ther 302(2):510–515. https://doi.org/10.1124/jpet.102.034140

Wang L, Chen J, Zeng Y, Wei J, Jing J, Li G, Su L, Tang X, Wu T, Zhou L (2016) Functional variant in the SLC22A3-LPAL2-LPA gene cluster contributes to the severity of coronary artery disease. Arterioscler Thromb Vasc Biol 36(9):1989–1996. https://doi.org/10.1161/ATVBAHA.116.307311

Wu X, George RL, Huang W, Wang H, Conway SJ, Leibach FH, Ganapathy V (2000) Structural and functional characteristics and tissue distribution pattern of rat OCTN1, an organic cation transporter, cloned from placenta. Biochim Biophys Acta 1466(1–2):315–327

Xia B, Heimbach T, He H, Lin TH (2012) Nilotinib preclinical pharmacokinetics and practical application toward clinical projections of oral absorption and systemic availability. Biopharm Drug Dispos 33(9):536–549. https://doi.org/10.1002/bdd.1821

Yamada S, Arrell DK, Martinez-Fernandez A, Behfar A, Kane GC, Perez-Terzic CM, Crespo-Diaz RJ, McDonald RJ, Wyles SP, Zlatkovic-Lindor J, Nelson TJ, Terzic A (2015) Regenerative therapy prevents heart failure progression in dyssynchronous nonischemic narrow QRS cardiomyopathy. J Am Heart Assoc 4(5). https://doi.org/10.1161/JAHA.114.001614

Yamaguchi H, Kobayashi M, Okada M, Takeuchi T, Unno M, Abe T, Goto J, Hishinuma T, Mano N (2008) Rapid screening of antineoplastic candidates for the human organic anion transporter OATP1B3 substrates using fluorescent probes. Cancer Lett 260(1–2):163–169. https://doi.org/10.1016/j.canlet.2007.10.040

Yan X, Maixner DW, Yadav R, Gao M, Li P, Bartlett MG, Weng HR (2015) Paclitaxel induces acute pain via directly activating toll like receptor 4. Mol Pain 11:10. https://doi.org/10.1186/s12990-015-0005-6

Yang X, Han L (2019) Roles of renal drug transporter in drug disposition and renal toxicity. Adv Exp Med Biol 1141:341–360. https://doi.org/10.1007/978-981-13-7647-4_7

Yao X, Panichpisal K, Kurtzman N, Nugent K (2007) Cisplatin nephrotoxicity: a review. Am J Med Sci 334(2):115–124. https://doi.org/10.1097/MAJ.0b013e31812dfe1e

Yokoo S, Masuda S, Yonezawa A, Terada T, Katsura T, Inui K (2008) Significance of organic cation transporter 3 (SLC22A3) expression for the cytotoxic effect of oxaliplatin in colorectal cancer. Drug Metab Dispos 36(11):2299–2306. https://doi.org/10.1124/dmd.108.023168

Zaher H, Meyer zu Schwabedissen HE, Tirona RG, Cox ML, Obert LA, Agrawal N, Palandra J, Stock JL, Kim RB, Ware JA (2008) Targeted disruption of murine organic anion-transporting polypeptide 1b2 (Oatp1b2/Slco1b2) significantly alters disposition of prototypical drug substrates pravastatin and rifampin. Mol Pharmacol 74(2):320–329. https://doi.org/10.1124/mol.108.046458

Zhang J, Zhou W (2012) Ameliorative effects of SLC22A2 gene polymorphism 808 G/T and cimetidine on cisplatin-induced nephrotoxicity in Chinese cancer patients. Food Chem Toxicol 50(7):2289–2293. https://doi.org/10.1016/j.fct.2012.03.077

Zhang S, Lovejoy KS, Shima JE, Lagpacan LL, Shu Y, Lapuk A, Chen Y, Komori T, Gray JW, Chen X, Lippard SJ, Giacomini KM (2006) Organic cation transporters are determinants of oxaliplatin cytotoxicity. Cancer Res 66(17):8847–8857. https://doi.org/10.1158/0008-5472.CAN-06-0769

Zhang B, Bolognia J, Marks P, Podoltsev N (2014) Enhanced skin toxicity associated with the combination of clofarabine plus cytarabine for the treatment of acute leukemia. Cancer Chemother Pharmacol 74(2):303–307. https://doi.org/10.1007/s00280-014-2504-y

Zhao M, Isami K, Nakamura S, Shirakawa H, Nakagawa T, Kaneko S (2012) Acute cold hypersensitivity characteristically induced by oxaliplatin is caused by the enhanced responsiveness of TRPA1 in mice. Mol Pain 8:55. https://doi.org/10.1186/1744-8069-8-55

Zhu HJ, Appel DI, Grundemann D, Richelson E, Markowitz JS (2012) Evaluation of organic cation transporter 3 (SLC22A3) inhibition as a potential mechanism of antidepressant action. Pharmacol Res 65(4):491–496. https://doi.org/10.1016/j.phrs.2012.01.008

Zimmerman EI, Hu S, Roberts JL, Gibson AA, Orwick SJ, Li L, Sparreboom A, Baker SD (2013) Contribution of OATP1B1 and OATP1B3 to the disposition of sorafenib and sorafenib-glucuronide. Clin Cancer Res 19(6):1458–1466. https://doi.org/10.1158/1078-0432.CCR-12-3306

Zimmerman EI, Gibson AA, Hu S, Vasilyeva A, Orwick SJ, Du G, Mascara GP, Ong SS, Chen T, Vogel P, Inaba H, Maitland ML, Sparreboom A, Baker SD (2016) Multikinase inhibitors induce cutaneous toxicity through OAT6-mediated uptake and MAP 3K7-driven cell death. Cancer Res 76(1):117–126. https://doi.org/10.1158/0008-5472.CAN-15-0694

Zong J, Borland J, Jerva F, Wynne B, Choukour M, Song I (2014) The effect of dolutegravir on the pharmacokinetics of metformin in healthy subjects. J Int AIDS Soc 17(4 Suppl 3):19584. https://doi.org/10.7448/IAS.17.4.19584

Zwart R, Verhaagh S, Buitelaar M, Popp-Snijders C, Barlow DP (2001) Impaired activity of the extraneuronal monoamine transporter system known as uptake-2 in Orct3/Slc22a3-deficient mice. Mol Cell Biol 21(13):4188–4196. https://doi.org/10.1128/MCB.21.13.4188-4196.2001

Rev Physiol Biochem Pharmacol (2022) 183: 217–250
https://doi.org/10.1007/112_2020_33

Published online: 1 August 2020

Ion Transport and Radioresistance

Bastian Roth and Stephan M. Huber

Contents

Abstract Neoplastic transformation is associated with alterations of the ion transports across plasma and intracellular membranes. These alterations are crucial elements of the phenotypical reprogramming of the transformed cells and may promote adaptation to hypoxia, malignant progression, tumor spreading and metastasis, as well as therapy resistance. The present review article focuses on ion transport processes in tumor cells that are induced by ionizing radiation and that

B. Roth
Department of Biology, Neurophysiology and Neurosensory Systems, Technische Universität Darmstadt, Darmstadt, Germany
e-mail: roth@bio.tu-darmstadt.de

S. M. Huber (✉)
Department of Radiation Oncology, University of Tübingen, Tübingen, Germany
e-mail: stephan.huber@uni-tuebingen.de

contribute to radioresistance and therapy failure. In particular, this article introduces radiogenic ion transports across plasma and mitochondrial membranes and discusses their functional significance for cell cycle control, DNA repair, accelerated repopulation, cell migration and metastasis, metabolic reprogramming, adaptation to hypoxia, and radiogenic formation of reactive oxygen species.

Keywords DNA damage response · Hypoxia adaptation · Ion channels · Ion transport · Oncochannels · Radiogenic reprogramming · Radioresistance

Abbreviations

$\Delta\Psi_m$	Inner mitochondrial membrane potential
$\bullet O_2^-$	Superoxide anion radical
AMPK	5′ Adenosine monophosphate-activated protein kinase
ATM	Protein kinase ataxia-telangiectasia mutated
CaMKIIs	Isoforms of the Ca^{2+}/calmodulin-dependent kinase-II
CLIC1	Chloride intracellular channel-1
CXCR4	C-X-C motif chemokine receptor-4
DSBs	DNA double-strand breaks
EGFR	Epidermal growth factor receptor
EMT	Epithelial–mesenchymal transition
ERK	Extracellular signal-regulated kinase
GMT	Glial-mesenchymal transition
GSCs	Glioblastoma stem cells
HIF-1α	Hypoxia-inducible factor-1α
K_{ATP}	ATP-sensitive potassium channel
K_{Ca}	Calcium-activated potassium channel
K_v	Voltage-gated potassium channel
MnSOD	Mitochondrial manganese superoxide dismutase (SOD2)
MPT	Mitochondrial membrane permeability transition pore
$_{mt}$	Mitochondrial
Na_v	Voltage-gated sodium channel
Pyk2	Focal adhesion kinase-2
ROS	Reactive oxygen species
SDF1	Stromal cell-derived factor-1 (CXCL12)
TRPM2/8	Member 2/8 of the melastatin family of transient receptor potential nonselective cation channels
TRPV1/5/6	Member 1/5/6 of the vanilloid family of transient receptor potential nonselective cation channels
UCP2/3	Uncoupling protein-2/3
VDAC1	Voltage-gated anion channel-1
V_m	Membrane potential

1 Introduction

By transporting ions between extracellular and/or intracellular compartments, ion channels and passive or primary active (pumps) transporters in plasma- or intracellular membrane(s) regulate membrane potentials, ion concentrations including Ca^{2+}-transients, osmotically obliged H_2O fluxes and cell volume, pH homeostasis, or co- and counter-transfer of nutrients, metabolites, or catabolites. Beyond this "operational transport processes," ion transports and, in particular, ion channel activities generate fast electro- and Ca^{2+} signals. Moreover, non-conductive conformational changes of ion channels have been identified to transduce signals within molecular complexes between, e.g., membrane receptors and downstream kinases. Importantly, these ion channel-mediated non-conductive, as well as the electro- and Ca^{2+} signals are integral modules of biochemical signal transduction pathways, of the redox or the pH signaling. In fact, ion transports across biological membranes interfere with virtually all cellular processes (for review, see (Huber 2013*)). As an example, ion transport-dependent plasma membrane potential and cell volume have been identified decades ago to act as superordinate regulators of cell proliferation or cell death (for review, see (Chen et al. 2008; Yang and Brackenbury 2013)).

Therefore, it is not surprising that neoplastically transformed cells express ion transports that differ profoundly from those of their healthy parental cells. Importantly, this aberrantly expressed ion transport toolkit has been demonstrated to contribute to malignant progression, metastasis, or therapy resistance of tumor cells. In particular, some ion channel types seem to have a high "oncogenic" potential since they are upregulated in several tumor entities, and thus, might be promising targets for future strategies of anti-cancer therapy (for review, see (Huber 2013*)).

About 50% of all cancer patients undergo radiotherapy in curative or palliative concepts. Radiotherapy applies ionizing radiation to a target tissue volume in order to induce lethal mutations in the targeted cells. DNA double-strand breaks (DSBs) are the most hazardous form of DNA damage for a proliferating cell since passing mitosis with unrepaired DSBs inevitably introduces chromosome aberrations. Such a cell is at high risk of running into mitotic catastrophe and to die by apoptotic or necrotic cell death (for review, see (Toulany 2019)). Irradiation of normoxic tissue with MV photons which is frequently applied by external beam radiotherapy causes, on average, about 20 nuclear DNA DSBs per 1 Gy (1 J/kg of tissue) of energy deposition (Matsuya et al. 2014). DSBs, in turn, trigger the cellular DNA damage response which comprises activation of the DNA repair machinery and in parallel deceleration of the cell cycle. More or less stringent DNA damage checkpoints in G_1, S, and G_2 phase of cell cycle trigger G_1 or G_2/M cell cycle arrest or slow-down of S progression (Chao et al. 2017) in order to provide the time required to accomplish DNA repair. Moreover, decatenation-, spindle assembly-, and post-mitotic tetraploidy checkpoint in late G_2, (pro)metaphase, and G_1, respectively, mount guard on the chromosome integrity (Brown and Geiger 2018). Whether or not ionizing radiation eradicates tumor cells does not only depend on cell cycle control and

DNA repair. Additionally, the O_2 tension of the tumor microenvironment and the cellular anti-oxidative defense determine the extent of DNA damage evoked by ionizing radiation. Both mutually depend on the bioenergetics that frequently is reprogrammed in tumor cells towards upregulated glycolysis and lactic acid formation on the expense of mitochondrial oxidative phosphorylation (for review, see (Eckert et al. 2019*)).

The target volume in radiotherapy covers the tumor (neoadjuvant and definitive radiotherapy) or, upon surgical resection the tumor bed (adjuvant radiotherapy), a safety margin around the gross tumor volume (with potentially disseminated tumor cells), and dependent on tumor type and stage, draining lymph nodes and/or other areas with a high risk of tumor relapse. Thereby, beyond improving local control, radiotherapy decreases the risk of lymph node or distant metastases. Fractionated regimens of radiotherapy apply several fractions of sublethal radiation doses.

The 5 *R*s of radiobiology (*r*epair, *r*epopulation, *r*edistribution, *r*eoxygenation, *r*adiosensitivity) describe parameters (classically only for tumor cells) that provide a framework to develop dose-fractionation regimens in radiotherapy that effectively eradicate tumor cells at maximal protection of normal tissue. The idea behind fractionated radiotherapy is to apply several sublethal radiation doses (e.g., 60 Gy in 30 fractions within 6 weeks for the radiotherapy of glioblastoma) that accumulate to a lethal dose in tumor cells but not in normal tissue in order to broaden the therapeutic window. The differing fractionated radiation response of tumor and normal tissue relies on differences in radiosensitivity, velocity and capacity of DNA repair, and radiation-induced redistribution into a more vulnerable phase of cell cycle between two radiation fractions. Moreover, reoxygenation of hypoxic tumor areas during fractionated radiotherapy increases the efficacy of tumor radiation. Finally, time periods of fractionation regimes are designed that way to utilize fractionated radiation-induced accelerated repopulation of early-responding normal tissue (which decreases toxicity) and/or to be completed before accelerated repopulation of tumors occurs. Because of DNA damage response, irradiated cells do not decay monoexponentially with increasing sublethal irradiation doses. Instead, the dependence of the survival fraction (S) and the radiation dose (D) follows a linear-quadratic function $\left(S = e^{-\left(\alpha D + \beta D^2\right)}\right)$ with α, β as tissue/tumor entity-specific parameters. Notably, the α/β ratio (unit [Gy]) can be used to predict the fractionation sensitivity of tissues/tumors and to develop effective and safe hypo-, hyper-, or normo-fractionation schedules.

Beyond widening the therapeutic window of radiotherapy, even optimized schedules of fractionated radiotherapy may induce accelerated repopulation, i.e., increased proliferation rates of the surviving clonogenic tumor cells (for review, see (Yom 2015)). Moreover, fractionated radiation may boost migration, spreading and metastasis of cancer cells (for review, see (Vilalta et al. 2016)). Both, accelerated repopulation and hypermigration, may contribute to therapy failure.

There is increasing evidence that ionizing radiation modulates ion transports in cancer cells and that these transports may confer radioresistance. The present review article aims to summarize our current knowledge on the function of ion transports in irradiated cells. Specifically, parts 2, 3, and 4 in this article summarize the data on the role of radiation-induced ion transport in DNA damage response, in accelerated repopulation and migration, and in metabolism and redox state, respectively. The first part introduces the methodology applied for the analysis of ion transports.

Worldwide, only few groups including our laboratories working on the physiology of irradiated cancer cells and many of the data discussed here were obtained in our current or former laboratories. We labeled our own publications by asterisk.

2 Methods to Study Ion Transports

Manual patch-clamp recording under visual microscopic control is suitable to characterize activities of individual types of ion channels or electrogenic transporters in defined morphological phenotypes such as epithelial-differentiated, migrating, dividing, apoptotic, senescent, or other cellular states. Automatized patch-clamp systems, by contrast, require detachment/isolation of cells and lose the visual information of the recorded cells. Extracellular fields recording and extracellular microelectrode arrays can give only spatial information about the overall electrical activity of cells. Therefore, almost half a century after the pioneering work of Bernd Sakmann and Erwin Neher (Neher and Sakmann 1976), manual patch-clamp recording is still the gold standard of analyzing ion channel activities.

Ion channel activities crosstalk with biochemical signal transduction pathways. Modulation of Ca^{2+} signaling and downstream Ca^{2+} effector proteins is a well-documented example for the integration of electrosignaling into biochemical signaling. Therefore, patch-clamp data are often combined with data on free Ca^{2+} concentrations obtained by ratiometric Ca^{2+} fluorescence imaging. Beyond that, fluorescence dyes specific for other ions, including protons, are on the market. For technical reasons, most patch-clamp and fluorescence imaging data are from conventional two-dimensional (2D) in vitro cultures. Solid tumors, however, grow three-dimensionally. Therefore, three-dimensional (3D) in vitro tumor cell cultures have been postulated to reflect better the in vivo situation than 2D cultures (Ravi et al. 2015). This, however, holds not always true. Glioblastoma cells, for instance, invade the brain parenchyma even mono-dimensionally (1D) by using vessels or axon bundles as tracks (Sontheimer 2008). Neoplastic transformation of renal, intestinal, prostate, etc. epithelial cells occurs in a 2D in vivo environment. Likewise, intra- and extravasation and circulation in lymph or blood vessels of tumor cells during distant metastasis (Klumpp et al. 2016b*) can hardly be studied in vitro in 3D tumor cell clusters. Moreover, due to the hypoxia-associated radioresistance, 3D cell clusters must not exceed a certain size for biochemical analysis of their radiobiology.

For most 3D tumor cell cultures, cells are embedded in an extracellular matrix (ECM), usually collagen or matrigel. This is a critical step since "unphysiological" matrices such as collagen for brain tumors (Ruoslahti 1996) may trigger a signaling that does not occur in the tumor in vivo and/or may hamper matrix invasion. As a matter of fact, matrigel (even growth factor-deriched preparations) triggers profound changes in gene expression and cancer stem cell phenotype (own unpublished data). Importantly, for analysis of, e.g., ion channel activity by the patch-clamp technique, matrix embedded cell clusters cannot be used directly since patch glass pipette once in contact with matrix material hardly forms giga-ohm seals with the plasma membrane which is a prerequisite for this technique.

Alternatively, specific media induce/select cancer stem cells that may form free-floating 3D spheroids (Klumpp et al. 2018*). Cells at the spheroid surface can be easily studied by patch-clamp recording. In addition, tumor slices from freshly resected specimens allow patch-clamp analysis of cells in a 3D context. In this ex vivo setting, however, the differentiation between cancer and stroma cells in the heterogeneous tumor requires further markers for cell type identification. Taken together, 3D cultures may better than conventional 2D cultures mirror the biology of solid tumors in most but not all oncological processes. Unfortunately, technical requirements of manual patch-clamp recording or fluorescence imaging limit the use of 3D cultures. Hence, most of our knowledge on ion channels in solid tumors relies on 2D in vitro data. The fact that many 2D in vitro observations could be confirmed in animal studies (e.g., (Steinle et al. 2011*; Edalat et al. 2016*)) suggests that 2D cultures of tumor cells are not a completely artificial model system.

3 DNA Damage Response

Conceptually, radiotherapy eradicates tumor cells by eliciting DNA DSBs to an extent that cannot be repaired by non-homologous end joining in G_1 phase and additional homologous recombination in S and G_2 phase of the cell cycle (for review, see (Valerie and Povirk 2003)). The capacity of DNA damage repair differs tremendously between tumor entities, within an entity between individual tumors, and within the heterogeneous tumor between subpopulations of tumor cells. Besides intrinsic cellular properties, factors of the tumor microenvironment (such as tumor hypoxia, as discussed later on in this article) determine the radioresistance of a tumor.

The intrinsic radioresistance of tumor cells may underlie therapy failure. Hence, interfering with cellular DNA damage response in combined therapy protocols is a strategy to radiosensitize tumors. As described in the following paragraphs, ionizing radiation has been demonstrated to stimulate ion transports which are pivotal elements of the cellular DNA damage response.

3.1 Radiogenic Modulation of Ion Channel Activity

Within 1–2 h, ionizing radiation induces activation of Ca^{2+}-dependent high conductance BK_{Ca} ($K_{Ca}1.1$, KCNMA1) (Steinle et al. 2011*) and intermediate conductance IK_{Ca} ($K_{Ca}3.1$, SK4, KCNN4) K^+ channels (Stegen et al. 2015*) in glioblastoma cells. Radiogenic IK_{Ca} activity has also been described in human lung adenocarcinoma (Gibhardt et al. 2015*; Roth et al. 2015*), in murine breast cancer cells (Mohr et al. 2019*), and in human T cell leukemia cells (Klumpp et al. 2016a*; Voos et al. 2018). Moreover, in chronic myeloid leukemia cells, ionizing radiation stimulates the activity of $K_v3.4$ (KCNC4) (Palme et al. 2013*) and hERG1 (KCNH2) voltage-gated K^+ channels (Palme et al. 2020*). In contrast to these fast stimulating effects, ionizing radiation has been demonstrated in animal models to inhibit BK_{Ca} in rat aortic smooth muscle cells at later time points (several days), resulting in a decreased vasorelaxant effect of these channels (Kyrychenko et al. 2012; Soloviev et al. 2009).

Upstream signaling of radiogenic channel modulation may involve the formation of reactive oxygen species (ROS) (Gibhardt et al. 2015*), lipid peroxidation, and subsequent activation of tyrosine kinases such as Src (Dittmann et al. 2009) and Pyk2 kinase (focal adhesion kinase-2) (Proudfoot et al. 2018). While Src stimulates the epidermal growth factor receptor (EGFR) and downstream pathways (Dittmann et al. 2009), Pyk2 kinase directly upregulates BK_{Ca} channel activity (Stegen et al. 2016*). The radiogenic inhibitory effect on BK_{Ca} in rat aortic smooth muscle cells, in contrast, is reportedly mediated by protein kinase C (Kizub et al. 2010).

In addition, radiogenic stabilization of the hypoxia-inducible factor-1α (HIF-1α) and upregulation of the HIF-1α target genes stromal cell-derived factor-1 (SDF1, CXCL12) and its C-X-C motif chemokine receptor-4 (CXCR4) stimulate auto-/paracrine chemokine signaling that contributes to radiogenic channel activation (Edalat et al. 2016*). Ligation of CXCR4 with SDF1 reportedly results in formation of inositol triphosphate and Ca^{2+} release from the stores (for review, see (Eckert et al. 2018*)). Along those lines, several studies reported radiogenic Ca^{2+} signaling (Kandasamy et al. 1991; Teshima et al. 2000; Todd and Mikkelsen 1994; Wojewodzka et al. 1994; Yoshida 1997; Gibhardt et al. 2015*; Stegen et al. 2015*; Klumpp et al. 2016a*; Edalat et al. 2016*; Stegen et al. 2016*; Klumpp et al. 2017*; Mohr et al. 2019*).

In particular, Ca^{2+}-permeable member 5/6 of the vanilloid (TRP*V*5/6) and members 2 and 8 of the melastatin (TRP*M*2/8) family of transient receptor potential (TRP) nonselective cation channels have been proposed to contribute to radiogenic Ca^{2+} entry into chronic myeloid leukemia (Heise et al. 2010*), in T cell leukemia (Klumpp et al. 2016a*), and glioblastoma cells (Klumpp et al. 2017*). One can assume that radiogenic TRPV or TRPM channel activation locally depolarizes the membrane potential and generates a profound increase in cytosolic free Ca^{2+} concentration directly beneath the channel. Membrane depolarization and Ca^{2+} rise activate voltage-gated K_v and Ca^{2+}-dependent K_{Ca} channels in close vicinity of the TRP channels (for review, see (Gueguinou et al. 1843)). K^+ channel activity, in turn, repolarizes the membrane potential, thereby stabilizing the inwardly directed electrochemical driving force for Ca^{2+} and regulating the activity of voltage-gated/regulated Ca^{2+} entry pathways. This scenario has been suggested for chronic myeloid (Heise et al. 2010*; Palme et al. 2013*) and T cell leukemia (Klumpp et al. 2016a*), in breast cancer cells as well as in glioblastoma cells (Klumpp et al. 2017*; Stegen et al. 2016*). Markedly, in the latter, radiogenic BK_{Ca} and IK_{Ca} K^+ channel activity have different, partly antagonizing effects on cytosolic free Ca^{2+}

concentration (Stegen et al. 2016*). Likewise, radiogenic $K_v3.4$ and hERG1 K^+ channel activities exert opposite effects on Ca^{2+} entry and steady-state cytosolic free Ca^{2+} concentration in chronic myeloid leukemia cells (Palme et al. 2013*, 2020*) suggesting that K^+ channels in concert with Ca^{2+} entry pathways generate complex radiogenic Ca^{2+} signaling. This signaling contributes to DNA damage response, as described in the next paragraphs.

3.2 Intrinsic Radioresistance Relying on Radiogenic Ion Channel Activity

Pharmacological inhibition or genetic knockdown of $K_v3.4$ (Palme et al. 2013) or hERG1 (Palme et al. 2020*) in chronic myeloid leukemia cells, of TRPM2 in T cell leukemia cells (Klumpp et al. 2016a*), of IK_{Ca} in breast cancer cells (Steudel et al. 2017*; Mohr et al. 2019*), and of TRPM8 (Klumpp et al. 2017*) or IK_{Ca} (Stegen et al. 2015*) in glioblastoma cells radiosensitizes these cells, clearly indicating a pivotal function of these channels for DNA damage response. The latter study also reported growth delay of ectopically transplanted glioblastoma *xeno*grafts in mice by pharmacological IK_{Ca} targeting concomitant to fractionated tumor irradiation (Stegen et al. 2015). In all four cell models, channel targeting impairs cell cycle control and isoforms of Ca^{2+}/calmodulin-dependent kinase-II (CaMKIIs) have been identified to play a key role herein. CaMKII kinases translate Ca^{2+} signals in long-lasting biochemical signaling. Upon CaM binding-triggered autophosphorylation, these kinases stay active independently of Ca^{2+} (Coultrap and Bayer 2012).

Nuclear isoforms of CaMKIIs reportedly regulate cell cycle in irradiated chronic myeloid (Heise et al. 2010*; Palme et al. 2013*, 2020*) and T cell leukemia (Klumpp et al. 2016a*), as well as glioblastoma cells (Klumpp et al. 2017*). Notably, radiogenic CaMKII activity depends on TRP and/or K^+ channel signaling. CaMKIIs, in turn, regulate isoforms of the cdc25 phosphatase by inhibitory phosphorylation, thereby preventing activation of phosphorylated cdc2 by cdc25. Phosphorylated cdc2 in complex with cyclin B arrests the cell cycle at the G_2/M boundary. Experimental interference with TRP or K^+ channel activity in irradiated glioblastoma and leukemia cells inhibits CaMKII activation, resulting in cdc25-mediated dephosphorylation and activation of cdc2 and, thus, in overriding of the cell cycle arrest (Klumpp et al. 2016a*, 2017*; Palme et al. 2013*; Stegen et al. 2015*). As a consequence, irradiated cells do not accomplish DNA DSB repair and enter mitosis with immature DNA repair which impairs clonogenic survival (Klumpp et al. 2017*; Stegen et al. 2015*).

Beyond cell cycle control, radiogenic Ca^{2+} signaling reportedly contributes to DNA repair. In A549 lung adenocarcinoma cells, inhibition or knockdown of TRPV1 or TRPM2 delays activation of the protein kinase ataxia-telangiectasia mutated (ATM), nuclear translocation of EGFR, and/or accumulation of p53-binding protein 1 (Masumoto et al. 2013; Nishino et al. 2016). ATM senses

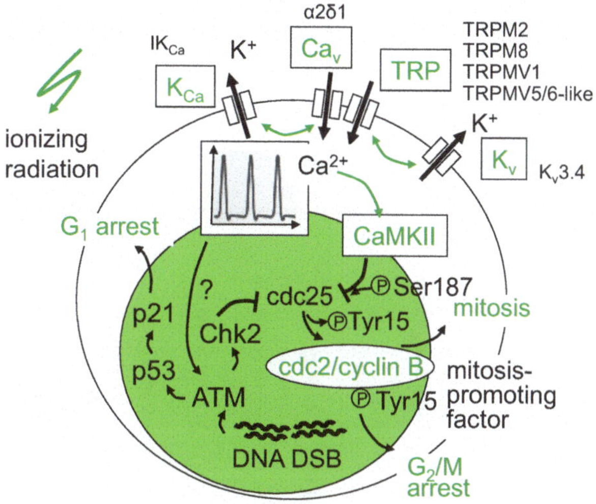

Fig. 1 Radiogenic activation of ion transports contributes to cell cycle control and DNA repair of irradiated cells (hypothetical model). Melastatin (M) and vanilloid (V) members of the transient receptor potential (TRP) nonselective cation channels in concert with voltage-gated Ca^{2+} (Ca_v), voltage-gated ($_tK_v$) K^+, and Ca^{2+}-activated (K_{Ca}) K^+ channels generate radiogenic Ca^{2+} signals that arrest the cell cycle via activation of Ca^{2+}/calmodulin kinase-II isoforms (CaMKII) and subsequent inhibition of the cdc25 phosphatase and the mitosis-promoting factor subunit cdc2. TRP-mediated signaling also regulates activation of protein kinase ataxia-telangiectasia mutated (ATM) which initiates DNA double-strand break (DSB) repair and cell cycle arrest (α2δ1: α2δ1subunit of the voltage-gated Ca^{2+} channel, CACNA2D1)

DNA DSBs and initiates DNA repair and cell cycle arrest (for review, see (Kim et al. 2019)). Accordingly, TRPV1 or TRPM2 inhibition/knockdown resulted in impaired early DNA damage responses and radiosensitization (Masumoto et al. 2013; Nishino et al. 2016). Likewise, in non-small cell lung cancer cells (NSCLC), the α2δ1 subunit of the voltage-gated Ca^{2+} channel (CACNA2D1) has been identified to contribute to ATM activation. Overexpression of CACNA2D1 in NSCLC accelerates DNA DSB repair and confers radioresistance (Sui et al. 2018) (Fig. 1; Table 1 provides a summary of the ion channels and transporters that have been identified to contribute to the radiobiology of cancer cells. In Table 1, also the (patho)physiological functions of these ion transports in normal tissue are described). In addition to DNA repair, radiogenic ion transports may contribute to accelerated repopulation, as discussed in the next paragraphs.

Table 1 Ion transports in radiobiology

Channel type	Expression normal tissue (selection)	(Patho-) physiology (selection)	Reference	Tumor entity	Function in radiobiology	References
IK_{Ca} ($K_{Ca}3.1$, SK4, KCNN4)	Blood cells including erythrocytes, smooth muscle cells, epithelia	Eryptosis, inflammation, atherosclerosis, blood pressure control, transepithelial transport	Foller et al (2010), Huber et al. (1999), Si et al. (2006), Zhou et al. (2015)	Human glioblastoma, human lung adenocarcinoma cells, murine breast cancer cells, human T cell leukemia cells	Radiogenic Ca^{2+} signaling, radiogenic cell cycle arrest, DNA repair, radioresistance	Stegen et al. (2015), Gibhardt et al. (2015), Roth et al. (2015), Mohr et al. (2019), Klumpp et al. (2016a), Voos et al. (2018)
BK_{Ca} ($K_{Ca}1.1$, KCNMA1)	Neurons, cardiomyocytes, epithelia	CNS development, hearing, motor control, colonic secretion	Sausbier et al. (2004), Soltysinska et al. (2014), Sorensen et al. (2008)	Glioblastoma	Radiogenic hypermigration	Steinle et al. (2011), Edalat et al. (2016), Birch et al. (2018)
$K_v3.4$ (KCNC4)	Neurons, skeletal muscle, pancreatic δ cells, arterial smooth muscle cells	Neuronal function, neurodegeneration, hormone release, hypoxic vasoconstriction	Kaczmarek and Zhang (2017), Vullhorst et al. (1998), Gopel et al. (2000), Guo et al. (2008)	Chronic myeloid leukemia cells	Radiogenic Ca^{2+} signaling, radiogenic cell cycle arrest, radioresistance	Palme et al. (2013)
hERG1 (Kv11.1, KCNH2)	Heart, retina, neurons	Repolarization of action potential, long QT syndrome	Sanguinetti (2010)	Chronic myeloid leukemia cells	Radiogenic Ca^{2+} signaling, radiogenic cell cycle arrest, radioresistance	Palme et al. (2020)
${}_{mt}K_{ATP}$ (Kir6.1/2-SUR1/2?)	Ubiquitous expressed K^+ channel of the inner mitochondrial membrane	Regulation of mitochondrial respiratory chain, cell protection	Cheng et al. (2010)	Glioblastoma	Radiogenic redox signaling?, radioresistance	Huang et al. (2015)

TRPM2	CNS; immune cells; pancreatic β cells	Core body temperature sensation, oxidative sensation, insulin secretion, inflammation	Huang et al. (2020)	Human T cell leukemia cells, lung adenocarcinoma cells	Radiogenic Ca^{2+} and redox signaling, DNA damage response, DNA repair, radioresistance	Klumpp et al. (2016a), Masumoto et al. (2013)
TRPM8	Sensory neurons, prostate	Cold sensation, pain?	Huang et al. (2020), Noyer et al. (2018)	Glioblastoma	Radiogenic Ca^{2+} signaling, radiogenic cell cycle arrest, DNA repair, radioresistance	Klumpp et al. (2017)
TRPV1	Sensory neurons, vascular endothelial cells, smooth muscle cells of various organs	Heat and inflammatory, pain sensation, blood pressure control, atherosclerosis	Zhang et al. (2020)	Lung adenocarcinoma cells	DNA damage response, DNA repair, radioresistance	Masumoto et al. (2013), Nishino et al. (2016)
TRPV5/6	Kidney, placenta, epididymis, pancreas, prostate, salivary gland, sweat gland	Intestinal and renal Ca^{2+} (re)absorption, maintenance of blood Ca^{2+} levels	Peng et al. (2018)	Chronic myeloid leukemia cells	Radiogenic Ca^{2+} signaling, radiogenic cell cycle arrest, radioresistance	Heise et al. (2010)
α2δ1 subunit Ca^{2+} channel (CACNA2D1)	Ubiquitously expressed, skeletal muscle	Epilepsy and neuropathic pain	Davies et al. (2007)	On-small lung cancer cells	DNA damage response, DNA repair, radioresistance	Sui et al. (2018)
ClC-3 (CLCN3)	Ubiquitously expressed mainly endosomal C^{-}/H^{-} exchanger	Acidification of endosomes and other compartments	Jentsch and Pusch (2018)	Plasma membrane of glioblastoma cells	Radiogenic hypermigration	Stegen et al. (2016)

(continued)

Table 1 (continued)

Channel type	Expression normal tissue (selection)	(Patho-) physiology (selection)	Reference	Tumor entity	Function in radiobiology	References
CLIC1	Ubiquitously intracellularly expressed (membrane integral and soluble)	Support of NADH oxidase in activated microglia (upon translocation to plasma membrane)	Averaimo et al. (2010), Littler et al. (2010)	Laryngeal cancer cells	Upregulation of radiogenic ROS formation	Kim et al. (2010)
VDAC1	Outer mitochondrial membrane	Mitochondrial gatekeeper	Arif et al. (2019)	Chronic myeloid leukemia cells	Downregulation of radiogenic mitochondrial ROS formation	Skonieczna et al. (2017)
MPT	Inner mitochondrial membrane, e.g., heart	Triggering of cell death	Bauer and Murphy (2020)		Spatial-temporally propagation of radiogenic mitochondrial ROS production	Leach et al. (2001)
Uncoupling proteins SLC25A8 (UCP2) SLC25A9 (UCP3)	Inner mitochondria membrane, spleen, kidney, thymus, pancreas, CNS, macrophages Skeletal muscle, BAT, heart	Fatty acid metabolism, mitochondrial ROS formation, cardioprotection	Cadenas (2018)	Renal clear cell carcinoma	Hypoxia and radioresistance	Braun et al. (2015)
SGLT1 (SLC5A1) SGLT2 (SLC5A2)	Small intestine, proximal tubule	Glucose (re)absorption by Na^{+}-coupled cotransport	Schumann et al. (2020)	Human lung adenocarcinoma, squamous head and neck carcinoma cells	Radiogenic glucose uptake, DNA decondensation, DNA repair, radioresistance	Dittmann et al. (2013), Huber et al. (2012), Peitzsch et al. (2014)

BAT brown adipose tissue, *ICC* interstitial cells of Cajal, *MPT* mitochondrial membrane permeability transition pore, *ROS* reactive oxygen species

4 Repopulation, Migration, and Invasiveness

Accelerated repopulation describes the phenomenon that ionizing radiation may boost the proliferation of the surviving clonogenic tumor cells. Normofractionated protocols frequently apply a daily dose of 2 Gy which is assumed to eradicate a certain low percentage of the tumor cells. Radiation-induced accelerated repopulation by the survivors may restore between two radiation fractions a part of the beforehand eradicated tumor mass and, thus, may lower local tumor control (Yom 2015) (Fig. 2).

4.1 Accelerated Repopulation

Cell proliferation requires cell cycle-associated changes in cell volume (Dupre and Hempling 1978) and membrane potential (V_m). Experimental depolarization of V_m may force G_0-arrested cells to enter cell cycling and, vice versa, hyperpolarization may arrest proliferating cells (Cone and Cone 1976; Cone and Tongier 1971) indicating a superior regulatory function of V_m. Similarly, experimental cell shrinkage may activate via serum and glucocorticoid-inducible kinase-1 Ca^{2+} oscillations that stimulate cell survival and cell proliferation (Lang et al. 2018). Changes in cell volume require fluxes of ions and osmotically obliged H_2O across the plasma membrane, changes in V_m, and ion channel activities.

The abundance of chloride channel accessory 1 (CLCA1) protein in the tumor specimens has been identified as an independent prognostic marker for local

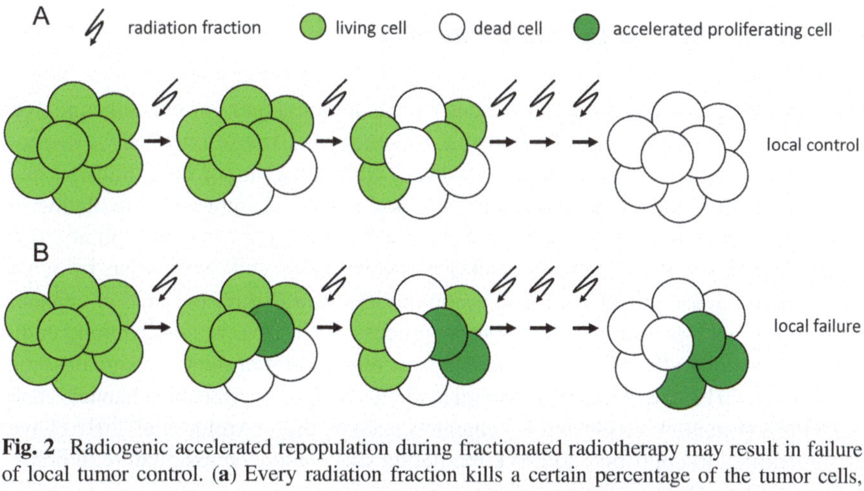

Fig. 2 Radiogenic accelerated repopulation during fractionated radiotherapy may result in failure of local tumor control. (**a**) Every radiation fraction kills a certain percentage of the tumor cells, resulting in the eradication of the tumor by the end of radiotherapy. (**b**) Radiation-induced accelerated proliferation may refill part of the eradicated tumor cell between two radiation fractions and may result in local failure

recurrence-free, metastasis-free, and disease-specific survival of patients with rectal cancer treated with concurrent radiochemotherapy (Chen et al. 2018). This channel regulator is being followed with great interest since it regulates the surface expression of the calcium-sensitive Cl^- channel TMEM16A (ANO1) (Sala-Rabanal et al. 2015). Cl^- fluxes across the plasma membrane are essential for osmo- and volume regulations which are especially critical processes in fast proliferating, migrating, and tissue-invading tumor cells. Consistently, several tumor entities show deregulation of ANO1 which leads to the assumption of a multifaceted role of this Cl^- channel in cancer (for review, see (Crottes and Jan 2019; Kunzelmann et al. 2019)). In breast (Britschgi et al. 2013) and prostate cancer cell lines (Liu et al. 2012), ANO1 reportedly promotes proliferation which may involve EGFR- and Ca^{2+}-dependent signaling. Although there are conflicting reports showing an inverse correlation between CLCA1 abundance of the tumor cells and radioresistance in pancreatic cancer cell lines (Ogawa et al. 2006) or prognosis of patients with colorectal cancer (Yang et al. 2015), a functional significance of the CLCA1/ANO1 axis in tumor cell proliferation has to be assumed.

Beyond cell proliferation, ionizing radiation may also accelerate migration and tissue invasion of tumor cells. Sublethal doses of ionizing radiation, as applied in fractionated radiotherapy, reportedly may induce a transition from a "grow" towards a "go" phenotype of the tumor cells and, in the worst case, may promote evasion of tumor cells from the target volume during therapy (for review, see (Vilalta et al. 2016)). Ion transports are indispensable for tumor cell migration, tissue infiltration, and metastasis by mediating signaling and cell volume changes (for review, see (Sontheimer 2008)). The next paragraphs introduce the involved ion channels and their modulation by ionizing radiation.

4.2 *Tumor Spreading and Metastasis*

To adopt a migratory and invasive phenotype, carcinoma and glioma cells undergo epithelial–mesenchymal (EMT) and glial-mesenchymal transition (GMT), respectively. For lymph node or distant metastasis, emigrating tumor cells transmigrate vessel walls and intravasate into lymph or blood vessels. Tumor cells circulating in the bloodstream have to survive large shear-stress forces in the narrow capillary bed. Upon adherence to the endothelium surface, metastasizing cells extravasate by again trans-migrating the vessel wall and accommodate in a distant organ. Ion channel activities are indispensable for all of these processes (for review, see (Klumpp et al. 2016b*)). Specifically, EMT is paralleled by a remodeling of the Ca^{2+} signalosome (Davis et al. 2012; Hu et al. 2011; Mahdi et al. 2015). Ca^{2+}-permeable channels such as TRPM8 (Liu et al. 2014) and K^+ channels such as IK_{Ca} (Arthur et al. 2015) have been identified to contribute to EMT-regulating Ca^{2+}- and electrosignaling in breast cancer and bronchial epithelial cells, respectively.

Likewise, IK_{Ca} is reportedly highly upregulated in the mesenchymal subpopulation of glioblastoma stem cells (GSCs) that are held responsible for brain infiltration

(Klumpp et al. 2018*). Functionally, IK_{Ca} is required for GSC migration/invasion (D'alessandro et al. 2013; Ruggieri et al. 2012). Beyond IK_{Ca}, high conductance Ca^{2+}-dependent BK_{Ca} K^+ channels (Weaver et al. 2006), Ca^{2+}-regulated ClC-3 (CLCN3) Cl^- channels (Cuddapah and Sontheimer 2010), and Ca^{2+}-permeable TRPM8 nonselective cation channels (Wondergem and Bartley 2009; Wondergem et al. 2008; Klumpp et al. 2017*) have been identified as key proteins involved in glioblastoma migration. This partial similarity in the ion channel toolkit of glioblastoma and metastasizing carcinoma cells might hint to analog processes in GMT and EMT.

Upon EMT, tissue infiltration has been demonstrated to crucially depend on voltage-gated Na^+ channels (Na_v) such as $Na_v1.5$ (SCN5A), $Na_v1.6$ (SCN8A), or $Na_v1.7$ (SCN9A) and a neonatal splice variant of $Na_v1.5$ upregulated in the lamellipodium of migrating glioma (Holley et al. 2014), melanoma, lung, breast cancer, and other carcinoma cells (for review, see (Djamgoz et al. 2019)). Patient data might suggest that aberrant expression of Na_v subunits in breast and colorectal cancer specimens associates with local control, lymph node metastasis, or patient survival (Fraser et al. 2005; Lin et al. 2019; Yang et al. 2012). Actually, genetic knockdown or pharmacological blockade of Na_v channels by antianginal or antiepileptic drugs or tetrodotoxin attenuates metastasis in preclinical rodent models of breast and prostate cancer (Driffort et al. 2014; Nelson et al. 2015a, b; Yildirim et al. 2012). Mechanistically, Na_v channels have been suggested to regulate Na^+/H^+ antiporters (NHEs, SLC9As) in the lamellipodium membrane by a yet ill-defined process. Proton extrusion by the antiporter acidifies locally extracellular peri-membrane sites which promotes digestion of the extracellular matrix by cysteine cathepsins and, thereby, tissue infiltration (Brisson et al. 2011, 2013; Gillet et al. 2009).

4.3 Ion Transports in Radiation-Induced Hypermigration and Metastasis

Whether or not fractionated radiotherapy may further boost tissue infiltration, tumor spreading, or metastasis of cancer cells is highly controversially discussed. Many preclinical studies could not demonstrate radiation-induced migration/invasion in 2D or 3D in vitro culture systems (e.g., (Eke et al. 2012)). In addition, there are only sparse clinical reports that hint to radiation-stimulated metastasis in head and neck squamous carcinoma and non-small cell lung cancer patients (Jesse and Lindberg 1975; Martin et al. 2014; Strong et al. 1978). The next paragraphs introduce in vitro and mouse studies that observed radiation-induced migration/invasion/metastasis of cancer cells and describe possible mechanisms.

Na_v channels in tumor cells are reportedly under the control of the epidermal growth factor receptor (EGFR) (Campbell et al. 2013; Gonzalez-Gonzalez et al. 2019; Mohammed et al. 2016; Uysal-Onganer and Djamgoz 2007; Zhang et al.

2019). Since ionizing radiation triggers plasmalemmal and nuclear EGFR signaling (Dittmann et al. 2008, 2009; Rodemann et al. 2007) it is tempting to speculate that fractionated radiotherapy stimulates pro-migratory Na_v channels. Evidence for an ionizing radiation-induced hypermigration/hyperinvasion of tumor cells came from several preclinical studies (for review, see (Vilalta et al. 2016)): These include experiments in glioblastoma that demonstrated ionizing radiation-induced migration in vitro (Steinle et al. 2011*; Klumpp et al. 2017*; D'alessandro et al. 2019) and ionizing radiation-stimulated brain infiltration in orthotopic glioblastoma mouse models (Edalat et al. 2016*; Birch et al. 2018). Along those lines, several studies hint to a radiation-induced distant metastasis of ectopically transplanted sarcoma (Krebs 1929; Yamamoto 1936) and breast cancer *xeno*grafts in mice (Bouchard et al. 2017; Kaplan and Murphy 1949; Sheldon and Fowler 1976).

In glioblastoma, programming and mechanics of radiogenic hypermigration/infiltration probably involve TRPM8- (Klumpp et al. 2017*), BK_{Ca}- (Edalat et al. 2016*; Steinle et al. 2011*), and IK_{Ca}-mediated (Stegen et al. 2016*; D'alessandro et al. 2019) Ca^{2+} signals and/or cell volume changes. Importantly, pharmacological blockade or molecular knockdown of these channels reverses the pro-migratory/invasive radiation effect in vitro and in vivo (Steinle et al. 2011*; Edalat et al. 2016*; Klumpp et al. 2017*). Mechanistically, CaMKII downstream of radiogenic BK channel activity probably directly activates ClC-3 Cl^- channel in the plasma membrane (Cuddapah et al. 2012, 2013; Cuddapah and Sontheimer 2010; Sontheimer 2008; Stegen et al. 2016*). ClC-3-mediated Cl^- efflux together with BK-mediated K^+ efflux and osmotically obliged efflux of H_2O results in local cell volume decrease which together with local volume acquisition at the cell lamellipodium is one process that motorizes migration of glioblastoma cells (for review, see (Huber 2013*); Fig. 3).

Taken together, preclinical data show that sublethal ionizing radiation may stimulate signaling that triggers migratory/invasive programs in tumor cells. This might apply only to certain tumor entities or even individual tumors. In addition, the clinical significance of these observations is still completely unknown. Nevertheless, fractionated radiotherapy should remain under suspicion of boosting spreading and metastasizing of tumors in the clinical situation. If so, the contributing ion channels are excellent targets for therapeutical intervention. Beyond channel activation, radiogenic HIF-1α and nuclear EGFR signaling regulate metabolic reprogramming of tumor cells. The involved ion transports will be discussed in the next chapter.

5 Metabolism and Redox State

Metabolic reprogramming of cancer cell is a dynamic process that evolves during malignant progression and that involves many metabolic pathways which are all interconnected and influence each other throughout the different cellular compartments (Faubert et al. 2020). The anaerobic generation of ATP by glycolysis and lactic acid fermentation even at O_2 partial pressures that are sufficient to run the

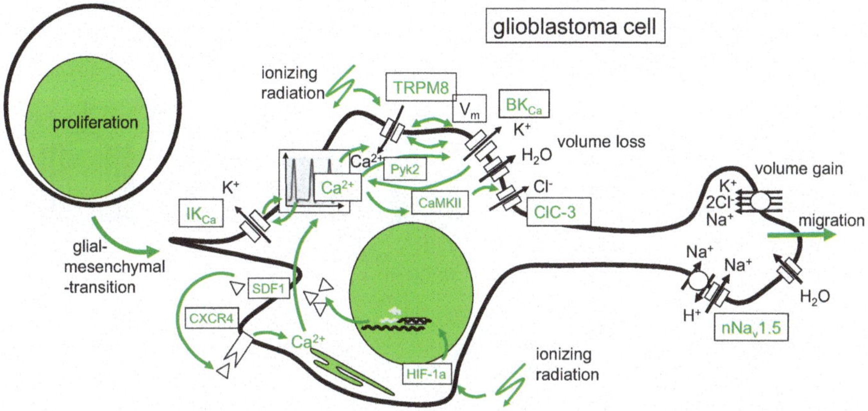

Fig. 3 Radiogenic ion transports foster cell migration of glioblastoma cells (hypothetical model). Volume increase of the lamellipodium (invadipodium) and volume loss by the cell rear motorizes directed migration. Local volume increase and decrease are accomplished by cotransport- and channel-mediated uptake and efflux of ions, respectively, followed by osmotically obliged water fluxes. Voltage-gated Na^+ channels (Na_v) in complex with Na^+/H^+ antiporters at the lamellipodium acidify locally extracellular peri-membrane sites, which promotes digestion of the extracellular matrix and directed infiltration of the brain parenchyma. For cell migration, glioblastoma cells have to undergo glial-mesenchymal transition which is accompanied by upregulation of IK_{Ca} K^+ channels. Radiogenic activity of melastatin member 8 of the TRP nonselective cation channels (TRPM8) and IK_{Ca} and BK_{Ca} Ca^{2+}-activated K^+ channels generates Ca^{2+} signals required for Ca^{2+}/calmodulin kinase-II isoforms (CaMKII) and subsequent activation of ClC-3 anion channels in the plasma membrane. Radiogenic stabilization of hypoxia-inducible factor-1α (HIF-1α) upregulates stromal cell-derived factor-1 (SDF1, CXCL12) and its C-X-C motif chemokine receptor-4 (CXCR4) which contribute via auto-/paracrine Ca^{2+} signaling to radiogenic IK_{Ca} and BK_{Ca} channel activation and which involve Ca^{2+}-dependent tyrosine kinase Pyk2

mitochondrial electron transport chain and to synthesize ATP by oxidative phosphorylation has been identified as a hallmark of cancer cells. This reprogrammed metabolism is (misleadingly) referred to as aerobic glycolysis and was proposed by Otto Warburg (Warburg 1956). Of note, ionizing radiation has been shown to downregulate oxidative phosphorylation and upregulate lactic acid formation, and nuclear EGFR (Dittmann et al. 2015*) and HIF-1α (Li et al. 2007) have been identified as regulators of this radiation-induced metabolic reprogramming. In aerobic glycolysis, bioenergetics pathways bypass the mitochondria at the expense of drastically reduced energy yield. Like nuclear power stations in our technological world, mitochondria, on the one hand, are highly efficient in ATP synthesis. On the other hand, mitochondria are at continuous risk of a "core melt accident," i.e., of forming superoxide anion ($\bullet O_2^-$) radicals. This may lead to structural disintegration of the mitochondria and eventually release of factors such as cytochrome C that trigger apoptotic cell death.

Therefore, one might speculate that a possible advantage of aerobic glycolysis might be – in addition to preserve O_2 and anabolic substrates for alternative use – a reduced risk of hazardous mitochondrial formation of $\bullet O_2^-$ and derived reactive

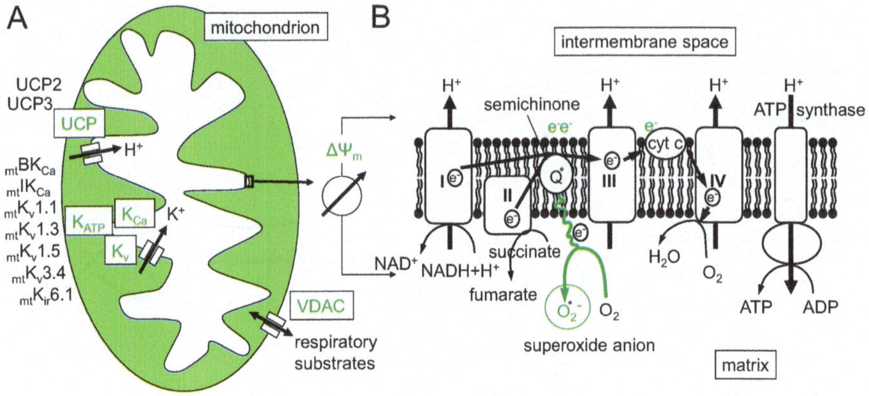

Fig. 4 Mitochondrial ion transports involved in metabolic reprogramming, radiogenic formation of superoxide anion ($\bullet O_2^-$), and adaptation to a radioresistance-conferring hypoxic tumor microenvironment (hypothetical model). (**a**) Voltage-gated anion channel (VDAC) in the outer mitochondrial membrane functions as a gatekeeper of the mitochondrial metabolism by fueling the mitochondria with respiratory substrates. Uncoupling proteins (UCPs), voltage-gated ($_{mt}K_v$) K^+, Ca^{2+}-activated ($_{mt}K_{Ca}$) K^+, and inwardly rectifying (K_{ir}) ATP-sensitive ($_{mt}K_{ATP}$) K^+ channels in inner mitochondrial membrane may modulate the inner mitochondrial membrane potential ($\Delta\Psi_m$). (**b**) $\bullet O_2^-$ formation by the mitochondrial electron transport chain. Complexes I–IV and the ATP synthase in the inner mitochondrial membrane are shown (Q: ubiquinone, cyt C: cytochrome C, e^-: electron)

oxygen species (ROS) that leads to oxidative cell damage (for review, see (Denko 2008)). In accordance with this assumption, decreasing mitochondrial anti-oxidative defense and increasing $\bullet O_2^-$ burden by heterozygous knockout of mitochondrial manganese superoxide dismutase (MnSOD, SOD2) reportedly induces compensatory uncoupling of the mitochondrial electron transport chain by uncoupling proteins (UCPs) and upregulation of aerobic glycolysis (Xu et al. 2015). Vice versa, cells pursuing oxidative glycolysis have been demonstrated to lower the activity of voltage-gated anion channel (VDAC1) in the outer mitochondrial membrane. VDAC "fuels" mitochondria with respiratory substrates, ADP, and P_i via this pathway and VDAC inactivation attenuates mitochondrial metabolism and ROS formation. Importantly, experimental disinhibition of VDAC increases mitochondrial metabolism resulting in decreased oxidative glycolysis and increased mitochondrial ROS formation (Maldonado 2017).

Beyond UCPs (Baffy 2017; Xu et al. 2015) and VDACs (Brahimi-Horn et al. 2012; Maldonado 2017) in inner and outer mitochondrial membrane, respectively, ATP-sensitive mitochondrial (mt) K_{ATP} ($_{mt}K_{ir}6.1$ (Kim et al. 2011)) (Fu et al. 2003) and Ca^{2+}-activated $_{mt}BK_{Ca}$ (Gu et al. 2014; Kulawiak et al. 2008) and $_{mt}IK_{Ca}$ (Kim et al. 2011; Leanza et al. 2014), as well as voltage-gated $_{mt}K_V1.1$ (KCNA1), $_{mt}K_V1.3$ (KCNA3) (Leanza et al. 2014), $_{mt}K_V1.5$ (KCNA5) (Archer et al. 2008), and $_{mt}K_V3.4$ (Song et al. 2017) K^+ channels have been identified in the inner mitochondrial membrane to regulate mitochondrial metabolism and/or ROS formation (Fig. 4). Since mitochondrial metabolism has been demonstrated to actively contribute to radiation damage (Richardson and Harper 2016) one can assume that mitochondrial transports may modulate the radiosensitivity of a given cell. The next paragraphs introduce into our current knowledge on the role of mitochondria in radiation-induced cell damage and the function of mitochondrial transports herein.

5.1 Mitochondria Contribute to Radiation-Induced DNA Damage

A central dogma of radiotherapy attributes its tumor-eradicating effect to the energy transfer of the radiation beam to the nucleus and subsequent damage of the nuclear DNA (Reisz et al. 2014). Evidence for such ionizing radiation-nucleus interaction as the principal mechanism of radiotherapy came from very early microbeam experiments that indicated higher radiation efficacy upon targeting the nucleus as compared to irradiation of the cytoplasm (Munro 1970). These ionizing radiation-induced direct and via radiolysis of H_20-mediated indirect damages of nuclear DNA occur instantaneously. Meanwhile, evidence accumulates that nuclear DNA damage is also caused by irradiation of the cytoplasm or neighboring cells (for review, see (Holley et al. 2014; Zhou et al. 2009)). Notably, this bystander effect may require ROS formation (Tartier et al. 2007) suggesting that ROS are spreading the absorbed radiation energy from the beam-targeted cell to the nuclear DNA of the unirradiated cells in the immediate proximity. Importantly, the electron transport chain of the mitochondria has been identified as a key player in this radiogenic bystander effect as deduced from the differences in DNA damage between cells with DNA-proficient (ρ^+) and -deficient (ρ^0) mitochondria (Cloos et al. 2009; Leach et al. 2001; Tartier et al. 2007; Yamazaki et al. 2008; Zhang et al. 2014).

Mitochondria-mediated DNA damage occurs late (minutes to hours after radiation) and might involve ionizing radiation-induced mitochondrial Ca^{2+} overflow (Leach et al. 2001) as a result of upregulation of Ca^{2+} fluxes across the plasma membrane through, e.g., TRP channels as demonstrated in several tumor entities (Heise et al. 2010*; Palme et al. 2013; Klumpp et al. 2016a*; Stegen et al. 2016*). With increasing cytosolic free Ca^{2+} concentration and decreasing ATP/ADP ratios (Dittmann et al. 2013*; Huber et al. 2012*), however, mitochondria increasingly form ROS (Leach et al. 2001) (for review, see (Huber et al. 2013*)). Mechanistically, low ATP/ADP ratios and high Ca^{2+} concentrations have been proposed to un-restrain the mitochondrial electron transport chain leading to hyperpolarization of the inner mitochondrial membrane ($\Delta\Psi_m$) potential and directly associated formation of $\bullet O_2^-$(for review, see (Huber et al. 2013*)) by slippage of single electrons from the mitochondrial electron transport chain to O_2 (Fig. 4). Along those lines, DNA damages that depend on mitochondrial ROS formation have been reported to increase with O_2 partial pressure (Richardson and Harper 2016). Ca^{2+}-mediated and O_2-dependent mitochondrial $\bullet O_2^-$ formation, in turn, may provoke mitochondrial disintegration upon opening of the mitochondrial membrane permeability transition

(MPT) pore and in the inner mitochondrial membrane dissipation of $\Delta\Psi_m$ (Vercesi et al. 1997).

As a matter of fact, ionizing radiation has been directly demonstrated in Jurkat T cell leukemia cells to activate Ca^{2+}-permeable TRPM2 channels in the plasma membrane resulting in cytosolic Ca^{2+} overflow and mitochondrial $\bullet O_2^-$ formation. The latter, which was apparent 6 h after irradiation, was associated with cell shrinkage and dissipation of $\Delta\Psi_m$ suggestive of triggering of the intrinsic apoptosis pathway. Markedly, inhibition of TRPM2 abolished the ionizing radiation-induced $\bullet O_2^-$ formation, cell shrinkage, and $\Delta\Psi_m$ dissipation (Klumpp et al. 2016a*). On the other hand, radiogenic TRPM2 activity also contributed to pro-survival signaling in these cells by inducing G_2/M cell cycle arrest. This pro-survival TRPM2 function prevailed in Bcl-2-transfected, apoptosis-resistant cells as compared to Mock-transfected Jurkat control cells. Bcl-2 overexpression was associated with higher radiogenic Ca^{2+} influx at lower mitochondrial $\bullet O_2^-$ formation and $\Delta\Psi_m$ dissipation (Klumpp et al. 2016a*). This might suggest that upon embanking the risk of mitochondria-triggered apoptosis by upregulation of anti-apoptotic proteins such as Bcl-2, tumor cells can afford higher cytosolic free Ca^{2+} concentrations and profit from the pro-survival function of the latter.

Of note, radiogenic mitochondrial ROS formation has been demonstrated to spatial-temporally propagate through the mitochondrial network: Radiogenic MPT of few injured mitochondria and subsequent local Ca^{2+} release from these affected mitochondria has been suggested to result in Ca^{2+}-overflow, ROS formation, and Ca^{2+} re-release of adjacent mitochondria (Leach et al. 2001). Consistently, pharmaceutical blockage of the MPT blunts radiogenic mitochondrial ROS formation (Leach et al. 2001) and induces radioresistance in some but not all cell lines (Anoopkumar-Dukie et al. 2009). Along those lines, chloride intracellular channel-1 (CLIC1) has been identified in laryngeal cancer cells to augment late (12 h post-irradiation) ionizing radiation-induced ROS formation (of undefined origin). Downregulation of CLIC1 was associated with radioresistance (Kim et al. 2010).

5.2 *Mitochondrial Transports and Radioresistance*

Differential analysis of ionizing radiation-induced changes in mRNA abundances between apoptosis-sensitive murine B cell lymphoma cells and Bcl2-overexpressing apoptosis-resistant sister clones disclosed the ionizing radiation (5 Gy)-stimulated upregulation of gene expression associated with anti-oxidative defense and mitochondrial transports. In particular, ionizing radiation induced within 1 h the upregulation of UCP2 in the inner and of VDAC1 (voltage-dependent anion channel) in the outer mitochondrial membrane notably solely in the apoptosis-sensitive cells (Voehringer et al. 2000). The authors of that study proposed a programmed mitochondrial disintegration mediated by radiogenic upregulation of VDAC1, a regulator of the MPT, and radiogenic UCP2-mediated uncoupling of $\Delta\Psi_m$. This mitochondrial disintegration, in turn, triggers mitochondrial cytochrome C release and the intrinsic apoptosis pathway.

In contrast to this assumption, inhibition of VDAC in K562 chronic myeloid leukemia cells reportedly augments ionizing radiation (4 Gy)-induced ROS formation, late DNA damage, and apoptotic cell death (Skonieczna et al. 2017) suggesting rather a pro-survival than a mitochondria-disintegrating function of VDAC. Along those lines, VDAC1 knockdown in *xeno*grafted human glioblastoma cells does neither alter apoptosis rate nor expression of apoptosis-regulating genes. Instead, VDAC1 knockdown induces profound metabolic and phenotypical reprogramming confirming VDAC1 as mitochondrial gatekeeper (Arif et al. 2019).

Likewise, hypoxia adaptation-associated upregulation of UCP3 in the inner mitochondrial membrane or renal proximal tubular cells (see below) has been shown to confer radioresistance (Braun et al. 2015*). Combined, this might suggest that the previously observed radiogenic upregulation of VDAC1 and UCP2 in the apoptosis-sensitive murine B cell lymphoma cells (Voehringer et al. 2000) rather protects than disintegrates mitochondria. One might further speculate that the Bcl2-overexpressing apoptosis-resistant lymphoma clones which have been documented in that previous study (Voehringer et al. 2000) to be endowed with a highly upregulated anti-oxidative defense were not at risk of oxidative injury by mitochondrial ROS formation and, thus, did not upregulate the proposed mitochondrial protection mechanisms seen in the apoptosis-sensitive clones. In addition to cell injury, ROS-induced ${}_{mt}K_{ATP}$ channel activity reportedly contributes to ERK (extracellular signal-regulated kinase)-induced radioresistance of glioma cells in vitro and in an ectopic glioblastoma mouse model in vivo (Huang et al. 2015) suggesting that mitochondrial ROS formation also triggers redox signaling.

In summary, a significant fraction of ionizing radiation-evoked cell damage depends on mitochondrial ROS formation. This is also illustrated by the fact that high mitochondrial anti-oxidative defense reportedly confers radioresistance (Qu et al. 2010). Moreover, oxidative glycolysis minimizes mitochondrial ROS formation in tumor cells and promotes radioresistance (for review, see (Cruz-Gregorio et al. 2019)). In addition to radiation damage, mitochondrial ROS formation has been identified to contribute to cell injury during reoxygenation after hypoxia. Chronic and intermittent hypoxia frequently occur in solid tumors and lower the efficiency of radiotherapy. Furthermore, hypoxia induces tumor cell transition towards a more malignant and therapy-resistant phenotype. Notably, mitochondrial channels/transporters have been reported to lower reoxygenation damages and to adapt to hypoxia and, hence, contribute to tumor radioresistance. These mitochondrial channels/transporters and their role in hypoxia adaptation will be described in the next paragraphs.

5.3 Tumor Hypoxia

Insufficient numbers and aberrant architecture of tumor vessels lacking normal walls in concert with high intra-tumoral pressures that may compress vessel lumina often restrict the perfusion of solid tumors (Nagy et al. 2009). In addition, O_2 diffusion distances may become too large in the expanding tumor mass. Chronic or intermittent hypoxia occurs whenever O_2 consumption of the tumor cells exceeds O_2 delivery by the tumor vasculature. Besides limiting the access of chemotherapeutics to hypoxia tumor areas and impairing the anti-tumor immune response, tumor hypoxia lowers the efficacy of radiotherapy (for review, see (Eckert et al. 2019*)). Accordingly, the prognosis of patients with hypoxic tumors is worse than that of patients with normoxic tumors, and, specifically, hypoxia is a negative predictive factor for the response to radiotherapy.

5.4 Radioresistance of Hypoxic Tumor Cells

Mechanistically, several processes have been proposed to contribute to hypoxia-associated failure of radiotherapy. A reducing redox state prevents the chemical O_2-dependent fixation of ionizing radiation-evoked DNA DSBs (Thoday and Read 1947). Moreover, DNA repair might also depend on O_2 tension (Ewing 1998). As discussed above in more detail, the late mitochondrial ROS formation-mediated damages in irradiated cells also rely on O_2 partial pressure. Finally, hypoxia reportedly is a key driver of malignant progression of tumor cells by reprogramming metabolism, inducing EMT, or increasing genetic instability, heterogeneity, and plasticity of the tumor (for review, see (Redfern et al. 2019; Terry et al. 2018; Trautmann et al. 2014)).

The O_2-responsive HIFs and the cellular energy-sensing 5′ adenosine monophosphate-activated protein kinase (AMPK) trigger metabolic reprogramming under hypoxia. This reprogramming comprises upregulation of nutrient import and oxidative glycolysis with lactic acid fermentation and downregulation of oxidative phosphorylation resulting in attenuated mitochondrial ROS formation (see above). In addition, hypoxia-induced reprogramming downregulates anabolic metabolism, adjusts the glutamine metabolism in order to maintain fueling of the citrate pool, alters the lipid metabolism, upregulates the anti-oxidative defense, and/or induces phenotypical cell transition (for review, see (Tyszka-Czochara et al. 2018; Xie and Simon 2017)).

5.5 Channel/Transporters Conferring Hypoxia-Associated Radioresistance

A hypoxic tumor microenvironment has been reported to induce/select cancer stem cells (for review, see (Terry et al. 2018)). Cancer stem cells have been suggested to upregulate anti-oxidative defense, DNA repair machinery, and anti-apoptotic pathways, rendering them less sensitive to radiotherapy (for review, see (Peitzsch et al. 2014)). Notably, the mesenchymal subpopulation of glioblastoma stem cells reportedly upregulates IK_{Ca} K^+ channels (Klumpp et al. 2018*) and the radioresistance of these cells critically depends on IK_{Ca} (Klumpp et al. 2018*; Stegen et al. 2015*).

Moreover, hypoxia-induced metabolic reprogramming involves Ca^{2+}- and electrosignaling and altered membrane transports. Among those are upregulated acid extrusion across the plasma membrane (Miranda-Goncalves et al. 2016), increased K^+ conductance of the inner mitochondrial (Gu et al. 2007, 2014), or truncation of VDAC1 in the outer mitochondrial membrane that confers apoptosis resistance by lowering mitochondrial cytochrome C release (for review, see (Mazure 2016)). Importantly, mitochondrial channels such as VDAC1 that attenuate hypoxia/reoxygenation damage confer at the same time radioresistance. Accordingly, pharmacological inhibition of VDAC1 results in radiosensitization, as reported in lymphoblastoid cells (Skonieczna et al. 2017). Similarly, highly O_2-dependent proximal tubular cells, when subjected to repetitive cycles of hypoxia/reoxygenation, have been shown to become insensitive to fluctuations of the O_2 partial pressure by upregulating anti-oxidative defense and by lowering reoxygenation-associated $\Delta\Psi_m$ hyperpolarization and mitochondrial ROS formation (Braun et al. 2015*). The latter two phenomena result from uncoupling of the electron transport chain achieved by upregulation of UCP2 and -3 in the inner mitochondrial membrane. As already discussed above, mitochondrial uncoupling also lowers the late mitochondria-mediated radiation damage and confers radioresistance in these cells (Braun et al. 2015*) (Fig. 4).

Combined, these reports disclose striking parallels between the metabolic stress responses to ionizing radiation and hypoxia/reoxygenation. Cellular insults of both ionizing radiation and hypoxia/reoxygenation are diminished in tumor cells with upregulated aerobic glycolysis and restrained mitochondrial electron transport chain. Ion channels in the inner or outer mitochondrial membrane seem to rule mitochondrial metabolism and, thus, mitochondrial response to ionizing radiation and hypoxia/reoxygenation stress. This suggests them as potential therapeutic targets to increase the mitochondria-mediated DNA damage and to radiosensitize tumor cells on the one hand and to prevent adaptation to hypoxia on the other. As a consequence of lowering energy-efficient mitochondrial bioenergetics, tumor cells with energy-inefficient aerobic glycolysis have to increase lactic acid extrusion and glucose and glutamine uptake pathways (for review, see (Meng et al. 2019)). In particular, this is true for irradiated cells as discussed in the next paragraph.

5.6 Irradiation-Induced Glucose Fueling

Besides overexpression of glucose uniporters (GLUTs, SLC2s (Meng et al. 2019)), several tumor entities reportedly harness the inwardly directed electrochemical driving force of Na^+ to take up glucose across the plasma membrane via Na^+/glucose (SGLT2, SLC5A2) and/or $2Na^+$/glucose (SGLT1, SLC5A1) cotransport (for review, see (Wright et al. 2017)). These cotransports guarantee glucose supply even at low extracellular glucose concentration that prevails in malperfused tumors. Importantly, the genotoxic stress response in irradiated A549 lung adenocarcinoma as well as in irradiated FaDu and SAS squamous head and neck carcinoma cells has been reported to demand even higher glucose fueling via SGLT (Huber et al. 2012*). Pharmaceutical blockade of SGLT radiosensitizes these tumor cells indicating the pivotal function of SGLT for the clonogenic survival. Mechanistically upregulated SGLT activity is needed to counterbalance increased ATP consumption and to provide carbohydrates for histone acetylation, with the latter being crucial for chromatin remodeling and DNA repair (Huber et al. 2012*; Dittmann et al. 2013*).

Electrogenic SGLT activity depolarizes the membrane potential (Huber et al. 2012*). This is in part counteracted by ionizing radiation-induced activation of IK_{Ca} channels in the plasma membrane, as shown in A549 cells that stabilizes the membrane potential and maintains the electrical driving force for Na^+-coupled cotransports (Huber et al. 2012*; Roth et al. 2015*). To sum up this part, the bioenergetics of irradiated tumor cells critically depends on increased glucose fueling which in case of electrogenic glucose uptake via Na^+-coupled cotransports involves upregulated K^+ channel activity in the plasma membrane (Fig. 5). To which

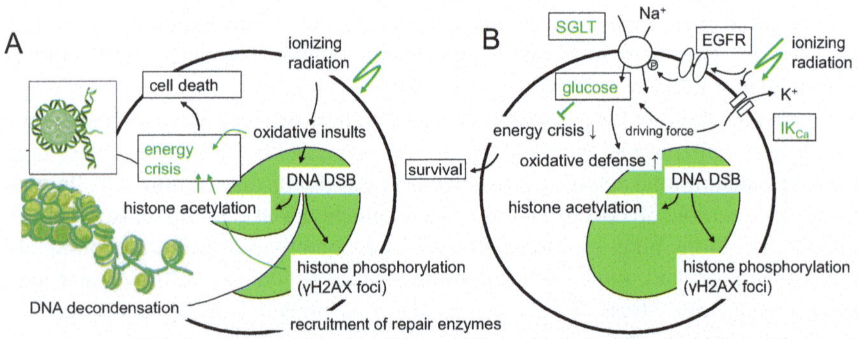

Fig. 5 Ion transports in radiogenic glucose fueling (hypothetical model). (**a**) Fixing of oxidative insults as well as histone acetylation and phosphorylation during DNA repair dissipate carbohydrates and ATP pools that result in an energy crisis of the irradiated cells. (**b**) Radiogenic activation of the epidermal growth factor receptor (EGFR) stimulates upregulation of Na^+/glucose cotransporters (SGLTs) in the plasma membrane that provide glucose for pentose phosphate cycle and glycolysis. The former replenishes NADPH+H^+ for glutathione reduction and antioxidative defense, while the latter provides pyruvate for carbohydrate refilling and ATP production by lactic acid formation. Parallel radiogenic activation of K^+ channels in the plasma membrane maintains the electrochemical driving force for Na^+/glucose cotransport (DNA DSB: DNA double-strand break)

extent pharmaceutical intervention with, e.g., FDA-approved SGLT2 inhibitors can be used to radiosensitize tumors in the clinical situation cannot be answered until data from preclinical experiments in tumor models are available.

6 Concluding Remarks

It is obvious that ionizing radiation induces in tumor cells adaptations of the cell physiology including radiogenic transmembrane ion transports that may contribute to the stress response and eventually survival of the irradiated cells. We are just at the very beginning of identifying the involved transport proteins and the underlying mechanisms. The limited data available, however, clearly show that interference with radiogenic ion transports may radiosensitize tumor cells either directly by inhibiting DNA damage response or indirectly by impairing survival in a radioresistance-conferring hypoxic microenvironment.

Hence, pharmacological targeting of ion transports concomitant to fractionated radiation seems to be a promising strategy for new anti-cancer therapy. Fortunately, a significant percentage of all approved drugs target ion transports. Those drugs comprise neuroleptics, antidepressants, analgesics, anticonvulsants, antiarrhythmics, antihypertensives, diuretics, antidiabetes, and much more (for review, see (Klumpp et al. 2016b*)). Moreover, an inhibitor (senicapoc) of the radioresistance-conferring IK_{Ca} K^+ channel that is upregulated in many tumor entities has been shown in clinical trials to be well tolerated (Ataga et al. 2008). As a consequence, many ion transports are druggable and the identification of pharmaceutical ion transport targets should rapidly become translatable into clinical trials. The available preclinical data, however, rely mostly on 2D in vitro cultures and very few animal studies and, thus, do neither allow any prognosis on the efficacy of channel/transporter targeting in a clinical setting nor on potential resistance mechanisms that might develop upon, e.g., pharmacological blockade of a certain ion channel type during fractionated radiotherapy. Therefore, much more basic research and, most importantly, preclinical animal studies, preferably in orthotopic tumor models, are needed to improve our very fragmentary knowledge on the function of ion transport for the radioresistance of tumor cells.

Acknowledgement This project was supported by a grant of the German Cancer Aid (70112872/70113144).

Conflict of Interest Both authors declare no competing interests.

References

Anoopkumar-Dukie S, Conere T, Sisk GD et al (2009) Mitochondrial modulation of oxygen-dependent radiosensitivity in some human tumour cell lines. Br J Radiol 82:847–854

Archer SL, Gomberg-Maitland M, Maitland ML et al (2008) Mitochondrial metabolism, redox signaling, and fusion: a mitochondria-ROS-HIF-1alpha-K_v1.5 O_2-sensing pathway at the intersection of pulmonary hypertension and cancer. Am J Physiol Heart Circ Physiol 294:H570–H578

Arif T, Stern O, Pittala S et al (2019) Rewiring of cancer cell metabolism by mitochondrial VDAC1 depletion results in time-dependent tumor reprogramming: Glioblastoma as a proof of concept. Cell 8:1330

Arthur GK, Duffy SM, Roach KM et al (2015) KCa3.1 K+ channel expression and function in human bronchial epithelial cells. PLoS One 10:e0145259

Ataga KI, Smith WR, De Castro LM et al (2008) Efficacy and safety of the Gardos channel blocker, senicapoc (ICA-17043), in patients with sickle cell anemia. Blood 111:3991–3997

Averaimo S, Milton RH, Duchen MR et al (2010) Chloride intracellular channel 1 (CLIC1): sensor and effector during oxidative stress. FEBS Lett 584:2076–2084

Baffy G (2017) Mitochondrial uncoupling in cancer cells: liabilities and opportunities. Biochim Biophys Acta Bioenerg 1858:655–664

Bauer TM, Murphy E (2020) Role of mitochondrial calcium and the permeability transition pore in regulating cell death. Circ Res 126:280–293

Birch JL, Strathdee K, Gilmour L et al (2018) A novel small-molecule inhibitor of MRCK prevents radiation-driven invasion in Glioblastoma. Cancer Res 78:6509–6522

Bouchard G, Therriault H, Geha S et al (2017) Radiation-induced lung metastasis development is MT1-MMP-dependent in a triple-negative breast cancer mouse model. Br J Cancer 116:479–488

Brahimi-Horn MC, Ben-Hail D, Ilie M et al (2012) Expression of a truncated active form of VDAC1 in lung cancer associates with hypoxic cell survival and correlates with progression to chemotherapy resistance. Cancer Res 72:2140–2150

Braun N, Klumpp D, Hennenlotter J et al (2015) UCP-3 uncoupling protein confers hypoxia resistance to renal epithelial cells and is upregulated in renal cell carcinoma. Sci Rep 5:13450

Brisson L, Gillet L, Calaghan S et al (2011) Na_V1.5 enhances breast cancer cell invasiveness by increasing NHE1-dependent H^+ efflux in caveolae. Oncogene 30:2070–2076

Brisson L, Driffort V, Benoist L et al (2013) Na_V1.5 Na^+ channels allosterically regulate the NHE-1 exchanger and promote the activity of breast cancer cell invadopodia. J Cell Sci 126:4835–4842

Britschgi A, Bill A, Brinkhaus H et al (2013) Calcium-activated chloride channel ANO1 promotes breast cancer progression by activating EGFR and CAMK signaling. Proc Natl Acad Sci U S A 110:E1026–E1034

Brown A, Geiger H (2018) Chromosome integrity checkpoints in stem and progenitor cells: transitions upon differentiation, pathogenesis, and aging. Cell Mol Life Sci 75:3771–3779

Cadenas S (2018) Mitochondrial uncoupling, ROS generation and cardioprotection. Biochim Biophys Acta Bioenerg 1859:940–950

Campbell TM, Main MJ, Fitzgerald EM (2013) Functional expression of the voltage-gated Na^+-channel Na_v1.7 is necessary for EGF-mediated invasion in human non-small cell lung cancer cells. J Cell Sci 126:4939–4949

Chao HX, Poovey CE, Privette AA et al (2017) Orchestration of DNA damage checkpoint dynamics across the human cell cycle. Cell Syst 5:445–459.e445

Chen JM, Sepramaniam S, Armugam A et al (2008) Water and ion channels: crucial in the initiation and progression of apoptosis in central nervous system? Curr Neuropharmacol 6:102–116

Chen TJ, He HL, Shiue YL et al (2018) High chloride channel accessory 1 expression predicts poor prognoses in patients with rectal cancer receiving chemoradiotherapy. Int J Med Sci 15:1171–1178

Cheng Y, Debska-Vielhaber G, Siemen D (2010) Interaction of mitochondrial potassium channels with the permeability transition pore. FEBS Lett 584:2005–2012

Cloos CR, Daniels DH, Kalen A et al (2009) Mitochondrial DNA depletion induces radioresistance by suppressing G_2 checkpoint activation in human pancreatic cancer cells. Radiat Res 171:581–587

Cone CD Jr, Cone CM (1976) Induction of mitosis in mature neurons in central nervous system by sustained depolarization. Science 192:155–158

Cone CD Jr, Tongier M Jr (1971) Control of somatic cell mitosis by simulated changes in the transmembrane potential level. Oncology 25:168–182

Coultrap SJ, Bayer KU (2012) CaMKII regulation in information processing and storage. Trends Neurosci 35:607–618

Crottes D, Jan LY (2019) The multifaceted role of TMEM16A in cancer. Cell Calcium 82:102050

Cruz-Gregorio A, Martinez-Ramirez I, Pedraza-Chaverri J et al (2019) Reprogramming of energy metabolism in response to radiotherapy in head and neck squamous cell carcinoma. Cancers (Basel) 11:182

Cuddapah VA, Sontheimer H (2010) Molecular interaction and functional regulation of ClC-3 by Ca^{2+}/calmodulin-dependent protein kinase II (CaMKII) in human malignant glioma. J Biol Chem 285:11188–11196

Cuddapah VA, Habela CW, Watkins S et al (2012) Kinase activation of ClC-3 accelerates cytoplasmic condensation during mitotic cell rounding. Am J Physiol Cell Physiol 302:C527–C538

Cuddapah VA, Turner KL, Seifert S et al (2013) Bradykinin-induced chemotaxis of human gliomas requires the activation of KCa3.1 and ClC-3. J Neurosci 33:1427–1440

D'alessandro G, Catalano M, Sciaccaluga M et al (2013) KCa3.1 channels are involved in the infiltrative behavior of glioblastoma in vivo. Cell Death Dis 4:e773

D'alessandro G, Monaco L, Catacuzzeno L et al (2019) Radiation increases functional KCa3.1 expression and invasiveness in glioblastoma. Cancers (Basel) 11:279

Davies A, Hendrich J, Van Minh AT et al (2007) Functional biology of the alpha$_2$delta subunits of voltage-gated calcium channels. Trends Pharmacol Sci 28:220–228

Davis FM, Peters AA, Grice DM et al (2012) Non-stimulated, agonist-stimulated and store-operated Ca^{2+} influx in MDA-MB-468 breast cancer cells and the effect of EGF-induced EMT on calcium entry. PLoS One 7:e36923

Denko NC (2008) Hypoxia, HIF1 and glucose metabolism in the solid tumour. Nat Rev Cancer 8:705–713

Dittmann K, Mayer C, Kehlbach R et al (2008) Radiation-induced caveolin-1 associated EGFR internalization is linked with nuclear EGFR transport and activation of DNA-PK. Mol Cancer 7:69

Dittmann K, Mayer C, Kehlbach R et al (2009) Radiation-induced lipid peroxidation activates SRC kinase and triggers nuclear EGFR transport. Radiother Oncol 92:379–382

Dittmann K, Mayer C, Rodemann HP et al (2013) EGFR cooperates with glucose transporter SGLT1 to enable chromatin remodeling in response to ionizing radiation. Radiother Oncol 107:247–251

Dittmann K, Mayer C, Paasch A et al (2015) Nuclear EGFR renders cells radio-resistant by binding mRNA species and triggering a metabolic switch to increase lactate production. Radiother Oncol 116:431–437

Djamgoz MBA, Fraser SP, Brackenbury WJ (2019) In vivo evidence for voltage-gated Sodium Channel expression in carcinomas and potentiation of metastasis. Cancers (Basel) 11:1675

Driffort V, Gillet L, Bon E et al (2014) Ranolazine inhibits $Na_V1.5$-mediated breast cancer cell invasiveness and lung colonization. Mol Cancer 13:264

Dupre AM, Hempling HG (1978) Osmotic properties of Ehrlich ascites tumor cells during the cell cycle. J Cell Physiol 97:381–396

Eckert F, Schilbach K, Klumpp L et al (2018) Potential role of CXCR4 targeting in the context of radiotherapy and immunotherapy of cancer. Front Immunol 9:3018

Eckert F, Zwirner K, Boeke S et al (2019) Rationale for combining radiotherapy and immune checkpoint inhibition for patients with hypoxic tumors. Front Immunol 10:407

Edalat L, Stegen B, Klumpp L et al (2016) BK K^+ channel blockade inhibits radiation-induced migration/brain infiltration of glioblastoma cells. Oncotarget 7:14259–14278

Eke I, Storch K, Kastner I et al (2012) Three-dimensional invasion of human glioblastoma cells remains unchanged by X-ray and carbon ion irradiation in vitro. Int J Radiat Oncol Biol Phys 84: e515–e523

Ewing D (1998) The oxygen fixation hypothesis: a reevaluation. Am J Clin Oncol 21:355–361

Faubert B, Solmonson A, Deberardinis RJ (2020) Metabolic reprogramming and cancer progression. Science 368(6487):eaaw5473

Foller M, Bobbala D, Koka S et al (2010) Functional significance of the intermediate conductance Ca^{2+}-activated K^+ channel for the short-term survival of injured erythrocytes. Pflugers Arch 460:1029–1044

Fraser SP, Diss JK, Chioni AM et al (2005) Voltage-gated sodium channel expression and potentiation of human breast cancer metastasis. Clin Cancer Res 11:5381–5389

Fu C, Cao CM, Xia Q et al (2003) Reactive oxygen species and mitochondrial KATP-sensitive channels mediated cardioprotection induced by TNF-alpha during hypoxia and reoxygenation. Sheng Li Xue Bao 55:284–289

Gibhardt CS, Roth B, Schroeder I et al (2015) X-ray irradiation activates K+ channels via H2O2 signaling. Sci Rep 5:13861

Gillet L, Roger S, Besson P et al (2009) Voltage-gated sodium channel activity promotes cysteine cathepsin-dependent invasiveness and colony growth of human cancer cells. J Biol Chem 284:8680–8691

Gonzalez-Gonzalez L, Gonzalez-Ramirez R, Flores A et al (2019) Epidermal growth factor potentiates migration of MDA-MB 231 breast cancer cells by increasing $Na_V1.5$ channel expression. Oncology 97:373–382

Gopel SO, Kanno T, Barg S et al (2000) Patch-clamp characterisation of somatostatin-secreting -cells in intact mouse pancreatic islets. J Physiol 528:497–507

Gu XQ, Siemen D, Parvez S et al (2007) Hypoxia increases BK channel activity in the inner mitochondrial membrane. Biochem Biophys Res Commun 358:311–316

Gu XQ, Pamenter ME, Siemen D et al (2014) Mitochondrial but not plasmalemmal BK channels are hypoxia-sensitive in human glioma. Glia 62:504–513

Gueguinou M, Chantome A, Fromont G et al (1843) K_{Ca} and Ca^{2+} channels: the complex thought. Biochim Biophys Acta:2322–2333

Guo L, Tang X, Tian H et al (2008) Subacute hypoxia suppresses $K_v3.4$ channel expression and whole-cell K^+ currents through endogenous 15-hydroxyeicosatetraenoic acid in pulmonary arterial smooth muscle cells. Eur J Pharmacol 587:187–195

Heise N, Palme D, Misovic M et al (2010) Non-selective cation channel-mediated Ca^{2+}-entry and activation of Ca2+/calmodulin-dependent kinase II contribute to G_2/M cell cycle arrest and survival of irradiated leukemia cells. Cell Physiol Biochem 26:597–608

Holley AK, Miao L, St Clair DK et al (2014) Redox-modulated phenomena and radiation therapy: the central role of superoxide dismutases. Antioxid Redox Signal 20:1567–1589

Hu J, Qin K, Zhang Y et al (2011) Downregulation of transcription factor Oct4 induces an epithelial-to-mesenchymal transition via enhancement of Ca^{2+} influx in breast cancer cells. Biochem Biophys Res Commun 411:786–791

Huang L, Li B, Tang S et al (2015) Mitochondrial KATP channels control glioma radioresistance by regulating ROS-induced ERK activation. Mol Neurobiol 52:626–637

Huang Y, Fliegert R, Guse AH et al (2020) A structural overview of the ion channels of the TRPM family. Cell Calcium 85:102111

Huber SM (2013) Oncochannels. Cell Calcium 53:241–255

Huber SM, Tschop J, Braun GS et al (1999) Bradykinin-stimulated Cl^- secretion in T84 cells. Role of Ca^{2+}-activated hSK4-like K^+ channels. Pflugers Arch 438:53–60

Huber SM, Misovic M, Mayer C et al (2012) EGFR-mediated stimulation of sodium/glucose cotransport promotes survival of irradiated human A549 lung adenocarcinoma cells. Radiother Oncol 103:373–379

Huber SM, Butz L, Stegen B et al (2013) Ionizing radiation, ion transports, and radioresistance of cancer cells. Front Physiol 4:212

Jentsch TJ, Pusch M (2018) CLC chloride channels and transporters: structure, function, physiology, and disease. Physiol Rev 98:1493–1590

Jesse RH, Lindberg RD (1975) The efficacy of combining radiation therapy with a surgical procedure in patients with cervical metastasis from squamous cancer of the oropharynx and hypopharynx. Cancer 35:1163–1166

Kaczmarek LK, Zhang Y (2017) Kv3 channels: enablers of rapid firing, neurotransmitter release, and neuronal endurance. Physiol Rev 97:1431–1468

Kandasamy SB, Howerton TC, Hunt WA (1991) Reductions in calcium uptake induced in rat brain synaptosomes by ionizing radiation. Radiat Res 125:158–162

Kaplan HS, Murphy ED (1949) The effect of local roentgen irradiation on the biological behavior of a transplantable mouse carcinoma; increased frequency of pulmonary metastasis. J Natl Cancer Inst 9:407–413

Kim JS, Chang JW, Yun HS et al (2010) Chloride intracellular channel 1 identified using proteomic analysis plays an important role in the radiosensitivity of HEp-2 cells via reactive oxygen species production. Proteomics 10:2589–2604

Kim MJ, Kim RK, Yoon CH et al (2011) Importance of PKC_{delta} signaling in fractionated-radiation-induced expansion of glioma-initiating cells and resistance to cancer treatment. J Cell Sci 124:3084–3094

Kim W, Lee S, Seo D et al (2019) Cellular stress responses in radiotherapy. Cell 8:1105

Kizub IV, Pavlova OO, Ivanova IV et al (2010) Protein kinase C-dependent inhibition of BK_{Ca} current in rat aorta smooth muscle cells following gamma-irradiation. Int J Radiat Biol 86:291–299

Klumpp D, Misovic M, Szteyn K et al (2016a) Targeting TRPM2 channels impairs radiation-induced cell cycle arrest and fosters cell death of T cell leukemia cells in a Bcl-2-dependent manner. Oxidative Med Cell Longev 2016:8026702

Klumpp L, Sezgin EC, Eckert F et al (2016b) Ion channels in brain metastasis. Int J Mol Sci 17:1513

Klumpp D, Frank SC, Klumpp L et al (2017) TRPM8 is required for survival and radioresistance of glioblastoma cells. Oncotarget 8:95896–95913

Klumpp L, Sezgin EC, Skardelly M et al (2018) KCa3.1 channels and glioblastoma: in vitro studies. Curr Neuropharmacol 16:627–635

Krebs C (1929) Effect of roentgen irradiation on the interrelation between malignant tumors and their hosts. Acta Radiol Supp 8:1–133

Kulawiak B, Kudin AP, Szewczyk A et al (2008) BK channel openers inhibit ROS production of isolated rat brain mitochondria. Exp Neurol 212:543–547

Kunzelmann K, Ousingsawat J, Benedetto R et al (2019) Contribution of Anoctamins to cell survival and cell death. Cancers (Basel) 11:382

Kyrychenko S, Tishkin S, Dosenko V et al (2012) The BK(Ca) channels deficiency as a possible reason for radiation-induced vascular hypercontractility. Vasc Pharmacol 56:142–149

Lang F, Pelzl L, Hauser S et al (2018) To die or not to die SGK1-sensitive ORAI/STIM in cell survival. Cell Calcium 74:29–34

Leach JK, Van Tuyle G, Lin PS et al (2001) Ionizing radiation-induced, mitochondria-dependent generation of reactive oxygen/nitrogen. Cancer Res 61:3894–3901

Leanza L, O'reilly P, Doyle A et al (2014) Correlation between potassium channel expression and sensitivity to drug-induced cell death in tumor cell lines. Curr Pharm Des 20:189–200

Li F, Sonveaux P, Rabbani ZN et al (2007) Regulation of HIF-1alpha stability through S-nitrosylation. Mol Cell 26:63–74

Lin S, Lv Y, Xu J et al (2019) Over-expression of $Na_v1.6$ channels is associated with lymph node metastases in colorectal cancer. World J Surg Oncol 17:175

Littler DR, Harrop SJ, Goodchild SC et al (2010) The enigma of the CLIC proteins: ion channels, redox proteins, enzymes, scaffolding proteins? FEBS Lett 584:2093–2101

Liu W, Lu M, Liu B et al (2012) Inhibition of Ca^{2+}-activated Cl^- channel ANO1/TMEM16A expression suppresses tumor growth and invasiveness in human prostate carcinoma. Cancer Lett 326:41–51
Liu J, Chen Y, Shuai S et al (2014) TRPM8 promotes aggressiveness of breast cancer cells by regulating EMT via activating AKT/GSK-3_{beta} pathway. Tumour Biol 35:8969–8977
Mahdi SH, Cheng H, Li J et al (2015) The effect of TGF-beta-induced epithelial-mesenchymal transition on the expression of intracellular calcium-handling proteins in T47D and MCF-7 human breast cancer cells. Arch Biochem Biophys 583:18–26
Maldonado EN (2017) VDAC-tubulin, an anti-Warburg pro-oxidant switch. Front Oncol 7:4
Martin OA, Anderson RL, Russell PA et al (2014) Mobilization of viable tumor cells into the circulation during radiation therapy. Int J Radiat Oncol Biol Phys 88:395–403
Masumoto K, Tsukimoto M, Kojima S (2013) Role of TRPM2 and TRPV1 cation channels in cellular responses to radiation-induced DNA damage. Biochim Biophys Acta 1830:3382–3390
Matsuya Y, Ohtsubo Y, Tsutsumi K et al (2014) Quantitative estimation of DNA damage by photon irradiation based on the microdosimetric-kinetic model. J Radiat Res 55:484–493
Mazure NM (2016) News about VDAC1 in hypoxia. Front Oncol 6:193
Meng Y, Xu X, Luan H et al (2019) The progress and development of GLUT1 inhibitors targeting cancer energy metabolism. Future Med Chem 11:2333–2352
Miranda-Goncalves V, Granja S, Martinho O et al (2016) Hypoxia-mediated upregulation of MCT1 expression supports the glycolytic phenotype of glioblastomas. Oncotarget 7:46335–46353
Mohammed FH, Khajah MA, Yang M et al (2016) Blockade of voltage-gated sodium channels inhibits invasion of endocrine-resistant breast cancer cells. Int J Oncol 48:73–83
Mohr CJ, Gross D, Sezgin EC et al (2019) $K_{Ca}3.1$ channels confer radioresistance to breast cancer cells. Cancers (Basel) 11:1285
Munro TR (1970) The relative radiosensitivity of the nucleus and cytoplasm of Chinese hamster fibroblasts. Radiat Res 42:451–470
Nagy JA, Chang SH, Dvorak AM et al (2009) Why are tumour blood vessels abnormal and why is it important to know? Br J Cancer 100:865–869
Neher E, Sakmann B (1976) Single-channel currents recorded from membrane of denervated frog muscle fibres. Nature 260:799–802
Nelson M, Yang M, Dowle AA et al (2015a) The sodium channel-blocking antiepileptic drug phenytoin inhibits breast tumour growth and metastasis. Mol Cancer 14:13
Nelson M, Yang M, Millican-Slater R et al (2015b) Nav1.5 regulates breast tumor growth and metastatic dissemination in vivo. Oncotarget 6:32914–32929
Nishino K, Tanamachi K, Nakanishi Y et al (2016) Radiosensitizing effect of TRPV1 channel inhibitors in cancer cells. Biol Pharm Bull 39:1224–1230
Noyer L, Grolez GP, Prevarskaya N et al (2018) TRPM8 and prostate: a cold case? Pflugers Arch 470:1419–1429
Ogawa K, Utsunomiya T, Mimori K et al (2006) Differential gene expression profiles of radioresistant pancreatic cancer cell lines established by fractionated irradiation. Int J Oncol 28:705–713
Palme D, Misovic M, Schmid E et al (2013) $K_v3.4$ potassium channel-mediated electrosignaling controls cell cycle and survival of irradiated leukemia cells. Pflugers Arch 465:1209–1221
Palme D, Misovic M, Ganser K et al (2020) hERG K^+ channels promote survival of irradiated leukemia cells. Front Pharmacol 11:489
Peitzsch C, Perrin R, Hill RP et al (2014) Hypoxia as a biomarker for radioresistant cancer stem cells. Int J Radiat Biol 90:636–652
Peng JB, Suzuki Y, Gyimesi G et al (2018) TRPV5 and TRPV6 calcium-selective channels. In: Kozak JA, Putney JW Jr (eds) Calcium entry channels in non-excitable cells. CRC Press, Taylor & Francis Group, Boca Raton, pp 241–274
Proudfoot JM, Murrey MW, Mclean S et al (2018) F2-isoprostanes affect macrophage migration and CSF-1 signalling. Free Radic Biol Med 126:142–152

Qu Y, Zhang H, Zhao S et al (2010) The effect on radioresistance of manganese superoxide dismutase in nasopharyngeal carcinoma. Oncol Rep 23:1005–1011

Ravi M, Paramesh V, Kaviya SR et al (2015) 3D cell culture systems: advantages and applications. J Cell Physiol 230:16–26

Redfern A, Agarwal V, Thompson EW (2019) Hypoxia as a signal for prison breakout in cancer. Curr Opin Clin Nutr Metab Care 22:250–263

Reisz JA, Bansal N, Qian J et al (2014) Effects of ionizing radiation on biological molecules--mechanisms of damage and emerging methods of detection. Antioxid Redox Signal 21:260–292

Richardson RB, Harper ME (2016) Mitochondrial stress controls the radiosensitivity of the oxygen effect: implications for radiotherapy. Oncotarget 7:21469–21483

Rodemann HP, Dittmann K, Toulany M (2007) Radiation-induced EGFR-signaling and control of DNA-damage repair. Int J Radiat Biol 83:781–791

Roth B, Gibhardt CS, Becker P et al (2015) Low-dose photon irradiation alters cell differentiation via activation of hIK channels. Pflugers Arch 467:1835–1849

Ruggieri P, Mangino G, Fioretti B et al (2012) The inhibition of $K_{Ca}3.1$ channels activity reduces cell motility in glioblastoma derived cancer stem cells. PLoS One 7:e47825

Ruoslahti E (1996) Brain extracellular matrix. Glycobiology 6:489–492

Sala-Rabanal M, Yurtsever Z, Nichols CG et al (2015) Secreted CLCA1 modulates TMEM16A to activate Ca^{2+}-dependent chloride currents in human cells. Elife 4:e05875

Sanguinetti MC (2010) HERG1 channelopathies. Pflugers Arch 460:265–276

Sausbier M, Hu H, Arntz C et al (2004) Cerebellar ataxia and Purkinje cell dysfunction caused by Ca^{2+}-activated K^+ channel deficiency. Proc Natl Acad Sci U S A 101:9474–9478

Schumann T, Konig J, Henke C et al (2020) Solute carrier transporters as potential targets for the treatment of metabolic disease. Pharmacol Rev 72:343–379

Sheldon PW, Fowler JF (1976) The effect of low-dose pre-operative X-irradiation of implanted mouse mammary carcinomas on local recurrence and metastasis. Br J Cancer 34:401–407

Si H, Heyken WT, Wolfle SE et al (2006) Impaired endothelium-derived hyperpolarizing factor-mediated dilations and increased blood pressure in mice deficient of the intermediate-conductance Ca^{2+}-activated K^+ channel. Circ Res 99:537–544

Skonieczna M, Cieslar-Pobuda A, Saenko Y et al (2017) The impact of DIDS-induced inhibition of voltage-dependent anion channels (VDAC) on cellular response of Lymphoblastoid cells to ionizing radiation. Med Chem 13:477–483

Soloviev A, Tishkin S, Ivanova I et al (2009) Functional and molecular consequences of ionizing irradiation on large conductance Ca^{2+}-activated K^+ channels in rat aortic smooth muscle cells. Life Sci 84:164–171

Soltysinska E, Bentzen BH, Barthmes M et al (2014) KCNMA1 encoded cardiac BK channels afford protection against ischemia-reperfusion injury. PLoS One 9:e103402

Song MS, Ryu PD, Lee SY (2017) $K_v3.4$ is modulated by HIF-1alpha to protect SH-SY5Y cells against oxidative stress-induced neural cell death. Sci Rep 7:2075

Sontheimer H (2008) An unexpected role for ion channels in brain tumor metastasis. Exp Biol Med (Maywood) 233:779–791

Sorensen MV, Matos JE, Sausbier M et al (2008) Aldosterone increases $K_{Ca}1.1$ (BK) channel mediated colonic K+ secretion. J Physiol 586:4251–4264

Stegen B, Butz L, Klumpp L et al (2015) Ca^{2+}-activated IK K^+ channel blockade radiosensitizes glioblastoma cells. Mol Cancer Res 13:1283–1295

Stegen B, Klumpp L, Misovic M et al (2016) K^+ channel signaling in irradiated tumor cells. Eur Biophys J 45:585–598

Steinle M, Palme D, Misovic M et al (2011) Ionizing radiation induces migration of glioblastoma cells by activating BK K^+ channels. Radiother Oncol 101:122–126

Steudel FA, Mohr CJ, Stegen B et al (2017) SK4 channels modulate Ca^{2+} signalling and cell cycle progression in murine breast cancer. Mol Oncol 11:1172–1188

Strong MS, Vaughan CW, Kayne HL et al (1978) A randomized trial of preoperative radiotherapy in cancer of the oropharynx and hypopharynx. Am J Surg 136:494–500

Sui X, Geng JH, Li YH et al (2018) Calcium channel alpha2delta1 subunit (CACNA2D1) enhances radioresistance in cancer stem-like cells in non-small cell lung cancer cell lines. Cancer Manag Res 10:5009–5018

Tartier L, Gilchrist S, Burdak-Rothkamm S et al (2007) Cytoplasmic irradiation induces mitochondrial-dependent 53BP1 protein relocalization in irradiated and bystander cells. Cancer Res 67:5872–5879

Terry S, Faouzi Zaarour R, Hassan Venkatesh G et al (2018) Role of hypoxic stress in regulating tumor immunogenicity, resistance and plasticity. Int J Mol Sci 19:3044

Teshima K, Yamamoto A, Yamaoka K et al (2000) Involvement of calcium ion in elevation of mRNA for gamma-glutamylcysteine synthetase (gamma-GCS) induced by low-dose gamma-rays. Int J Radiat Biol 76:1631–1639

Thoday JM, Read J (1947) Effect of oxygen on the frequency of chromosome aberrations produced by X-rays. Nature 160:608

Todd DG, Mikkelsen RB (1994) Ionizing radiation induces a transient increase in cytosolic free [Ca^{2+}] in human epithelial tumor cells. Cancer Res 54:5224–5230

Toulany M (2019) Targeting DNA double-strand break repair pathways to improve radiotherapy response. Genes (Basel) 10:25

Trautmann F, Cojoc M, Kurth I et al (2014) CXCR4 as biomarker for radioresistant cancer stem cells. Int J Radiat Biol 90:687–699

Tyszka-Czochara M, Konieczny P, Majka M (2018) Recent advances in the role of AMP-activated protein kinase in metabolic reprogramming of metastatic cancer cells: targeting cellular bioenergetics and biosynthetic pathways for anti-tumor treatment. J Physiol Pharmacol:69

Uysal-Onganer P, Djamgoz MB (2007) Epidermal growth factor potentiates in vitro metastatic behaviour of human prostate cancer PC-3M cells: involvement of voltage-gated sodium channel. Mol Cancer 6:76

Valerie K, Povirk LF (2003) Regulation and mechanisms of mammalian double-strand break repair. Oncogene 22:5792–5812

Vercesi AE, Kowaltowski AJ, Grijalba MT et al (1997) The role of reactive oxygen species in mitochondrial permeability transition. Biosci Rep 17:43–52

Vilalta M, Rafat M, Graves EE (2016) Effects of radiation on metastasis and tumor cell migration. Cell Mol Life Sci 73:2999–3007

Voehringer DW, Hirschberg DL, Xiao J et al (2000) Gene microarray identification of redox and mitochondrial elements that control resistance or sensitivity to apoptosis. Proc Natl Acad Sci U S A 97:2680–2685

Voos P, Fuck S, Weipert F et al (2018) Ionizing radiation induces morphological changes and immunological modulation of Jurkat cells. Front Immunol 9:922

Vullhorst D, Klocke R, Bartsch JW et al (1998) Expression of the potassium channel $K_V3.4$ in mouse skeletal muscle parallels fiber type maturation and depends on excitation pattern. FEBS Lett 421:259–262

Warburg O (1956) On respiratory impairment in cancer cells. Science 124:269–270

Weaver AK, Bomben VC, Sontheimer H (2006) Expression and function of calcium-activated potassium channels in human glioma cells. Glia 54:223–233

Wojewodzka M, Walicka M, Sochanowicz B et al (1994) Calcium antagonist, TMB-8, prevents the induction of adaptive response by hydrogen peroxide or X-rays in human lymphocytes. Int J Radiat Biol 66:99–109

Wondergem R, Bartley JW (2009) Menthol increases human glioblastoma intracellular Ca2+, BK channel activity and cell migration. J Biomed Sci 16:90

Wondergem R, Ecay TW, Mahieu F et al (2008) HGF/SF and menthol increase human glioblastoma cell calcium and migration. Biochem Biophys Res Commun 372:210–215

Wright EM, Ghezzi C, Loo DDF (2017) Novel and unexpected functions of SGLTs. Physiology (Bethesda) 32:435–443

Xie H, Simon MC (2017) Oxygen availability and metabolic reprogramming in cancer. J Biol Chem 292:16825–16832

Xu Y, Miriyala S, Fang F et al (2015) Manganese superoxide dismutase deficiency triggers mitochondrial uncoupling and the Warburg effect. Oncogene 34:4229–4237

Yamamoto T (1936) Experimental study on effect of x-ray on metastasis of malignant tumor, especially in bones. Jpn J Obst Gynec 19:559–569

Yamazaki H, Yoshida K, Yoshioka Y et al (2008) Impact of mitochondrial DNA on hypoxic radiation sensitivity in human fibroblast cells and osteosarcoma cell lines. Oncol Rep 19:1545–1549

Yang M, Brackenbury WJ (2013) Membrane potential and cancer progression. Front Physiol 4:185

Yang M, Kozminski DJ, Wold LA et al (2012) Therapeutic potential for phenytoin: targeting $Na_v1.5$ sodium channels to reduce migration and invasion in metastatic breast cancer. Breast Cancer Res Treat 134:603–615

Yang B, Cao L, Liu J et al (2015) Low expression of chloride channel accessory 1 predicts a poor prognosis in colorectal cancer. Cancer 121:1570–1580

Yildirim S, Altun S, Gumushan H et al (2012) Voltage-gated sodium channel activity promotes prostate cancer metastasis in vivo. Cancer Lett 323:58–61

Yom SS (2015) Accelerated repopulation as a cause of radiation treatment failure in non-small cell lung cancer: review of current data and future clinical strategies. Semin Radiat Oncol 25:93–99

Yoshida S (1997) Effects of X-irradiation on the calcium channel of the mouse oocyte. Life Sci 60:1377–1383

Zhang B, Davidson MM, Hei TK (2014) Mitochondria regulate DNA damage and genomic instability induced by high LET radiation. Life Sci Space Res 1:80–88

Zhang J, Mao W, Dai Y et al (2019) Voltage-gated sodium channel $Na_v1.5$ promotes proliferation, migration and invasion of oral squamous cell carcinoma. Acta Biochim Biophys Sin Shanghai 51:562–570

Zhang C, Ye L, Zhang Q et al (2020) The role of TRPV1 channels in atherosclerosis. Channels (Austin) 14:141–150

Zhou H, Hong M, Chai Y et al (2009) Consequences of cytoplasmic irradiation: studies from microbeam. J Radiat Res 50(Suppl A):A59–A65

Zhou XB, Feng YX, Sun Q et al (2015) Nucleoside diphosphate kinase B-activated intermediate conductance potassium channels are critical for neointima formation in mouse carotid arteries. Arterioscler Thromb Vasc Biol 35:1852–1861

Rev Physiol Biochem Pharmacol (2022) 183: 251–278
https://doi.org/10.1007/112_2021_66

Published online: 12 January 2022

Ion Transporting Proteins and Cancer: Progress and Perspectives

Mustafa B. A. Djamgoz

Contents

M. B. A. Djamgoz (✉)
Department of Life Sciences, Imperial College London, London, UK

Biotechnology Research Centre, Cyprus International University, Nicosia, Mersin, Turkey
e-mail: m.djamgoz@imperial.ac.uk

Abstract Ion transporting proteins (ITPs) comprise a wide range of ion channels, exchangers, pumps and ionotropic receptors many of which are expressed in tumours and contribute dynamically to the different components and stages of the complex cancer process, from initiation to metastasis. In this promising major field of biomedical research, several candidate ITPs have emerged as clinically viable. Here, we consider a series of general issues concerning the oncological potential of ITPs focusing on voltage-gated sodium channels as a 'case study'. First, we outline some key properties of 'cancer' as a whole. These include epigenetics, stemness, metastasis, heterogeneity, neuronal characteristics and bioelectricity. Cancer specificity of ITP expression is evaluated in relation to tissue restriction, splice variance, functional specificity and macro-molecular complexing. As regards clinical potential, diagnostics is covered with emphasis on enabling early detection. For therapeutics, we deal with molecular approaches, drug repurposing and combinations. Importantly, we emphasise the need for carefully designed clinical trials. We highlight also the area of 'social responsibility' and the need to involve the public (cancer patients and healthy individuals) in the work of cancer research professionals as well as clinicians. In advising patients how best to manage cancer, and live with it, we offer the following four principles: Awareness and prevention, early detection, specialist, integrated care, and psychological support. Finally, we highlight four key prerequisites for commercialisation of ITP-based technologies against cancer. We conclude that ITPs offer significant potential as regards both understanding the intricacies of the complex process of cancer and for developing much needed novel therapies.

Keywords Clinical trial · Diet and lifestyle · Drug repurposing · Metastasis · Social responsibility · Voltage-gated sodium channel

Abbreviations

cfDNA	Cell-free DNA
ctDNA	Circulating tumour DNA
EGF	Epidermal growth factor
EMT	Epithelial-mesenchymal transition
HIF	Hypoxia-inducible factor
IC	Immune cell
ITP	Ion transporting protein
K_{Ca}	Calcium-activated potassium channel
lncRNA	Long, noncoding RNA
miRNA	microRNA

NCX	Sodium-calcium exchanger
NE	Neural element
NHE	Sodium-hydrogen exchanger
NRSF	Neuron restrictive silencer factor
NSCLC	Non-small-cell lung cancer
PKA	Protein kinase A
REST	RE1-silencing transcription factor
siRNA	Small interfering RNA
TME	Tumour micro-environment
TTX	Tetrodotoxin
VGIC	Voltage-gated ion channel
VGSC	Voltage-gated sodium channel

1 Introduction

Cancer has been around a long time and will remain a part of modern life for the foreseeable future. In fact, current rates are expected to increase by some 70% in the next 20 years (Bray et al. 2018). Major reasons for this are increasing life expectancy, environmental conditions and worsening lifestyles leading also to associated adverse conditions like obesity and type 2 diabetes. Although major advances are continuously being made in the clinical management of cancer and the potential of 'integrated management' is increasingly being recognised, many problems remain in both diagnostic and therapeutic areas. These include lack of functional biomarkers resulting in uncertainty of even some of the most commonly used diagnostic methods, side effects of therapy, drug resistance and, ultimately, cost.

Research over the last several decades has increasingly been suggesting that ion transporting proteins (ITPs) – ion channels, transporters (exchangers and pumps) and ionotropic receptors – offer distinct advantages as novel predictive biomarkers and mechanistic targets that could overcome at least some of the current clinical limitations (e.g. Arcangeli and Becchetti 2010; Djamgoz et al. 2014; Lastraioli et al. 2015a). This may not be so surprising since as small, charged entities ions can get around the body very effectively and influence a wide range of intracellular activities and whole-cell behaviours. Collectively, ion channels and ion exchangers are encoded by numerous multi-gene families, comprising ca. 4% of the protein-coding genes in the human genome, with 406 genes encoding ion channels and 883 encoding a broad variety of transporters, of which 350 are intracellular (Huang et al. 2004; Venter et al. 2001).

Here, we present a perspective of recent developments involving functional expression of ITPs in cancers and suggest ways in which the field could further become clinically viable. First, we highlight some relevant general properties of cancer in relation to intrinsic expression and pathophysiological role of ITPs. Our aim is not to deal exhaustively with individual ITPs or specific cancers.

Nevertheless, as a case study, we focus on voltage-gated sodium channel (VGSC) expression as a pan-carcinoma phenomenon (Horne et al. 2021; Djamgoz et al. 2019).

1.1 Epigenetics

In the first instance, cancer can be said to be an epigenetic rather than a genetic disease, i.e. it results from qualitatively and/or quantitatively aberrant expression of otherwise normal genes with given amino acid sequences (Darwiche 2020; Ortiz-Barahona et al. 2020). The primary consequence of this is disruption of cells' normal, well-balanced functioning and loss of tissue homeostasis, which can avalanche into waves of further disruption. Importantly, epigenetic changes can be reversible and, accordingly, some 40% of cancer cases are thought to be due to modifiable factors (Islami et al. 2018; Jentzsch et al. 2020). This is 'good news' as it means the cancer machinery may thus be switched off or put into reverse. Indeed, the Nobel Prize winning scientist Jennifer Doudna recently predicted that CRISPR will be used in the future '*. . . not to edit genomes, or at least not to make permanent changes to genomes, but instead to regulate them, to control levels of human proteins that are produced from different genes*' (Mullin 2020). This emphasises the importance of managing disease, including cancer, through modification of gene expression rather than fixing 'faulty' genes. Nevertheless, protein-coding genes are controlled intimately by the noncoding components of the genome, e.g. long-noncoding RNAs (lncRNAs), microRNAs (miRNAs), which themselves are prone to mutations (Pfister and Ashworth 2017). Ultimately, therefore, cancer can be considered to be a 'genomic' disease. Overall, both genetic and epigenetic mechanisms are increasingly and promisingly being exploited clinically against cancer with emphasis on personalised treatment, i.e. 'precision medicine' (Jebelli et al. 2021; Zhang et al. 2020).

1.2 Stemness

Cancer tissues demonstrate varying degrees of dedifferentiation. Consistent with this, cancer cells, at least subpopulations, express properties characteristics of stem cells (Reddy 2020). This is important clinically since it is the stemness ('plasticity') of cancer cells that will enable them to survive by adapting to varying conditions, some adverse. Such conditions could change markedly in space and time especially during the progression of cancer (leading to metastasis) as well as during therapy. In fact, it is by changing their essential characteristics during treatment that cancer stem cells (CSCs) can give rise to 'drug resistance'. On the whole, the more dedifferentiated a cancer, the worse the prognosis. Accordingly, much effort is being exerted to find ways of selectively eliminating the CSC population of tumours (e.g. Sridharan et al. 2019). Importantly, a part of the stemness is manifested in the expression in cancer of some genes, including those for ITPs, in their embryonic forms, a phenomenon called 'oncofoetal' (Zaidi et al. 2017). This property has significant implications for cancer management and can be exploited in both diagnosis and therapy.

1.3 Subtypes and Hallmarks

Cancer is a multi-stage group of many different diseases or disease states, perhaps even hundreds (Weinberg 2014). It can arise in any organ of the body, most commonly in epithelial tissues giving rise to carcinomas which comprise some 90% of all cancers. However, whilst several types of carcinoma may share some common characteristics, there are also significant differences (e.g. hormone sensitivity). Such differences may directly affect the preferred treatment modalities, in a personalised setting, including in relation to ITP expression. Overall, 10 'hallmarks' have been associated with the cancer process – 6 classical, 2 enabling and 2 emerging (Hanahan and Weinberg 2000, 2011). In a seminal study, Prevarskaya et al. (2018) linked various ITPs to each of these hallmarks, and more associations are likely to follow.

1.4 Tumour Micro-environment and Cellular Heterogeneity

Tumours are heterogenous in their cellular make-up. Importantly, alongside cancer cells are immune cells (ICs), fibroblasts, endothelial cells, neurones and specialised CSCs, which collectively drive the cancer process. The whole ensemble is embedded within the extracellular matrix of the tumour micro-environment (TME) and is highly interactive through both gap junctions, tumour nano/microtubes, vesicles and chemical synapses. Most recently, local microbiota has also been found to make up an important part of the TME in many types of cancer, especially in cancers arising from mucosal sites, including lung, skin and gastrointestinal tract (Wong-Rolle et al. 2020). Interestingly, within the microbiome, bacteria also possess ITPs and remarkable bioelectric properties (Das et al. 2018; Martinez-Corral et al. 2019). Altogether, consequently, the cellular interactions and the metabolic biochemicals released into the narrow spaces of the TME can impact significantly upon the cancer process. Another aspect of the 'heterogeneity' of the cancer process is its spatial and temporal dynamism, including in relation to regulated functional ITP expression (Djamgoz 2013; Djamgoz et al. 2014).

1.5 Immune Component

As already noted, ICs are an integral part of the TME. A functional association between the immune system and cancer has been known since the work of William Coley in the nineteenth century (Richardson et al. 1999). This has been formalised as an 'emerging hallmark' of cancer highlighting the avoidance by cancer cells of immune destruction with increasing evidence for the underlying mechanisms (Hanahan and Weinberg 2011; Li and Fei 2020). Increasing evidence also suggests that ITPs play a significant role in the functioning of several types of IC, including lymphocytes, natural killer cells and macrophages (Feske et al. 2015, 2019; Firmenich and Djamgoz 2020). Furthermore, functional ITP expression in ICs can be exploited in immunotherapy, either individually or in combination with other modalities (Djamgoz and Firmenich 2021). This area would be extremely worthwhile cultivating further.

1.6 Metabolism

It is well known that cancer cells' metabolism is fundamentally different from normal cells, exemplified by the Warburg effect (e.g. Pascale et al. 2020). The metabolic alterations in tumours are non-homogeneous, complex and dynamic, reflecting the intricate make-up of the TME. In particular, cancer cells upregulate their expression of the glucose transporter GLUT-1. Thus, cancer cells use glucose as their primary source of ATP rather than produce it by oxidative phosphorylation in mitochondria. The switch can be triggered irreversibly by hypoxia and results in generation of H^+ which is extruded by NHE1 and facilitates extracellular proteolysis and invasion during metastasis. The sodium-dependent glucose transporter SGLT-1 can also contribute to development of cancer (Yamazaki et al. 2018). It would be interesting to determine if the sodium dependence of these transporters associates with Na^+-permeating channels. Clearly, ITPs are inherent to the Warburg effect and worthy of more in-depth study (Iorio et al. 2019).

1.7 Metastasis

Even an individual cancer could involve multiple pathologies rolled into one 'disease'. Overall, primary tumourigenesis ('proliferation') and secondary tumourigenesis ('metastasis') are controlled differently, even partially independently (e.g. Corsini et al. 2020). In most cases, it is metastasis that is the main cause of death from cancer. Although metastasis is a complex process, it can be considered in a reductionist approach to comprise a series of 'basic' cellular behaviours, including attachment/detachment, secretion, motility/invasion, even gene expression and apoptosis. It is well known that ITPs control such cellular behaviours. The same is true for proliferative activity. As already noted, a common feature of growing tumours is development of inner hypoxia which can have a significant impact on tumour progression via expression of hypoxia-inducible 'transcription' factors (HIFs) (Montenegro and Indraccolo 2020). Hypoxia can also impact upon ITP expression/activity (e.g. Guzel et al. 2019).

1.8 Association with 'Mainstream' Cancer Mechanisms

Importantly, functional ITP expression is controlled by 'mainstream' mechanisms of cancer, especially growth factors and steroid hormones (Fraser et al. 2014). For example, Campbell et al. (2013) showed that the pro-invasive effect of epidermal growth factor (EGF) on non-small-cell lung cancer (NSCLC) cells was blocked by tetrodotoxin (TTX), consistent with the notion that VGSC expression/activity mediates the cancer-promoting effect of EGF. A similar phenomenon of functional EGF-VGSC association was shown for human prostate cancer cells by Uysal-Onganer and Djamgoz (2007). Furthermore, Cardone et al. (2015) demonstrated for pancreatic cancer cells that EGF promoted invasion and ECM proteolysis and this was dependent on NHE1. It is also interesting that for the most common hormone-sensitive cancers (e.g. breast and prostate), it is hormone *insensitivity* that correlates positively with functional VGSC expression (Grimes et al. 1995; Laniado et al. 1997; Fraser et al. 2005). Indeed, VGSC expression could be upregulated by silencing oestrogen receptor expression in MCF-7 cells (Mohammed et al. 2016). This association is significant for two reasons. First, it demonstrates, again, that ITPs are integral to the cancer process. Second, it raises possibilities of 'combination' treatments and ITP-based therapies that could overcome drug resistance in hormone-sensitive cancers (Sect. 5.2.3; also, Cardone et al. 2015).

1.9 Neuronal Characteristics of Carcinomas

A remarkable property of carcinomas is their intrinsic acquisition of characteristics normally associated with neurones. Recently, even a new term has emerged to define this concept – *cancer neuroscience* (Monje et al. 2020; Demir et al. 2020). The more aggressive a carcinoma, the more 'neuronal' it appears to be. This is seen most clearly in the case of small-cell lung cancer (SCLC), a carcinoma which expresses several 'neuronal' biomarkers and is so aggressive that it can spread without forming an identifiable primary tumour (Onganer et al. 2005; Jung et al. 2020). Several neuronal biomarkers are also expressed in SCLC, including antineuronal nuclear antibody, Hu antigen proteins, neurone-specific enolase, aromatic l-amino acid decarboxylase, 'neuronal' cell adhesion molecules (NCAMs), neurone-restrictive silencer factor (NRSF/REST), voltage-activated ion channels and neurotransmitter receptors (Onganer et al. 2005). Initially identified as a transcriptional repressor of neuronal genes in non-neuronal cells, REST/NRSF also functions as a tumour suppressor (Coulson 2005). Another neuronal facet of cancer is the occurrence of *Paraneoplastic Neurological Syndrome* where the presence of some neuronal antigen in tumour can stimulate the immune system to produce antibodies which exert neuromuscular effects (Joubert and Honnorat 2015) (see also Sect. 5.1). Breast cancer cells metastasising to brain express components of GABAergic signalling apparently to mimic neurone-like behaviour (Neman et al. 2014). An even more remarkable neuronal property to emerge more recently is the extrinsic physical innervation of tumours and the role that this plays in cancer progression (Hutchings et al. 2020). Tumours can exploit the nerve connection in a variety of ways. For example, as shown recently by Banh et al. (2020), pancreatic cancer cells can use the nerve connection as a supply of nutrients to regulate protein synthesis. This is reminiscent of how tumours can also cleverly exploit (avoid) immune surveillance. In fact, all three cellular elements – nerves, cancer cells and ICs – are highly interactive in the TME (Yang et al. 2020a, b).

1.10 Bioelectricity and Cellular Excitability

Along with their innervation and in line with some of the acquired neuronal characteristics, cancer cells (and all other cells within the TME) possess a range of inherent bioelectric properties. These properties include membrane potentials, expression of voltage- and ligand-gated ion channels, electrogenic ion exchangers and pumps and ion-binding proteins (Djamgoz 2011). In addition, cancer especially carcinoma cells are subject to local (trans-cellular) field potentials that occur naturally in epithelial structures.

Cancer cells maintain a plasma membrane potential (V_m) which, in common with other proliferative cell types, is depolarised relative to normal cells. Such potentials are typically −20 to −30 mV (Levin 2007; Yang and Brackenbury 2013). The membrane potential is associated with a range of ionic mechanisms. For example, in addition to K^+ channels that commonly control V_m, both VGSCs and epithelial sodium channels (ENaCs) are expressed in a range of cancers and could contribute to the depolarised nature of the V_m (Ware et al. 2021; Yang et al. 2020a, b). In parallel, the rise in the intracellular Na^+ concentration resulting from the Na^+ influx that such channels enable is regulated by the Na^+/K^+-ATPase pump. Importantly, the cells use the trans-membrane Na^+ concentration gradient to regulate the intracellular concentrations of a number of key ions such as Ca^{2+} (via Na^+- Ca^{2+} exchange) and pH (via $Na^+–H^+$ exchange) and essential amino acids like glutamate and gamma-aminobutyric acid. Such scenarios can readily be made for any of the common inorganic ions and related to the various components/stages of the cancer process. In short, whilst exceptions and variations will occur, it would be safe, at least tempting, to state sweepingly that all ion channels are expressed in all cancers. Thus, cancer, a pathological universe, is like the brain, a biological universe!

The bioelectric characteristics of carcinomas were formulated into the 'Celex Hypothesis' which states that it is the electrical excitability of carcinomas, driven by VGSC activity, that makes them aggressive, invasive and, ultimately, metastatic (Djamgoz 2013). Indeed, minimally disruptive microelectrode array recordings showed that human breast cancer cells at rest are spontaneously active with action potentials that can be blocked reversibly by TTX (Ribeiro et al. 2020). Also, using chronic, differential recording techniques, McCallum et al. (2020) provided evidence that electrical activity is present within mammary tumours induced in mice in vivo (although the possibility that a part of the activity could be coming from the extrinsic nerve input could not be eliminated). Further work involving recordings from naturally occurring tumours would be greatly welcome!

The aim of this perspective is to highlight some key issues involving cancer cells' ITPs and bioelectricity. This promises to improve both our understanding of the complex process of cancer and, ultimately, its clinical management.

2 Voltage-Gated Ion Channel Expression in Relation to Metastasis

With the exception of gliomas, primary tumours are not usually fatal, especially if detected early and cleanly removed by surgery. Even then, however, micrometastases, if present, can be a major problem and removal of the primary tumour can somehow stimulate them to start growing! Thus, metastatic disease remains the main problem in cancer management, so the question comes up: What are the ion channels driving the metastatic process in carcinomas? We approached this problem directly by focusing on *voltage-gated* ion channels (VGICs), motivated by the huge impact of the V_m (equivalent to some 10^7 V/m) on intrinsic membrane proteins and cellular functioning as a whole. The natural technique to recording the electrical activity of cells is patch clamp recording. We applied this technique to carcinomas of strong and weak metastatic potential in a comparative approach analogous to 'subtractive hybridisation' in molecular biology. The first application was to rat prostate cancer isogenic cell lines and revealed that strong metastatic potential was associated with functional VGSC expression (Grimes et al. 1995). Further work on a wide range of carcinomas demonstrated the same phenomenon (Djamgoz et al. 2019). In all cases, treating the cells with VGSC blockers such as TTX, various pharmacological blockers (e.g. anti-epileptics), gene silencing or an antibody suppressed cellular invasiveness in vitro. Importantly, also, silencing the VGSC (by siRNA) prior to inoculation or injecting TTX into the primary tumour suppressed metastasis in vivo and even prolonged life expectancy (Driffort et al. 2014; Nelson et al. 2015; Yildirim et al. 2012). Thus, the VGSC was established as a pro-metastatic mechanism. However, in order to exploit the VGSC as a clinical target in human patients, cancer-specific properties have to be determined. In fact, such properties are applicable to ITP expression in cancers generally.

3 Cancer-Specificity of ITP Expression

Ion channels and transporters are expressed widely in the human body and are involved in every aspect of cellular functioning. Accordingly, an important question in the possible exploitation of ITPs in clinical management of cancer is their qualitative and/or quantitative specificity for the cancer process. This is possible in several different ways, as follows.

3.1 Splice Variance

It is well known that specific splice variants of various genes are expressed in cancers (e.g. Eymin 2020; Lee et al. 2020). A particularly promising class of genes are those that are developmentally regulated, resulting in expression of '*neonatal*' genes and their products in *adult* cancer tissues. This has been demonstrated most effectively for Nav1.5 (gene: *SCN5A*). Exon 6 of this gene is alternatively spliced such that the 3′ form is expressed in the adult whilst the 5′ form occurs primarily in the neonate. The splicing affects mainly the DI:S3/S4 loop region of the channel protein, the two splice variants differing by 6 amino acids in an extracellular part of the protein. 5'-Nav1.5 (nNav1.5) was discovered as a molecular entity in a neuroblastoma cell line and shown to be developmentally regulated (Ou et al. 2005). Functional expression of nNav1.5 also occurs in cancers of breast, colon, melanoma, astrocytoma, and possibly ovarian cancer (Djamgoz et al. 2019).

3.2 Restricted Tissue Expression

In some cases, although the ITP in question is the product of a normal adult gene, it has a restricted tissue expression in the adult body. A clear example is Kv10.1 which is highly expressed in a range of cancers but amongst normal tissues it is restricted to brain (Cázares-Ordoñez and Pardo 2017). This limitation coupled with the physically restricted access to brain, enabled by the blood-brain barrier, makes systemic use of Kv10.1 blockers as potential anti-cancer agents feasible. Also, nNav1.5 protein has been shown to be absent amongst a wide range of adult human tissues but expressed in several cancers, as listed above (Yamaci et al. 2017; Djamgoz et al. 2019).

3.3 Macro-molecular Complexing

A 'normal' ITP may by be complexed with another protein such that the combination becomes cancer specific. This is seen clearly in the case of hERG1 + beta-integrin complexing which occurs only in cancer cells and this is 'non-conducting' (Becchetti et al. 2017, 2019). In contrast, in normal physiological situations, as in heart muscle, hERG1 expression occurs by itself and functions in conducting mode. Another example is Nav1.5 + NHE1 complexing (protein–protein interaction) that has been demonstrated in breast cancer cells (Brisson et al. 2013). Such complexing raises the possibility of developing bi-specific antibodies as anti-cancer agents (e.g. Duranti and Arcangeli 2019). Recent evidence has shown that nNav1.5 is pharmacologically distinct (Fraser et al. 2021).

3.4 Functional State

In this situation, a normal ITP may generate a novel signal under conditions that are associated particularly with tumours. A clear example of this is the 'persistent current' component (I_{NaP}) of the VGSC which develops under hypoxic conditions, inherent to growing tumours (Djamgoz and Onkal 2013). Another example, also noted above, is the cancer-specific functioning of hERG1 (complexed with beta-integrin) in a non-conducting mode of the channel (Becchetti et al. 2017).

4 Systems Pathophysiology

Cancer ITPs do not impact upon our health alone! Just as in the brain, cancer ITPs and other associated signalling mechanisms work in concert. In the first instance, functional expression of ITPs is under the control of growth factors and hormones (Fraser et al. 2014). Downstream is a variety of signalling cascades that ultimately generate the relevant cancer cell behaviour. In an important study, House et al. (2010) showed that that VGSC (Nav1.5) expression associates with a number of canonical genes driving colon cancer invasiveness. These include MAPK, Wnt, Ca^{2+} signalling as well as genes for membrane remodelling (secretion and proteolysis). In the other well-characterised case of strongly metastatic human breast cancer cells, VGSC activity stimulates NHE1 which leads to H^+ extrusion from the cells and pericellular acidification (Brisson et al. 2011). In turn, the latter turns on a proteolytic enzyme which degrades the extracellular matrix thus opening the way for cancer cells to invade their surroundings and once they hit upon a blood vessel and enter the circulation, metastasis will have started. There is probably a similar scenario involving Na^+ flux and NCX (Chovancova et al. 2020; Rodrigues et al. 2019). The latter is likely to have a major impact on our understanding of cancer mechanisms due to the pivotal role of Ca^{2+} in cell physiology and pathophysiology, including cancer (e.g. Izquierdo-Torres et al. 2020; Monteith et al. 2017). Other VGSC-associated signalling mechanisms include protein kinase A (PKA) (Brackenbury and Djamgoz 2006; Chioni et al. 2010); Rho-associated protein kinase (ROCK) (Poisson et al. 2020); Rac1 (Yang et al. 2020a, b); beta subunits (Haworth and Brackenbury 2019); and salt-inducible kinase-1 (Gradek et al. 2019).

Characterisation of ITP-specific signalling mechanisms and associated cancer cell behaviour will not only improve our understanding of the cancer, especially the metastatic, process, but also generate possibilities of combination therapies across such signalling cascades.

There are other aspects of cancer's 'systems pathophysiology' that need in-depth consideration. First, the TME in which cancer cell interactions involving ITPs occur also contain ICs and neural elements (NEs) comprising intrinsic neurones and/or extrinsic nerve inputs. Both ICs and NEs also express ITPs which bring up two further considerations: First, we need to understand the role of these ITPs and integrate them into the complex cancer process. Second, effects of intended anti-cancer systemic treatments should take into account these additional ITPs as well. This is important in order to avoid generating possible 'antagonistic' effects in therapy.

The neuronal input to cancer has particular therapeutic potential since there is evidence that the sympathetic and parasympathetic nervous systems exert significant influence upon the cancer process through their neurotransmitters, adrenaline and acetylcholine, respectively (Hutchings et al. 2020). As well as neurochemically, it would be possible to influence the NE input to tumours using implantable devices and optogenetics amongst other techniques (Phillips et al. 2021).

5 Clinical Potential

The clinical potential of ITP expression in cancer can be considered in both diagnostic and therapeutic settings.

5.1 *Diagnosis*

A clinically viable cancer diagnostic biomarker should have two important characteristics. First, it should be expressed early in the cancer/metastatic process. Second, it should be predictive or functional, i.e. provide a clue for possible subsequent treatment. These criteria are met by the VGSC expression. Indeed, in the evidence presented by House et al. (2010) the association of VGSC (Nav1.5) expression with MAPK, Wnt, Ca^{2+} signalling, etc. was such that the VGSC appeared to turn on first amongst the genes driving colon cancer invasiveness. Consistent with this, Gradek et al. (2019) showed in breast cancer cells that VGSC (Nav1.5) controls epithelial-to-mesenchymal transition (EMT), an early event in invasiveness. Regarding the second criterion, it is well known that inhibitors of the VGSC(s) expressed can suppress metastasis (Driffort et al. 2014; Nelson et al. 2015; Djamgoz et al. 2019). There is also substantial evidence showing clearly that Kv11.1 expression correlates with a range of clinicopathological characteristics of several cancers (Lastraioli et al. 2015b). Another manifestation of ITP expression with diagnostic potential is in relation to autoantibody (autoAb) generation. A well-known example of this is the Lambert-Eaton Myasthenic Syndrome in SCLC where the immune system generates autoantibodies to voltage-gated sodium and calcium channels (Kesner et al. 2018). More recently, a similar phenomenon has been demonstrated for breast cancer where autoAb's to nNav1.5 were detected in serum of patients at levels that fell following treatment (Rajaratinam et al. 2021). This phenomenon is most likely to be due to the highly antigenic nature of the spliced region of nNav1.5 (Djamgoz and Firmenich 2021). Since this autoimmune response would be expected to be very early (whilst the metastatic cells are still confined to the primary tumour), autoAb's can serve as ideal biomarkers of disease state and expected progression and thus facilitate treatment decisions.

5.2 *Therapy*

A major advantage of targeting ITPs as anti-cancer mechanisms is the availability of a wide range of drugs as well as molecular tools. Thus, both pharmacologic inhibitors and gene silencing of VGSCs have been shown to significantly inhibit metastasis in in vivo animal models (Driffort et al. 2014; Nelson et al. 2015; Djamgoz et al. 2019).

5.2.1 Molecular Approaches

Gene silencing (siRNA) techniques targeting ITPs have been used in vitro and in vivo to demonstrate a range of anti-cancer including anti-metastatic effects (e.g. Driffort et al. 2014; Nelson et al. 2015). Such RNA-based approaches are increasingly becoming clinically viable (e.g. Balwani et al. 2020; also, https://www.nature.com/articles/s41587-020-0494-3). Another promising approach is application of antibodies capable of blocking cancer-promoting ITPs (e.g. Chioni et al. 2005; Duranti and Arcangeli 2019; Djamgoz and Firmenich 2021). Such molecular techniques could rapidly enable the progression of ITP targeting from the laboratory to the cancer clinic.

5.2.2 Repurposing

Several types of existing 'bioelectric' pharmacological agents have already been shown to produce anti-metastatic effects in vivo. Driffort et al. (2014) used the anti-angina drug ranolazine to demonstrate suppression of lung metastasis in the tail-vein model of breast cancer. Similarly, Nelson et al. (2015) applied the anti-convulsant drug phenytoin systemically to an orthotopic model of breast cancer to show suppression of metastasis to several organs. Accordingly, a large number of such available agents can be ‘repurposed’ as anti-cancer/metastatic drugs (Capatina et al. 2020).

5.2.3 Combinations and Drug Resistance

Another approach is to combine ITP-based agents with cancer treatments, especially in relation to overcoming chemotherapeutic resistance (Kischel et al. 2019; Almasi and El Hiani 2020). Thus, Pillozzi et al. (2018) showed that combined application of riluzole (a K_{Ca} channel opener) and inhibition of Kv11.1/hERG1 currents could overcome cisplatin resistance in colorectal cancer cells. Also, NHE1 could synergise with paclitaxel therapy of triple-negative breast cancer cells (Amith et al. 2015). A further combination would be to co-target ITPs in cancer and/or ICs with immunotherapy (Djamgoz and Firmenich 2021). A particularly interesting possibility is to combine checkpoint inhibition with VGSC blockage since it has been shown (1) that EMT is controlled by VGSC activity (Gradek et al. 2019) and (2) that checkpoint (anti-CTLA4) immunotherapy works much better for ‘epithelial’ versus ‘mesenchymal’ tumours (Dongre et al. 2017). Such combination treatments have enormous potential, and this could be enhanced by application of modern techniques of artificial intelligence to seek out novel inhibitors (including repurposed drugs) and their combinations (Liang et al. 2020).

Both ‘repurposing’ and ‘combinations’ could be beneficial. First, by enhancing drug efficacy, dosage could be reduced which could result in decreased chance of drug resistance developing. Second, side effects of the given therapies would be reduced which would be welcome by patients since cancer drugs often manifest undesirable side effects. These would be serious considerations especially if one of the drugs in the combination is chemotherapeutic since chemotherapy can make cancer more aggressive in the long term (e.g. Shnaider et al. 2020). Indeed, there is emerging evidence that ITP modulators can be coupled effectively with chemotherapy (Capatina et al. 2020).

6 Patient Needs: Social Responsibility

The intricacies of cancer, and problems in clinical management, including those covered here, are not always known amongst the general public. Even the average oncologist may not be aware of a relatively recently emerging area, like ITPs in cancer! So, cancer-ITP researchers have significant social responsibility in disseminating their knowledge to the wider world and connect it with the needs of patients who always appreciate learning about promising new therapies. Patients and their families/associates also have responsibility in reverse. In many countries, cancer research is funded by charitable foundations which depend fundamentally on public support.

6.1 Diet and Lifestyle

Unfortunately, the current clinical care does not always result in a lasting cure. Remission is achievable but a complete 'cure' is much harder, and cancer often recurs, even after several years of patients being given the 'all clear'. Importantly, some 40% of cancers are due to modifiable factors and, hence, are preventable (e.g. Britt et al. 2020). Amongst the preventable factors are dietary and nutraceutical agents and lifestyle factors. Interestingly, ITPs can be included in such preventable effects. An example is the body's pH which generally promotes the metastatic dissemination of cancer cells. Accordingly, alkaline diet can produce anti-cancer effects including during treatment (Jentzsch et al. 2020). A recent study reported that the antacid sodium bicarbonate can metabolically reprogram T-cells in acute myeloid leukaemia patients to resist the immune-suppressing effects of cancer cells that can drive leukaemia relapse after stem cell transplants (Uhl et al. 2020). Also, anti-cancer agents like resveratrol, curcumin, capsaicin, genistein, ginseng, omega-3 polyunsaturated fatty acids, vitamin D, epigallocatechin gallate, capsaicin, even salt target ITPs as their targets (Djamgoz and Isbilen 2006; Lopez-Charcas et al. 2021). These are complemented by non-dietary natural products (Bajaj et al. 2020; Jentzsch et al. 2020).

An important aspect of the impact of diet on cancer is 'redox homeostasis' essential for a healthy physiological steady state. This includes oxidation–reduction reactions and their overall balance. The role of dietary and nutraceutical agents with antioxidant properties is hotly debated (e.g. Jentzsch et al. 2020). In an interesting recent study, Senyuk et al. (2021) showed that potassium channel Kv11.1 activity significantly altered expression of genes controlling the production of reactive oxygen species (ROS). An adaptive response to the potentially lethal augmentation of ROS was identified involving increased Nrf2-dependent transcription of antioxidant genes. Indeed, Nrf2 promoted survival in breast cancer, whereas knockdown of Nrf2 led to Kv11.1-induced cell death.

Further work in this area would be extremely worthwhile and could further encourage cancer patients (as well as healthy people) to moderate their diets so as to reduce the risk of cancer progressing, recurring or occurring in the first place.

6.2 *Patient Involvement and Care*

As noted in the Introduction, cancer now is a part of modern living and is here to stay for the foreseeable future. In advising patients how best to manage cancer, and live with it, we can offer patients and those at risk the following four principles.

6.2.1 Awareness and Prevention

As with any ailment, prevention is best. This starts with awareness which, in turn, depends on knowledge and education. Thus, being aware of risk factors, including genetic, and taking necessary measures such as maintaining a healthy diet and lifestyle can go a long way towards suppressing cancer. Such issues are even more important in the context of a relatively new field like ITPs.

6.2.2 Early Detection

Early detection greatly increases the chance of surviving cancer after initial treatment. Here, again, being vigilant and doing self-examination (for breast cancer) or watching bowel or urinary movements (for colorectal and urological cancers, respectively) can save lives. 'Early' means detecting cancer at a pre-metastatic stage and/or whilst the primary tumour which, in turn, makes treatment (including surgery) easier, including as regards any undesirable side effect of the treatment. A clear example of an ITP (VGSC) enabling early detection of pre-invasive colon cancer was demonstrated by House et al. (2010).

6.2.3 Specialist, Integrated Care

Cancer has an extremely complex pathophysiology, changing in space and time. Accordingly, it is imperative that it is treated by specialist, ideally research-active medical oncologists. Furthermore, due to the current limitations of the cancer therapies, it is best to integrate these with *evidence-based* complementary measures comprising dietary and nutraceutical agents and lifestyle. This area was comprehensively reviewed recently in relation to one of the most difficult-to-treat cancers, pancreatic cancer (Jentzsch et al. 2020).

6.2.4 Psychological Support

As already noted, most tumours are innervated, and the nerve input exerts significant influence upon cancer progression. Evidence also suggests that 'optimism' can significantly boost general physical and mental health (Lee et al. 2019).

Furthermore, immune tissues are also innervated, and cells of the immune system possess ITPs including neurotransmitter receptors (Hutchings et al. 2020). Indeed, a remarkable experimental study showed that stimulation of the ventral tegmental area (VTA) of the brain would suppress the growth of human lung cancer and melanoma implanted in mouse models by 40–50%, possibly by boosting anti-tumour immunity (Ben-Shaanan et al. 2018). VTA is a part of the brain's 'reward system' and uses dopamine as its primary messenger. Thus, nerve input provides a mechanistic basis for the modulatory role of mental state and psychological support in helping cancer patients and their carers (Hutchings et al. 2020). As with dietary issues, the impact of 'psychology' on cancer is not readily accepted by clinical oncologists, mainly for historic reasons. Nevertheless, this is destined to change as the evidence for mechanistic insights, including epigenetic, for the role of nerves in the cancer process continues to gather momentum.

7 Clinical Trials

The evidence has grown steadily over the years in support of functional expression of ITPs in cancer. There is no doubt, indeed, that ITPs are now viable targets for anti-cancer clinical applications. Furthermore, because of their active contribution to the cancer process, ITPs can serve as novel markers for patient stratification. So, the scene is set for clinical trials. However, an inherent problem is the fact that ITPs are expressed widely in the human body so manipulating them, say with drugs, can generate complicated, even seriously undesirable, side effects. This problem can be solved by exploiting the cancer-specific properties of ITPs (Sect. 3). Initial trials can readily involve 'repurposed drugs' with known safety profiles and dosage regimes (Sect. 5.2.2). Thus, in many instances, direct phase II testing may be possible. Unfortunately, clinical trials on metastatic disease historically have been less favoured by the field, especially large pharmaceutical companies, due to (1) the relatively long time metastasis takes to develop and (2) the difficulty in monitoring, unlike, say, primary tumours that can simply be measured. Nevertheless, such problems can be overcome by carefully designed clinical trials involving viable biomarkers. For example, liquid biopsy techniques and biomarkers like CTCs, exosomes, miRNAs, cfDNA, ctDNA/tfDNA can go a long way to circumvent the latter problem (e.g. Liu et al. 2020; Marczynski et al. 2020; Palazzolo et al. 2020; Wen et al. 2020). Also, recent developments in non-invasive imaging techniques and parameters (e.g. ^{23}Na-MRI, Hyperpolarised MRI) could speed up progress in this area and make clinical trials on metastasis much more readily manageable.

8 Commercial Opportunism

We anticipate that commercial opportunities will follow successful clinical trials or may even precede them. Already several biotechnology companies are involved in applications of drugs targeting ITPs, especially ion channels. Research scientists, even clinicians, are often so engrossed in their technology that they neglect to see how it will work in the 'real world' or who will be prepared to pay for its commercialisation. Importantly, successful commercialisation necessitates a number of key attributes, as follows: (https://www.nature.com/articles/d41586-020-02040-x?utm_source=Nature+Briefing&utm_campaign=e1a2d18570-briefing-dy-20201221&utm_medium=email&utm_term=0_c9dfd39373-e1a2d18570-44492645).

8.1 *Clinical Need and Competition*

Obviously, the first prerequisite is that there has to be some need in the clinic (served by the pharmaceutical industry) for a new idea or drug to be desirable. This is clearly the case for cancer. As highlighted in the Introduction, significant limitations remain in clinical management of cancer as regards both definitive diagnoses and long-term effective therapies. Having met the clinical need, it is also important to demonstrate that the proposed approach is better than its potential competitors. It is important, therefore, to be aware of the competition in the field and to be able to argue that one has the better prospect. Furthermore, what may seem like competition can also be turned into an attractive attribute. By comparison, for example, annual sales of the breast cancer drug Herceptin reached $6.08 billion in 2019 (https://www.fiercepharma.com/special-report/top-20-drugs-by-global-sales-2019-herceptin#:~:text=Herceptin%20hit%20%246.08%20billion%20in,with%20copycats%20on%20the%20market). So, an alternative, including an ITP-based drug, could well be positioned in the market so as to generate comparable profits!

8.2 *Technology and Protection of Intellectual Property*

As expected, since research on ITPs is invariably multi-disciplinary, ranging in approaches from molecular biology to biophysics and bioengineering, much novel mechanistic knowledge and associated technologies are steadily accumulating. Assuming that the 'technology' in question is viable in meeting clinical needs and being supported by independent evidence that has stood the test time, it is crucial to for the 'intellectual property' (IP) to be protected. This brings up additional needs, e.g. engaging a reputable IP attorney, having the finance to support the drawing up (as well the upkeep) of the patents and ensuring an efficient and fruitful interaction between the lawyers and the inventors.

8.3 People and Teamwork

Making a fantastic discovery is only the first step in its potential commercialisation. For investors, the management team is at least as essential as the new technology. In particular, a CEO with first-class leadership skills and expertise in seeing an early discovery all the way to the marketplace is essential. A successful team, led by the CEO, should be multi-talented, inspiring and able to pull things together and influence people. It needs to project confidence, be good at presenting ideas and building trust and, ultimately, be able to raise money to get the job done. All this also requires an enormous degree of patience because the time scales can be long (several months to years).

8.4 Finance and Investment

Finding investment for the kind of early-stage company based upon an ITP as a target may not be easy, even after the previous three conditions are met. Even when potential investors say they would invest in an 'early-stage' venture, their definition/expectation of 'early' is often not less than the drug having been tested successfully in man, i.e. subject to at least a phase II trial. There is no magical rule for successfully funding a company, apart from trying every possibility from grants to various kinds of investment, ranging from venture capital to angel investors. During this period, it is essential that the foundations of the company, i.e. the technology and its IP remain strong and protected. Thus, one hidden advantage of this interim period could be the demonstration that the company's foundations have stood the test of time.

9 Conclusion

In overall conclusion, the case is strong for adoption of various ITPs as novel cancer mechanisms and targets with significant diagnostic and therapeutic potential. The case is particularly strong for nNav1.5 expression in several carcinomas. Many other ITPs are also being actively evaluated and we can confidently predict that the numbering of ITPs satisfying clinical expectations will increase steadily in time. These include ligand-gated (ionotropic) ion channels for which, again, much non-cytotoxic pharmacology is already available and can readily be 'repurposed'. All such ITPs and associated signalling mechanisms and drugs can be exploited alone as well in combination with existing therapies to improve treatment efficacy.

This would be possible even for the most advanced therapies, such as immunotherapy (Geng et al. 2012; Pilon-Thomas et al. 2016; Uhl et al. 2020; Djamgoz and Firmenich 2021). It would seem only a matter of time, therefore, before ITPs become a reality in clinical oncology.

Acknowledgements Our research has been supported over many years by numerous granting agencies and first-class research students and associates (too many to list individually). We are particularly indebted to the Trustees of the Pro Cancer Research Fund (PCRF) for unfailing support of some 20 years.

Conflict of Interest Statement MBAD is involved in a small biotech company aiming broadly to realise the clinical potential of VGSC expression in cancer.

References

Almasi S, El Hiani Y (2020) Exploring the therapeutic potential of membrane transport proteins: focus on cancer and chemoresistance. Cancers (Basel) 12(6):1624. https://doi.org/10.3390/cancers12061624

Amith SR, Wilkinson JM, Baksh S, Fliegel L (2015) The Na^+/H^+ exchanger (NHE1) as a novel co-adjuvant target in paclitaxel therapy of triple-negative breast cancer cells. Oncotarget 6:1262–1275. https://doi.org/10.18632/oncotarget.2860

Arcangeli A, Becchetti A (2010) New trends in cancer therapy: targeting ion channels and transporters. Pharmaceuticals (Basel) 3:1202–1224. https://doi.org/10.3390/ph3041202

Bajaj S, Ong ST, Chandy KG (2020) Contributions of natural products to ion channel pharmacology. Nat Prod Rep 37:703–716. https://doi.org/10.1039/c9np00056a

Balwani M, Sardh E, Ventura P, Peiró PA, Rees DC, Stölzel U, Bissell DM, Bonkovsky HL, Windyga J, Anderson KE, Parker C, Silver SM, Keel SB, Wang JD, Stein PE, Harper P, Vassiliou D, Wang B, Phillips J, Ivanova A, Langendonk JG, Kauppinen R, Minder E, Horie Y, Penz C, Chen J, Liu S, Ko JJ, Sweetser MT, Garg P, Vaishnaw A, Kim JB, Simon AR, Gouya L, Investigators ENVISION (2020) Phase 3 trial of RNAi therapeutic Givosiran for acute intermittent porphyria. N Engl J Med 382:2289–2301. https://doi.org/10.1056/NEJMoa1913147

Banh RS, Biancur DE, Yamamoto K, Sohn ASW, Walters B, Kuljanin M, Gikandi A, Wang H, Mancias JD, Schneider RJ, Pacold ME, Kimmelman AC (2020) Neurons release serine to support mrna translation in pancreatic cancer. Cell 183:1202–1218.e25. https://doi.org/10.1016/j.cell.2020.10.016

Becchetti A, Crescioli S, Zanieri F, Petroni G, Mercatelli R, Coppola S, Gasparoli L, D'Amico M, Pillozzi S, Crociani O, Stefanini M, Fiore A, Carraresi L, Morello V, Manoli S, Brizzi MF, Ricci D, Rinaldi M, Masi A, Schmidt T, Quercioli F, Defilippi P, Arcangeli A (2017) The conformational state of hERG1 channels determines integrin association, downstream signaling, and cancer progression. Sci Signal 10(473):eaaf3236. https://doi.org/10.1126/scisignal.aaf3236

Becchetti A, Petroni G, Arcangeli A (2019) Ion channel conformations regulate integrin-dependent signaling. Trends Cell Biol 29:298–307. https://doi.org/10.1016/j.tcb.2018.12.005

Ben-Shaanan TL, Schiller M, Azulay-Debby H, Korin B, Boshnak N, Koren T, Krot M, Shakya J, Rahat MA, Hakim F, Rolls A (2018) Modulation of anti-tumor immunity by the brain's reward system. Nat Commun 9:2723. https://doi.org/10.1038/s41467-018-05283-5

Brackenbury WJ, Djamgoz MBA (2006) Activity-dependent regulation of voltage-gated Na+ channel expression in Mat-LyLu rat prostate cancer cell line. J Physiol 573:343–356. https://doi.org/10.1113/jphysiol.2006.106906

Bray F, Ferlay J, Soerjomataram I, Siegel RL, Torre LA, Jemal A (2018) Global cancer statistics 2018: GLOBOCAN estimates of incidence and mortality worldwide for 36 cancers in 185 countries. CA Cancer J Clin 68:394–424. https://doi.org/10.3322/caac.21492

Brisson L, Gillet L, Calaghan S, Besson P, Le Guennec JY, Roger S, Gore J (2011) Na(V)1.5 enhances breast cancer cell invasiveness by increasing NHE1-dependent H(+) efflux in caveolae. Oncogene 30:2070–2076. https://doi.org/10.1038/onc.2010.574

Brisson L, Driffort V, Benoist L, Poet M, Counillon L, Antelmi E, Rubino R, Besson P, Labbal F, Chevalier S, Reshkin SJ, Gore J, Roger S (2013) NaV1.5 Na^+ channels allosterically regulate the NHE-1 exchanger and promote the activity of breast cancer cell invadopodia. J Cell Sci 126:4835–4842. https://doi.org/10.1242/jcs.123901

Britt KL, Cuzick J, Phillips KA (2020) Key steps for effective breast cancer prevention. Nat Rev Cancer 20:417–436. https://doi.org/10.1038/s41568-020-0266-x

Campbell TM, Main MJ, Fitzgerald EM (2013) Functional expression of the voltage-gated Na^+-channel Nav1.7 is necessary for EGF-mediated invasion in human non-small cell lung cancer cells. J Cell Sci 126:4939–4949. https://doi.org/10.1242/jcs.130013

Capatina AL, Lagos D, Brackenbury WJ (2020) Targeting ion channels for cancer treatment: current progress and future challenges. Rev Physiol Biochem Pharmacol. https://doi.org/10.1007/112_2020_46

Cardone RA, Greco MR, Zeeberg K, Zaccagnino A, Saccomano M, Bellizzi A, Bruns P, Menga M, Pilarsky C, Schwab A, Alves F, Kalthoff H, Casavola V, Reshkin SJ (2015) A novel NHE1-centered signaling cassette drives epidermal growth factor receptor-dependent pancreatic tumor metastasis and is a target for combination therapy. Neoplasia 17:155–166. https://doi.org/10.1016/j.neo.2014.12.003

Cázares-Ordoñez V, Pardo LA (2017) Kv10.1 potassium channel: from the brain to the tumors. Biochem Cell Biol 95:531–536. https://doi.org/10.1139/bcb-2017-0062

Chioni AM, Fraser SP, Pani F, Foran P, Wilkin GP, Diss JK, Djamgoz MBA (2005) A novel polyclonal antibody specific for the Na(v)1.5 voltage-gated Na(+) channel 'neonatal' splice form. J Neurosci Methods 147:88–98. https://doi.org/10.1016/j.jneumeth.2005.03.010

Chioni AM, Shao D, Grose R, Djamgoz MBA (2010) Protein kinase A and regulation of neonatal Nav1.5 expression in human breast cancer cells: activity-dependent positive feedback and cellular migration. Int J Biochem Cell Biol 42:346–358. https://doi.org/10.1016/j.biocel.2009.11.021

Chovancova B, Liskova V, Babula P, Krizanova O (2020) Role of sodium/calcium exchangers in tumours. Biomol Ther 10(9):1257. https://doi.org/10.3390/biom10091257

Corsini EM, Wang J, Wu CC, Fujimoto J, Negrao MV, Chen R, Quek K, Mitchell KG, Chow CB, Little L, Gumbs C, Song X, Behrens C, Correa AM, Antonoff MB, Swisher SG, Heymach JV, Zhang J, Wistuba II, Futreal PA, Sepesi B, Zhang J (2020) Genomic assessment distinguishes intrapulmonary metastases from synchronous primary lung cancers. J Thorac Dis 12:1952–1959. https://doi.org/10.21037/jtd-20-1

Coulson JM (2005) Transcriptional regulation: cancer, neurons and the REST. Curr Biol 15:R665–R668. https://doi.org/10.1016/j.cub.2005.08.032

Darwiche N (2020) Epigenetic mechanisms and the hallmarks of cancer: an intimate affair. Am J Cancer Res 10:1954–1978

Das S, Jayaratne R, Barrett KE (2018) The role of ion transporters in the pathophysiology of infectious diarrhea. Cell Mol Gastroenterol Hepatol 6:33–45. https://doi.org/10.1016/j.jcmgh.2018.02.009

Demir IE, Reyes CM, Alrawashdeh W, Ceyhan GO, Deborde S, Friess H, Görgülü K, Istvanffy R, Jungwirth D, Kuner R, Maryanovich M, Na'ara S, Renders S, Saloman JL, Scheff NN, Steenfadt H, Stupakov P, Thiel V, Verma D, Yilmaz BS, White RA, Wang TC, Wong RJ, Frenette PS, Gil Z (2020) Neural influences in cancer (NIC) international research consortium, Davis BM. Future directions in preclinical and translational cancer neuroscience research. Nat Cancer. 1:1027–1031. https://doi.org/10.1038/s43018-020-00146-9

Djamgoz MBA (2011) Bioelectricity of cancer: voltage-gated ion channels and direct-current electric fields. In: Pullar C (ed) The physiology of bioelectricity in development, tissue regeneration and cancer. Taylor & Francis, London, pp 269–294

Djamgoz MBA (2013) Biophysics of cancer: cellular excitability ("CELEX") hypothesis of metastasis. J Clin Exp Oncol S1:005. https://doi.org/10.4172/2324-9110.S1-005

Djamgoz MBA, Firmenich L (2021) Novel immunotherapeutic approaches to cancer: voltage-gated sodium channel expression in immune cells and tumors. In: Amiji M (ed) Cancer immunology and immunotherapy, vol 1. Elsevier. In press

Djamgoz MBA, Isbilen B (2006) Dietary compounds as anti-cancer agents: a preliminary evaluation of ion channels and membrane excitability as possible target mechanisms. Turk J Biochem 31:57–68

Djamgoz MBA, Onkal R (2013) Persistent current blockers of voltage-gated sodium channels: a clinical opportunity for controlling metastatic disease. Recent Pat Anticancer Drug Discov 8:66–84. https://doi.org/10.2174/15748928130107

Djamgoz MBA, Coombes RC, Schwab A (2014) Ion transport and cancer: from initiation to metastasis. Philos Trans R Soc B Biol Sci 369:20130092. https://doi.org/10.1098/rstb.2013.0092

Djamgoz MBA, Fraser SP, Brackenbury WJ (2019) In vivo evidence for voltage-gated sodium channel expression in carcinomas and potentiation of metastasis. Cancers (Basel) 11:1675. https://doi.org/10.3390/cancers11111675

Dongre A, Rashidian M, Reinhardt F, Bagnato A, Keckesova Z, Ploegh HL, Weinberg RA (2017) Epithelial-to-mesenchymal transition contributes to immunosuppression in breast carcinomas. Cancer Res 77:3982–3989. https://doi.org/10.1136/jitc-2020-SITC2020.0232

Driffort V, Gillet L, Bon E, Marionneau-Lambot S, Oullier T, Joulin V, Collin C, Pagès JC, Jourdan ML, Chevalier S, Bougnoux P, Le Guennec JY, Besson P, Roger S (2014) Ranolazine inhibits NaV1.5-mediated breast cancer cell invasiveness and lung colonization. Mol Cancer 13:264. https://doi.org/10.1186/1476-4598-13-264

Duranti C, Arcangeli A (2019) Ion channel targeting with antibodies and antibody fragments for cancer diagnosis. Antibodies (Basel) 8(2):33. https://doi.org/10.3390/antib8020033

Eymin B (2020) Targeting the spliceosome machinery: a new therapeutic axis in cancer? Biochem Pharmacol 189:114039. https://doi.org/10.1016/j.bcp.2020.114039

Feske S, Wulff H, Skolnik EY (2015) Ion channels in innate and adaptive immunity. Annu Rev Immunol 33:291–353. https://doi.org/10.1146/annurev-immunol-032414-112212

Feske S, Concepcion AR, Coetzee WA (2019) Eye on ion channels in immune cells. Sci Signal 12 (572):eaaw8014. https://doi.org/10.1126/scisignal.aaw8014

Firmenich L, Djamgoz MBA (2020) Meeting report: society of general physiologists symposium on "Ion Channels and Transporters in Immunity, Inflammation and Antitumor Immunity". Bioelectricity 2:418–423. https://doi.org/10.1089/bioe.2020.0045

Fraser SP, Diss JK, Chioni AM, Mycielska ME, Pan H, Yamaci RF, Pani F, Siwy Z, Krasowska M, Grzywna Z, Brackenbury WJ, Theodorou D, Koyutürk M, Kaya H, Battaloglu E, De Bella MT, Slade MJ, Tolhurst R, Palmieri C, Jiang J, Latchman DS, Coombes RC, Djamgoz MBA (2005) Voltage-gated sodium channel expression and potentiation of human breast cancer metastasis. Clin Cancer Res 11:5381–5389. https://doi.org/10.1158/1078-0432.CCR-05-0327

Fraser SP, Ozerlat-Gunduz I, Brackenbury WJ, Fitzgerald EM, Campbell TM, Coombes RC, Djamgoz MBA (2014) Regulation of voltage-gated sodium channel expression in cancer: hormones, growth factors and auto-regulation. Philos Trans R Soc Lond B Biol Sci 369 (1638):20130105. https://doi.org/10.1098/rstb.2013.0105

Fraser SP, Onkal R, Theys M, Bosmans F, Djamgoz MBA (2021) Neonatal NaV1.5: pharmacological distinctiveness of a cancer-related voltage-gated sodium channel splice variant. Br J Pharmacol (in press)

Geng S, Zhong Y, Wang S, Liu H, Zou Q, Xie X, Li C, Yu Q, He Z, Wang B (2012) Amiloride enhances antigen specific CTL by facilitating HBV DNA vaccine entry into cells. PLoS One 7 (3):e33015. https://doi.org/10.1371/journal.pone.0033015

Gradek F, Lopez-Charcas O, Chadet S, Poisson L, Ouldamer L, Goupille C, Jourdan ML, Chevalier S, Moussata D, Besson P, Roger S (2019) Sodium channel Nav1.5 controls epithelial-to-mesenchymal transition and invasiveness in breast cancer cells through its regulation by the salt-inducible kinase-1. Sci Rep 9(1):18652. https://doi.org/10.1038/s41598-019-55197-5

Grimes JA, Fraser SP, Stephens GJ, Downing JE, Laniado ME, Foster CS, Abel PD, Djamgoz MBA (1995) Differential expression of voltage-activated Na+ currents in two prostatic tumour cell lines: contribution to invasiveness in vitro. FEBS Lett 369:290–294. https://doi.org/10.1016/0014-5793(95)00772-2

Guzel RM, Ogmen K, Ilieva KM, Fraser SP, Djamgoz MBA (2019) Colorectal cancer invasiveness in vitro: predominant contribution of neonatal Nav1.5 under normoxia and hypoxia. J Cell Physiol 234(5):6582–6593. doi: https://doi.org/10.1002/jcp.27399

Hanahan D, Weinberg RA (2000) The hallmarks of cancer. Cell 100:57–70. https://doi.org/10.1016/s0092-8674(00)81683-9

Hanahan D, Weinberg RA (2011) Hallmarks of cancer: the next generation. Cell 144:646–674. https://doi.org/10.1016/j.cell.2011.02.013

Haworth AS, Brackenbury WJ (2019) Emerging roles for multifunctional ion channel auxiliary subunits in cancer. Cell Calcium 80:125–140. https://doi.org/10.1016/j.ceca.2019.04.005

Horne J, Mansur S, Bao Y (2021) Sodium ion channels as potential therapeutic targets for cancer metastasis. Drug Discov Today 26(5):1136–1147. https://doi.org/10.1016/j.drudis.2021.01.026

House CD, Vaske CJ, Schwartz AM, Obias V, Frank B, Luu T, Sarvazyan N, Irby R, Strausberg RL, Hales TG, Stuart JM, Lee NH (2010) Voltage-gated Na+ channel SCN5A is a key regulator of a gene transcriptional network that controls colon cancer invasion. Cancer Res 70:6957–6967. https://doi.org/10.1158/0008-5472.CAN-10-1169

Huang Y, Anderle P, Bussey KJ, Barbacioru C, Shankavaram U, Dai Z, Reinhold WC, Papp A, Weinstein JN, Sadée W (2004) Membrane transporters and channels: role of the transportome in cancer chemosensitivity and chemoresistance. Cancer Res 64:4294–4301. https://doi.org/10.1158/0008-5472.CAN-03-3884

Hutchings C, Phillips J, Djamgoz MBA (2020) Nerve input to tumours: pathophysiological consequences of a dynamic relationship. Biochim Biophys Acta Rev Cancer 1874:188411. https://doi.org/10.1016/j.bbcan.2020.188411

Iorio J, Petroni G, Duranti C, Lastraioli E (2019) Potassium and sodium channels and the Warburg effect: biophysical regulation of cancer metabolism. Bioelectricity 1:188–200. https://doi.org/10.1089/bioe.2019.0017

Islami F, Goding Sauer A, Miller KD, Siegel RL, Fedewa SA, Jacobs EJ, McCullough ML, Patel AV, Ma J, Soerjomataram I, Flanders WD, Brawley OW, Gapstur SM, Jemal A (2018) Proportion and number of cancer cases and deaths attributable to potentially modifiable risk factors in the United States. CA Cancer J Clin 68:31–54. https://doi.org/10.3322/caac.21440

Izquierdo-Torres E, Hernández-Oliveras A, Fuentes-García G, Zarain-Herzberg Á (2020) Calcium signaling and epigenetics: a key point to understand carcinogenesis. Cell Calcium 91:102285. https://doi.org/10.1016/j.ceca.2020.102285

Jebelli A, Baradaran B, Mosafer J, Baghbanzadeh A, Mokhtarzadeh A, Tayebi L (2021) Recent developments in targeting genes and pathways by RNAi-based approaches in colorectal cancer. Med Res Rev 41:395–434. https://doi.org/10.1002/med.21735

Jentzsch V, Davis JAA, Djamgoz MBA (2020) Pancreatic Cancer (PDAC): introduction of evidence-based complementary measures into integrative clinical management. Cancers (Basel) 12:E3096. https://doi.org/10.3390/cancers12113096

Joubert B, Honnorat J (2015) Autoimmune channelopathies in paraneoplastic neurological syndromes. Biochim Biophys Acta 1848(10 Pt B):2665–2676. https://doi.org/10.1016/j.bbamem.2015.04.003

Jung E, Alfonso J, Monyer H, Wick W, Winkler F (2020) Neuronal signatures in cancer. Int J Cancer 147:3281–3291. https://doi.org/10.1002/ijc.33138

Kesner VG, Oh SJ, Dimachkie MM, Barohn RJ (2018) Lambert-Eaton myasthenic syndrome. Neurol Clin 36:379–394. https://doi.org/10.1016/j.ncl.2018.01.008

Kischel P, Girault A, Rodat-Despoix L, Chamlali M, Radoslavova S, Abou Daya H, Lefebvre T, Foulon A, Rybarczyk P, Hague F, Dhennin-Duthille I, Gautier M, Ouadid-Ahidouch H (2019)

Ion channels: new actors playing in chemotherapeutic resistance. Cancers (Basel) 11(3):376. https://doi.org/10.3390/cancers11030376

Laniado ME, Lalani EN, Fraser SP, Grimes JA, Bhangal G, Djamgoz MBA, Abel PD (1997) Expression and functional analysis of voltage-activated Na+ channels in human prostate cancer cell lines and their contribution to invasion in vitro. Am J Pathol 150:1213–1221

Lastraioli E, Iorio J, Arcangeli A (2015a) Ion channel expression as promising cancer biomarker. Biochim Biophys Acta 1848(10 Pt B):2685–2702. https://doi.org/10.1016/j.bbamem.2014.12.016

Lastraioli E, Lottini T, Bencini L, Bernini M, Arcangeli A (2015b) hERG1 potassium channels: novel biomarkers in human solid cancers. Biomed Res Int:896432. https://doi.org/10.1155/2015/896432

Lee LO, James P, Zevon ES, Kim ES, Trudel-Fitzgerald C, Spiro A 3rd, Grodstein F, Kubzansky LD (2019) Optimism is associated with exceptional longevity in 2 epidemiologic cohorts of men and women. Proc Natl Acad Sci U S A 116:18357–18362. https://doi.org/10.1073/pnas.1900712116

Lee SE, Alcedo KP, Kim HJ, Snider NT (2020) Alternative splicing in hepatocellular carcinoma. Cell Mol Gastroenterol Hepatol 10(4):699–712. https://doi.org/10.1016/j.jcmgh.2020.04.018

Levin M (2007) Large-scale biophysics: ion flows and regeneration. Trends Cell Biol 17:261–270. https://doi.org/10.1016/j.tcb.2007.04.007

Li Z, Fei T (2020) Improving cancer immunotherapy with CRISPR-based technology. Adv Biosyst 4(11):e1900253. https://doi.org/10.1002/adbi.201900253

Liang G, Fan W, Luo H, Zhu X (2020) The emerging roles of artificial intelligence in cancer drug development and precision therapy. Biomed Pharmacother 128:110255. https://doi.org/10.1016/j.biopha.2020.110255

Liu HE, Vuppalapaty M, Wilkerson C, Renier C, Chiu M, Lemaire C, Che J, Matsumoto M, Carroll J, Crouse S, Hanft VR, Jeffrey SS, Di Carlo D, Garon EB, Goldman J, Sollier E (2020) Detection of EGFR mutations in cfDNA and CTCs, and comparison to tumor tissue in non-small-cell-lung-cancer (NSCLC) patients. Front Oncol 10:572895. https://doi.org/10.3389/fonc.2020.572895

Lopez-Charcas O, Pukkanasut P, Velu SE, Brackenbury WJ, Hales TG, Besson P, Gomora JC, Roger S (2021) Pharmacological and nutritional targeting of voltage-gated sodium channels in the treatment of cancers. iScience 24:102270. https://doi.org/10.1016/j.isci.2021.102270

Marczynski GT, Laus AC, Dos Reis MB, Reis RM, Vazquez VL (2020) Circulating tumor DNA (ctDNA) detection is associated with shorter progression-free survival in advanced melanoma patients. Sci Rep 10(1):18682. https://doi.org/10.1038/s41598-020-75792-1

Martinez-Corral R, Liu J, Prindle A, Süel GM, Garcia-Ojalvo J (2019) Metabolic basis of brain-like electrical signalling in bacterial communities. Philos Trans R Soc Lond B Biol Sci 374:20180382. https://doi.org/10.1098/rstb.2018.0382

McCallum GA, Shiralkar J, Suciu D, Covarrubias G, Yu JS, Karathanasis E, Durand DM (2020) Chronic neural activity recorded within breast tumors. Sci Rep 10(1):14824. https://doi.org/10.1038/s41598-020-71670-y

Mohammed FH, Khajah MA, Yang M, Brackenbury WJ, Luqmani YA (2016) Blockade of voltage-gated sodium channels inhibits invasion of endocrine-resistant breast cancer cells. Int J Oncol 48:73–83. https://doi.org/10.3892/ijo.2015.3239

Monje M, Borniger JC, D'Silva NJ, Deneen B, Dirks PB, Fattahi F, Frenette PS, Garzia L, Gutmann DH, Hanahan D, Hervey-Jumper SL, Hondermarck H, Hurov JB, Kepecs A, Knox SM, Lloyd AC, Magnon C, Saloman JL, Segal RA, Sloan EK, Sun X, Taylor MD, Tracey KJ, Trotman LC, Tuveson DA, Wang TC, White RA, Winkler F (2020) Roadmap for the emerging field of cancer neuroscience. Cell 181(2):219–222. https://doi.org/10.1016/j.cell.2020.03.034

Monteith GR, Prevarskaya N, Roberts-Thomson SJ (2017) The calcium-cancer signalling nexus. Nat Rev Cancer 17:367–380. https://doi.org/10.1038/nrc.2017.18

Montenegro F, Indraccolo S (2020) Metabolism in the tumor microenvironment. Adv Exp Med Biol 1263:1–11. https://doi.org/10.1007/978-3-030-44518-8_1

Mullin E (2020) Fresh Off Her Nobel Prize Win, Jennifer Doudna Predicts What's Next for CRISPR. https://futurehuman.medium.com/fresh-off-her-nobel-prize-win-jennifer-doudna-predicts-whats-next-for-crispr-1fea0225c41d

Nelson M, Yang M, Millican-Slater R, Brackenbury WJ (2015) Nav1.5 regulates breast tumor growth and metastatic dissemination in vivo. Oncotarget 6(32):32914–32929. https://doi.org/10.18632/oncotarget.5441

Neman J, Termini J, Wilczynski S, Vaidehi N, Choy C, Kowolik CM, Li H, Hambrecht AC, Roberts E, Jandial R (2014) Human breast cancer metastases to the brain display GABAergic properties in the neural niche. Proc Natl Acad Sci U S A 111:984–989. https://doi.org/10.1073/pnas.1322098111

Onganer PU, Seckl MJ, Djamgoz MBA (2005) Neuronal characteristics of small-cell lung cancer. Br J Cancer 93:1197–1201. https://doi.org/10.1038/sj.bjc.6602857

Ortiz-Barahona V, Joshi RS, Esteller M (2020) Use of DNA methylation profiling in translational oncology. Semin Cancer Biol 19.:S1044-579X(20)30271-6. https://doi.org/10.1016/j.semcancer.2020.12.011

Ou SW, Kameyama A, Hao LY, Horiuchi M, Minobe E, Wang WY, Makita N, Kameyama M (2005) Tetrodotoxin-resistant Na+ channels in human neuroblastoma cells are encoded by new variants of Nav1.5/SCN5A. Eur J Neurosci 22:793–801. https://doi.org/10.1111/j.1460-9568.2005.04280.x

Palazzolo S, Memeo L, Hadla M, Duzagac F, Steffan A, Perin T, Canzonieri V, Tuccinardi T, Caligiuri I, Rizzolio F (2020) Cancer extracellular vesicles: next-generation diagnostic and drug delivery nanotools. Cancers (Basel) 12(11):E3165. https://doi.org/10.3390/cancers12113165

Pascale RM, Calvisi DF, Simile MM, Feo CF, Feo F (2020) The Warburg effect 97 years after its discovery. Cancers (Basel) 12(10):2819. https://doi.org/10.3390/cancers12102819

Pfister SX, Ashworth A (2017) Marked for death: targeting epigenetic changes in cancer. Nat Rev Drug Discov 16:241–263. https://doi.org/10.1038/nrd.2016.256

Phillips JA, Hutchinson C, Djamgoz MBA (2021) Clinical potential of nerve input to tumors: a bioelectricity perspective. Bioelectricity 3. https://doi.org/10.1089/bioe.2020.0051

Pillozzi S, D'Amico M, Bartoli G, Gasparoli L, Petroni G, Crociani O, Marzo T, Guerriero A, Messori L, Severi M, Udisti R, Wulff H, Chandy KG, Becchetti A, Arcangeli A (2018) The combined activation of KCa3.1 and inhibition of Kv11.1/hERG1 currents contribute to overcome cisplatin resistance in colorectal cancer cells. Br J Cancer 118:200–212. https://doi.org/10.1038/bjc.2017.392

Pilon-Thomas S, Kodumudi KN, El-Kenawi AE, Russell S, Weber AM, Luddy K, Damaghi M, Wojtkowiak JW, Mulé JJ, Ibrahim-Hashim A, Gillies RJ (2016) Neutralization of tumor acidity improves antitumor responses to immunotherapy. Cancer Res 76:1381–1390. https://doi.org/10.1158/0008-5472.CAN-15-1743

Poisson L, Lopez-Charcas O, Chadet S, Bon E, Lemoine R, Brisson L, Ouaissi M, Baron C, Besson P, Roger S, Moussata D (2020) Rock inhibition promotes NaV1.5 sodium channel-dependent SW620 colon cancer cell invasiveness. Sci Rep 10(1):13350. https://doi.org/10.1038/s41598-020-70378-3

Prevarskaya N, Skryma R, Shuba Y (2018) Ion channels in cancer: are cancer hallmarks oncochannelopathies? Physiol Rev 98:559–621. https://doi.org/10.1152/physrev.00044.2016

Rajaratinam H, Rasudin NS, Al Astani TAD, Mokhtar NF, Yahya MM, Zain WZW, Asma-Abdullah N, Fuad WEM (2021) Breast cancer therapy affects the expression of antineonatal Nav1.5 antibodies in the serum of patients with breast cancer. Oncol Lett 21(2):108. https://doi.org/10.3892/ol.2020.12369

Reddy KB (2020) Stem cells: current status and therapeutic implications. Genes (Basel) 11(11):1372. https://doi.org/10.3390/genes11111372

Ribeiro M, Elghajiji A, Fraser SP, Burke ZD, Tosh D, Djamgoz MBA, Rocha PRF (2020) Human breast cancer cells demonstrate electrical excitability. Front Neurosci 14:404. https://doi.org/10.3389/fnins.2020.00404

Richardson MA, Ramirez T, Russell NC, Moye LA (1999) Coley toxins immunotherapy: a retrospective review. Altern Ther Health Med 5:42–47

Rodrigues T, Estevez GNN, Tersariol ILDS (2019) Na+/Ca2+ exchangers: unexploited opportunities for cancer therapy? Biochem Pharmacol 163:357–361. https://doi.org/10.1016/j.bcp.2019.02.032

Senyuk V, Eskandari N, Jiang Y, Garcia-Varela R, Sundstrom R, Leanza L, Peruzzo R, Burkard M, Minshall RD, Gentile S (2021) Compensatory expression of NRF2-dependent antioxidant genes is required to overcome the lethal effects of Kv11.1 activation in breast cancer cells and PDOs. Redox Biol 45:102030. https://doi.org/10.1016/j.redox.2021.102030

Shnaider PV, Ivanova OM, Malyants IK, Anufrieva KS, Semenov IA, Pavlyukov MS, Lagarkova MA, Govorun VM, Shender VO (2020) New insights into therapy-induced progression of cancer. Int J Mol Sci 21(21):7872. https://doi.org/10.3390/ijms21217872

Sridharan S, Howard CM, Tilley AMC, Subramaniyan B, Tiwari AK, Ruch RJ, Raman D (2019) Novel and alternative targets against breast cancer stemness to combat chemoresistance. Front Oncol 9:1003. https://doi.org/10.3389/fonc.2019.01003

Uhl FM, Chen S, O'Sullivan D, Edwards-Hicks J, Richter G, Haring E, Andrieux G, Halbach S, Apostolova P, Büscher J, Duquesne S, Melchinger W, Sauer B, Shoumariyeh K, Schmitt-Graeff A, Kreutz M, Lübbert M, Duyster J, Brummer T, Boerries M, Madl T, Blazar BR, Groß O, Pearce EL, Zeiser R (2020) Metabolic reprogramming of donor T cells enhances graft-versus-leukemia effects in mice and humans. Sci Transl Med 12(567):eabb8969. https://doi.org/10.1126/scitranslmed.abb8969

Uysal-Onganer P, Djamgoz MBA (2007) Epidermal growth factor potentiates in vitro metastatic behaviour of human prostate cancer PC-3M cells: involvement of voltage-gated sodium channel. Mol Cancer 6:76. https://doi.org/10.1186/1476-4598-6-76

Venter JC, Adams MD, Myers EW, Li PW, Mural RJ, Sutton GG, Smith HO, Yandell M, Evans CA, Holt RA, Gocayne JD, Amanatides P, Ballew RM, Huson DH, Wortman JR, Zhang Q, Kodira CD, Zheng XH, Chen L, Skupski M, Subramanian G, Thomas PD, Zhang J, Gabor Miklos GL, Nelson C, Broder S, Clark AG, Nadeau J, McKusick VA, Zinder N, Levine AJ, Roberts RJ, Simon M, Slayman C, Hunkapiller M, Bolanos R, Delcher A, Dew I, Fasulo D, Flanigan M, Florea L, Halpern A, Hannenhalli S, Kravitz S, Levy S, Mobarry C, Reinert K, Remington K, Abu-Threideh J, Beasley E, Biddick K, Bonazzi V, Brandon R, Cargill M, Chandramouliswaran I, Charlab R, Chaturvedi K, Deng Z, Di Francesco V, Dunn P, Eilbeck K, Evangelista C, Gabrielian AE, Gan W, Ge W, Gong F, Gu Z, Guan P, Heiman TJ, Higgins ME, Ji RR, Ke Z, Ketchum KA, Lai Z, Lei Y, Li Z, Li J, Liang Y, Lin X, Lu F, Merkulov GV, Milshina N, Moore HM, Naik AK, Narayan VA, Neelam B, Nusskern D, Rusch DB, Salzberg S, Shao W, Shue B, Sun J, Wang Z, Wang A, Wang X, Wang J, Wei M, Wides R, Xiao C, Yan C, Yao A, Ye J, Zhan M, Zhang W, Zhang H, Zhao Q, Zheng L, Zhong F, Zhong W, Zhu S, Zhao S, Gilbert D, Baumhueter S, Spier G, Carter C, Cravchik A, Woodage T, Ali F, An H, Awe A, Baldwin D, Baden H, Barnstead M, Barrow I, Beeson K, Busam D, Carver A, Center A, Cheng ML, Curry L, Danaher S, Davenport L, Desilets R, Dietz S, Dodson K, Doup L, Ferriera S, Garg N, Gluecksmann A, Hart B, Haynes J, Haynes C, Heiner C, Hladun S, Hostin D, Houck J, Howland T, Ibegwam C, Johnson J, Kalush F, Kline L, Koduru S, Love A, Mann F, May D, McCawley S, McIntosh T, McMullen I, Moy M, Moy L, Murphy B, Nelson K, Pfannkoch C, Pratts E, Puri V, Qureshi H, Reardon M, Rodriguez R, Rogers YH, Romblad D, Ruhfel B, Scott R, Sitter C, Smallwood M, Stewart E, Strong R, Suh E, Thomas R, Tint NN, Tse S, Vech C, Wang G, Wetter J, Williams S, Williams M, Windsor S, Winn-Deen E, Wolfe K, Zaveri J, Zaveri K, Abril JF, Guigó R, Campbell MJ, Sjolander KV, Karlak B, Kejariwal A, Mi H, Lazareva B, Hatton T, Narechania A, Diemer K, Muruganujan A, Guo N, Sato S, Bafna V, Istrail S, Lippert R, Schwartz R, Walenz B, Yooseph S, Allen D, Basu A, Baxendale J, Blick L, Caminha M, Carnes-Stine J, Caulk P, Chiang YH, Coyne M, Dahlke C, Mays A, Dombroski M, Donnelly M, Ely D, Esparham S, Fosler C, Gire H, Glanowski S, Glasser K, Glodek A, Gorokhov M, Graham K, Gropman B, Harris M, Heil J, Henderson S, Hoover J, Jennings D, Jordan C, Jordan J, Kasha J, Kagan L, Kraft C, Levitsky A,

Lewis M, Liu X, Lopez J, Ma D, Majoros W, McDaniel J, Murphy S, Newman M, Nguyen T, Nguyen N, Nodell M, Pan S, Peck J, Peterson M, Rowe W, Sanders R, Scott J, Simpson M, Smith T, Sprague A, Stockwell T, Turner R, Venter E, Wang M, Wen M, Wu D, Wu M, Xia A, Zandieh A, Zhu X (2001) The sequence of the human genome. Science 291:1304–1351. https://doi.org/10.1126/science.1058040. Erratum in: Science 2001 Jun 5;292(5523):1838

Ware AW, Harris JJ, Slatter TL, Cunliffe HE, McDonald FJ (2021) The epithelial sodium channel has a role in breast cancer cell proliferation. Breast Cancer Res Treat. https://doi.org/10.1007/s10549-021-06133-7

Weinberg RA (2014) The biology of cancer, 2nd edn. Garland Science, New York

Wen G, Zhou T, Gu W (2020) The potential of using blood circular RNA as liquid biopsy biomarker for human diseases. Protein Cell. https://doi.org/10.1007/s13238-020-00799-3

Wong-Rolle A, Wei HK, Zhao C, Jin C (2020) Unexpected guests in the tumor microenvironment: microbiome in cancer. Protein Cell. https://doi.org/10.1007/s13238-020-00813-8

Yamaci RF, Fraser SP, Battaloglu E, Kaya H, Erguler K, Foster CS, Djamgoz MBA (2017) Neonatal Nav1.5 protein expression in normal adult human tissues and breast cancer. Pathol Res Pract 213:900–907. https://doi.org/10.1016/j.prp.2017.06.003

Yamazaki Y, Harada S, Tokuyama S (2018) Sodium-glucose transporter as a novel therapeutic target in disease. Eur J Pharmacol 822:25–31. https://doi.org/10.1016/j.ejphar.2018.01.003

Yang M, Brackenbury WJ (2013) Membrane potential and cancer progression. Front Physiol 4:185. https://doi.org/10.3389/fphys.2013.00185

Yang M-W, Tao L-Y, Jiang Y-S, Yang J-Y, Huo Y-M, Liu D-J, Li J, Xue-Liang F, He R, Lin C, Liu W, Zhang J-F, Hua R, Li Q, Jiang S-H, Hu L-P, Tian G-A, Zhang X-X, Niu N, Lu P, Shi J, Xiao GG, Wang L-W, Xue J, Zhang Z-G, Sun Y-W (2020a) Perineural invasion reprograms the immune microenvironment through cholinergic signaling in pancreatic ductal adenocarcinoma. Cancer Res 80:1991–2003. https://doi.org/10.1158/0008-5472.CAN-19-2689

Yang M, James AD, Suman R, Kasprowicz R, Nelson M, O'Toole PJ, Brackenbury WJ (2020b) Voltage-dependent activation of Rac1 by Nav 1.5 channels promotes cell migration. J Cell Physiol 235:3950–3972. https://doi.org/10.1002/jcp.29290

Yildirim S, Altun S, Gumushan H, Patel A, Djamgoz MBA (2012) Voltage-gated sodium channel activity promotes prostate cancer metastasis in vivo. Cancer Lett 323:58–61. https://doi.org/10.1016/j.canlet.2012.03.036

Zaidi SK, Frietze SE, Gordon JA, Heath JL, Messier T, Hong D, Boyd JR, Kang M, Imbalzano AN, Lian JB, Stein JL, Stein GS (2017) Bivalent epigenetic control of oncofetal gene expression in cancer. Mol Cell Biol 37:e00352–e00317. https://doi.org/10.1128/MCB.00352-17

Zhang S, Gong Y, Li C, Yang W, Li L (2020) Beyond regulations at DNA levels: a review of epigenetic therapeutics targeting cancer stem cells. Cell Prolif:e12963. https://doi.org/10.1111/cpr.12963

The manufacturer's authorised representative in the EU is Springer Nature Customer Service Centre GmbH, Europaplatz 3, 69115 Heidelberg, Germany. If you have any concerns regarding our products, please contact ProductSafety@springernature.com

Printed and bound by CPI Group (UK) Ltd, Croydon, CR0 4YY
30/09/2026
02225498-0011